Introduction to Functional Data Analysis

CHAPMAN & HALL/CRC
Texts in Statistical Science Series

Series Editors

Francesca Dominici, *Harvard School of Public Health, USA*
Julian J. Faraway, *University of Bath, UK*
Martin Tanner, *Northwestern University, USA*
Jim Zidek, *University of British Columbia, Canada*

Extending the Linear Model with R:
Generalized Linear, Mixed Effects and
Nonparametric Regression Models, Second
Edition
J.J. Faraway

Linear Models with R, Second Edition
J.J. Faraway

A Course in Large Sample Theory
T.S. Ferguson

Multivariate Statistics: A Practical
Approach
B. Flury and H. Riedwyl

Readings in Decision Analysis
S. French

Discrete Data Analysis with R: Visualization
and Modeling Techniques for Categorical and
Count Data
M. Friendly and D. Meyer

Markov Chain Monte Carlo:
Stochastic Simulation for Bayesian Inference,
Second Edition
D. Gamerman and H.F. Lopes

Bayesian Data Analysis, Third Edition
A. Gelman, J.B. Carlin, H.S. Stern, D.B. Dunson,
A. Vehtari, and D.B. Rubin

Multivariate Analysis of Variance and
Repeated Measures: A Practical Approach for
Behavioural Scientists
D.J. Hand and C.C. Taylor

Practical Longitudinal Data Analysis
D.J. Hand and M. Crowder

Logistic Regression Models
J.M. Hilbe

Richly Parameterized Linear Models:
Additive, Time Series, and Spatial Models
Using Random Effects
J.S. Hodges

Statistics for Epidemiology
N.P. Jewell

Stochastic Processes: An Introduction,
Second Edition
P.W. Jones and P. Smith

The Theory of Linear Models
B. Jørgensen

Pragmatics of Uncertainty
J.B. Kadane

Principles of Uncertainty
J.B. Kadane

Graphics for Statistics and Data Analysis with R
K.J. Keen

Mathematical Statistics
K. Knight

Introduction to Functional Data Analysis
P. Kokoszka and M. Reimherr

Introduction to Multivariate Analysis:
Linear and Nonlinear Modeling
S. Konishi

Nonparametric Methods in Statistics with SAS
Applications
O. Korosteleva

Modeling and Analysis of Stochastic Systems,
Second Edition
V.G. Kulkarni

Exercises and Solutions in Biostatistical Theory
L.L. Kupper, B.H. Neelon, and S.M. O'Brien

Exercises and Solutions in Statistical Theory
L.L. Kupper, B.H. Neelon, and S.M. O'Brien

Design and Analysis of Experiments with R
J. Lawson

Design and Analysis of Experiments with SAS
J. Lawson

A Course in Categorical Data Analysis
T. Leonard

Statistics for Accountants
S. Letchford

Introduction to the Theory of Statistical
Inference
H. Liero and S. Zwanzig

Statistical Theory, Fourth Edition
B.W. Lindgren

Stationary Stochastic Processes: Theory and
Applications
G. Lindgren

Statistics for Finance
E. Lindström, H. Madsen, and J. N. Nielsen

The BUGS Book: A Practical Introduction to
Bayesian Analysis
D. Lunn, C. Jackson, N. Best, A. Thomas, and
D. Spiegelhalter

Introduction to General and Generalized
Linear Models
H. Madsen and P. Thyregod

Time Series Analysis
H. Madsen

Pólya Urn Models
H. Mahmoud

Texts in Statistical Science

Introduction to Functional Data Analysis

Piotr Kokoszka

Colorado State University
Ft. Collins, Colorado

Matthew Reimherr

The Pennsylvania State University
University Park, Pennsylvania

CRC Press
Taylor & Francis Group
Boca Raton London New York

CRC Press is an imprint of the
Taylor & Francis Group an **informa** business

A CHAPMAN & HALL BOOK

CRC Press
Taylor & Francis Group
6000 Broken Sound Parkway NW, Suite 300
Boca Raton, FL 33487-2742

First issued in paperback 2021

© 2017 by Taylor & Francis Group, LLC
CRC Press is an imprint of Taylor & Francis Group, an Informa business

No claim to original U.S. Government works

ISBN-13: 978-1-03-209659-9 (pbk)
ISBN-13: 978-1-4987-4634-2 (hbk)

Visit the Taylor & Francis Web site at
http://www.taylorandfrancis.com

and the CRC Press Web site at
http://www.crcpress.com

To Gudrun and Vanessa
  Piotr

To Zsuzsanna and William, Elliott, and Alice
  Matthew

Contents

Preface

Audience and scope

This book provides a concise introduction to the field of functional data analysis (FDA). It can be used as a textbook for a semester long course on FDA for advanced undergraduate or MS statistics majors, as well as for MS and PhD students in other disciplines, including applied mathematics, environmental science, public health, medical research, geophysical sciences, and economics. It can also be used for self–study and as a reference for researchers in those fields who wish to acquire a solid understanding of FDA methodology and practical guidance for its implementation. Each chapter contains problems and plentiful examples of relevant R code.

The field of FDA has seen rapid development over the last two decades. At present, FDA can be seen as a subfield of statistics that has reached a certain maturity with its central ideas and methods crystalized and generally viewed as fundamental to the subject. At the same time, its methods have been applied to quite broadly in medicine, science, business, and engineering. While new theoretical and methodological developments, and new applications, are still being reported at a fair rate, an introductory account will be useful to students and researchers seeking an accessible and sufficiently comprehensive introduction to the subject. Several FDA monographs exist, but they are either older or cover very specific topics, and none of them is written in the style of a textbook, with problems that can be assigned as homework or used as part of examinations. Our objective is to furnish a textbook that provides an accessible introduction to the field rather than a monograph that explores cutting edge developments. The book assumes a solid background in calculus, linear algebra, distributional probability theory, foundations of statistical inference, and some familiarity with R programming. Such a background is acquired by US senior and MS students majoring in statistics or mathematics with a statistics emphasis, and by European third year students with similar specializations. We do not assume background in nonparametric statistics or advanced regression methods. The required concepts are explained in scalar settings before the related functional concepts are developed. References to more advanced research are provided for those who wish to gain a more in-depth understanding of a specific topic. Each chapter ends with problems that fall into two categories: 1) theoretical problems, mostly simple exercises intended to solidify definitions and concepts, 2) R based data analytic problems.

There are a number of very good books on FDA. The best known is the monograph of Ramsay and Silverman (2005), which is the second edition of a book originally published in 1997. The first edition is largely credited with solidifying FDA as an official subbranch of statistics. Their work provides a more detailed treatment of many topics only highlighted in our book, including

computational aspects of smoothing, smoothing under positivity and monotonicity constraints, various special cases of the functional linear model, and
the dynamical systems approach to FDA. The companion book, Ramsay and
Silverman (2002), presents a more detailed analyses of several data sets. Another companion book, Ramsay *et al.* (2009) explains how the methodology of
Ramsay and Silverman (2005) is implemented in the R package fda. Almost
twenty years have passed since the publication of the monograph of Ramsay
and Silverman (2005) in which there have been several notable developments
that deserve a place in a textbook exposition. These include methodology for
sparsely observed functions and generalized linear models together with their
R implementations, as well as methodology for dependent functional data and
its implementation in several recently developed R packages. While Ramsay
and Silverman (2005) focus on methodology, the book of Hsing and Eubank
(2015) contains a rigorous and comprehensive exposition of the Hilbert space
framework for functional data. Functional linear time series models within an
even more general Banach space framework are studied in Bosq (2000). Our
book contains several chapters that present the most fundamental results of
Hilbert space theory for functional data. The book of Ferraty and Vieu (2006)
presents mathematical theory behind nonlinear functional regression, a topic
which is only briefly mentioned in our book. Shi and Choi (2011) develop a
Bayesian Gaussian process framework for functional regression. The book of
Horváth and Kokoszka (2012) presents a general introduction to the mathematical FDA framework and then branches into several directions with the
most novel exposition pertaining to functional data which exhibit dependence
over time or space. The mathematical level of Horváth and Kokoszka (2012)
is significantly higher than of this book. The collections Ferraty and Romain
(2011) and Ferraty (2011) contain contributions by leading researchers in the
field which summarize a number of recent developments in FDA and point
towards future research directions.

Outline of the contents

The book consists of 12 chapters that can be divided into several groups.
Chapters 1 and 2 introduce the most fundamental concepts of FDA including
basis expansions, mean and covariance functions, functional principal components and penalized smoothing. Chapter 3 gives an introduction to the
Hilbert space framework for modeling and inference of functional data. This
is chapter is designed to be very broadly accessible, focusing on mathematical
concepts. The objective is to provide the reader with sufficient background to
understand the mathematical concepts and formulations used in subsequent
chapters. Those interested in a more mathematically rigorous foundation can
replace Chapter 3 with Chapters 10 and 11. Chapters 4, 5, and 6 focus on
functional regression models. Chapter 4 is concerned with scalar–on–function
regression. In this simplest, but perhaps most important, setting, we discuss
the differences between the functional and traditional regression, and explain

how difficulties and advantages specific to functional settings are addressed. In Chapter 5, we turn to regression models with functional responses. Chapter 6 is dedicated to a functional version of generalized linear models, which has found many applications in medical and biology research. Chapter 7 provides an introduction to the analysis of sparsely observed functions. Such data arise often in longitudinal medical studies where a small number of measurements, often characterized by a nonnegligible observation error, are available for each subject. Chapters 8 and 9 introduce dependent functional data structures, for time series and spatial data, respectively. Chapters 10 and 11 provide a self–contained and rigorous introduction to the mathematical theory underlying FDA. Generally, only references to complex proofs are given, but the exposition is rigorous and can serve as a starting point for further advanced research. We conclude with Chapter 12 which introduces more advanced inferential tools for functional data, focusing on hypothesis testing and confidence bands. If Chapters 10 and 11 are used as a substitution for Chapter 3, then this chapter can follow immediately after.

Chapters 6, 7, 8, 9, and 12 introduce relatively recent research, which has not yet been presented elsewhere in a systematic textbook exposition. There are important, dynamically developing areas of FDA which are not included in this textbook. These include classification and clustering, the interface between functional data and differential equations, and functional data defined on manifolds. Some references are given in Section 2.4.

Acknowledgements

John Kimmel provided useful editorial guidance during the preparation of this book. The reviews he obtained helped us reshape and enhance the book at several stages of preparation; we thank the anonymous reviewers. Several researchers provided substantive help in the preparation of this book. Sections that contain R code are often based on the code provided by researchers involved in the preparation of the specific packages. Jeff Goldsmith provided the code presented in Section 4.7, Phil Reiss and Fabian Scheipl in Section 5.4, and Giles Hooker in Section 5.5. Fabian Scheipl also contributed to the code in Section 6.4. The final version of the code presented in Chapter 8 was prepared by Han Lin Shang, with contributions from Alexander Aue, Siegfried Hörmann and Gregory Rice. Robertas Gabrys prepared the code presented in Section 8.7. Pedro Delicado wrote the code shown in Chapter 9 and helped us improve other sections of that chapter. Some problems were suggested by PhD students at Colorado State University: Cody Alsaker, Ran Fu, Aaron Nielsen and Zach Weller. Hyunphil Choi, Johannes Klepsch, Neda Mohammadi Jouzdani, Stathis Paparoditis, and Ben Zheng found many typos and recommended improvements to the exposition. We thank them all for their kind help. We acknowledge generous support from the United States National Science Foundation and the National Security Agency.

Piotr Kokoszka and Matthew Reimherr

1

First steps in the analysis of functional data

Functional data analysis, FDA, has expanded rapidly in the years leading up to this text. The wide breadth of applications and tools make a precise definition of FDA somewhat difficult. At a high level, one should think of FDA as arising when one of the variables or units of interest in a data set can be naturally viewed as a smooth curve or function. FDA can then be thought of as the statistical analysis of samples of curves (possibly combined with vectors or scalars as well).

This chapter serves the dual purpose of introducing basic concepts and discussing the most important R functions for manipulating functional data. Ramsay *et al.* (2009) provide a much more comprehensive and detailed description of the R and MATLAB tools used for analysis of functional data. Our objective is to enable the reader to perform simple analyses in R and gain some working understanding of FDA. The scope of this chapter is therefore limited, many more developments will be presented in the following chapters. In particular, we restrict ourselves to the analysis of random samples of densely observed curves. The monograph of Ramsay and Silverman (2005) elaborates on many topics touched upon in this chapter; further detailed examples are presented in Ramsay and Silverman (2002).

The simplest data set encountered in FDA is a sample of the form

$$x_n(t_{j,n}) \in \mathbb{R}, \quad t_{j,n} \in [T_1, T_2], \quad n = 1, 2, \ldots, N, \quad j = 1, \ldots, J_n. \tag{1.1}$$

By this we mean that N curves are observed on a common interval $[T_1, T_2]$. The values of the curves are never known at all points $t \in [T_1, T_2]$, they are available only at some specific points $t_{j,n}$ which can be different for different curves x_n. Many important applications of FDA deal with situations where the number of points, $\{t_{j,n}\}$, per curve is small, for example a single digit number. Such cases require custom methodology which will be presented in Chapter 7. In this chapter, we focus on situations where the number of points per curve is large. In either case, a fundamental idea of FDA is that the objects we wish to study are smooth curves

$$\{x_n(t) : t \in [T_1, T_2], \ n = 1, 2, \ldots, N\}, \tag{1.2}$$

for which the values $x_n(t)$ exist at any point t, but are observed only at selected points $t_{j,n}$. For example, in a medical treatment, one may be interested in the

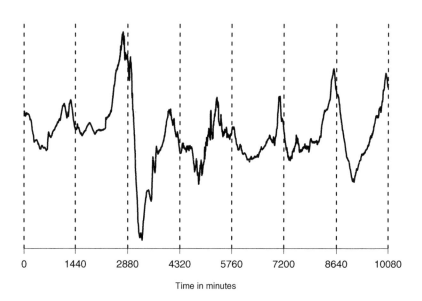

FIGURE 1.1: The horizontal component of the magnetic field measured at Honolulu magnetic observatory from 1/1/2001 00:00 UT to 1/7/2001 24:00 UT. The vertical dashed lines separate 24h days. Each daily curve is viewed as a single functional observation.

concentration $x_n(t)$ of a certain protein in the blood of patient n at time t. The number $x_n(t)$ exists at any time t, but its value is measured only at those times when the patients blood is tested, possibly twice a year. Functional data for which only a few observations are available for every curve are referred to as *sparse*. At the other end of the spectrum, we may consider the strength, $x_n(t)$, of the magnetic field measured at some location, for example at the Honolulu magnetic observatory, at time t on day n, see Figure 1.1. Again, $x_n(t)$ exists at any time, but its value is available only at specific times. In this case, a digital magnetometer records the values every five seconds, resulting in 17,280 observations per day. Such functional data are referred to as *densely* observed. Of course, there is no obvious cut off point between sparsely and densely observed curves, and there are many situations which are difficult to classify in this way. Still another interesting example is provided by high frequency financial data. Figure 1.2 shows the values of Microsoft stock minute by minute. In fact, shares of large companies are traded up to a thousand times every second, so such functions can be very densely observed. Conceptually, it is not clear if the price exists between trades, but the usual mathematical framework for price data uses continuous time stochastic differential equations, so the price function is defined for every t.

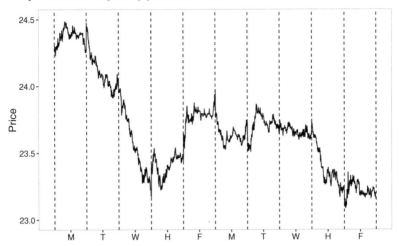

FIGURE 1.2: Microsoft stock prices in one-minute resolution, May 1-5, 8-12, 2006. The closing price on day n is not the same as the opening price on day $n + 1$. The displayed data can be viewed as a sample of 10 functional observations.

A major focus in FDA is the shape of the observed functions or of functions which summarize the properties of the data in some specific way. For example, it may be of interest how a certain treatment affects the mean protein level of patients. One would be interested in constructing a curve which describes how this level changes over time in a group of patients. For magnetometer data, space physics researchers are not interested in the value of the field every five seconds, but in the shape of the curve over several days. The shapes of such curves at many locations allow researchers to make inferences about physical processes occurring in near Earth space where no instruments are placed.

1.1 Basis expansions

Typically, the first step in working with functional data of the form (1.1) is to express them by means of a *basis expansion*

$$X_n(t) \approx \sum_{m=1}^{M} c_{nm} B_m(t), \quad 1 \leq n \leq N. \tag{1.3}$$

In (1.3), the B_m are some standard collection of *basis functions*, like splines wavelets, or sine and cosine functions; we will discuss them in the following. Expansion (1.3) reflects the intuition that the data (1.1) are observations from

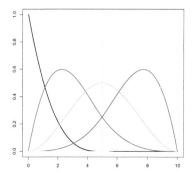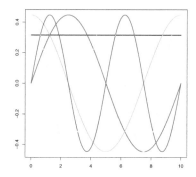

FIGURE 1.3: Plots of the first five B–splines (left) and Fourier (right) basis functions.

smooth functions that share some shape properties, and so can be approximated as linear combinations of some M basic shapes B_m, with M being typically smaller than the number of observed time points, J_n. If the number of points t_{jn} is very large, as for the magnetometer or high frequency financial data, expansion (1.3) serves the practical purpose of replacing the original scalar data $X_n(t_{jn})$ by a smaller collection of the coefficients c_{nm}. If the time points t_{jn} differ between subjects, the expansion puts the curves into a common domain, so that they are more readily comparable. For each n, the curve x_n is represented by the column vector $\mathbf{c}_n = [c_{n1}, c_{n2}, \ldots, c_{nM}]^T$ of dimension M. Expansion (1.3) can also be viewed as a preliminary smoothing of the curves; if the basis functions are smooth, as is the case for most commonly used basis systems, then the right–hand side of (1.3) will inherit this smoothness. However, additional and more targeted smoothing, say for parameter estimates, can be done later on.

We now discuss R implementation. The fda package in R was originally designed to accompany the book of Ramsay and Silverman (2005). In it lie the building blocks to carry out nearly all of the methods discussed in this book. In addition to the fda package, the refund package has emerged with a large collection of flexible tools which we will discuss later on. After loading the package fda using the RGui interface, use the following code

```
spline.basis=create.bspline.basis(rangeval=c(0,10), nbasis=5)
plot(spline.basis, lty=1, lwd=2)
```

to produce Figure 1.3 whose left panel shows five B-spline basis functions defined on the interval $[0, 10]$. There are additional arguments which specify the degree of smoothness and the position of the basis functions, we use default values. The code

```
fourier.basis=create.fourier.basis(rangeval=c(0,10), nbasis=5)
```

```
plot(fourier.basis, lty=1, lwd=2)
```

produces the right panel of Figure 1.3 which shows the first five Fourier basis functions. The first function in the Fourier basis is the constant function, then there are sine and cosine functions of period equal to the length of the interval. These are followed by sine and cosine functions of decreasing periods or, equivalently increasing frequencies. An odd number of basis functions is used: the constant function and sine/cosine pairs. It is important to keep in mind that the Fourier system is usually only suitable for functions which have approximately the same values at the beginning and the end of the interval; this is discussed in greater detail in Section 10.2.

The following example illustrates the construction of a basis expansion using simulated data. This allows us to focus on essential aspects without going into preprocessing of real data. Data examples are presented later in this chapter.

EXAMPLE 1.1.1 [B-spline expansion of the Wiener process] The Wiener process, also known as the Brownian motion, is defined in Section 11.4. It is an appropriate limit of the random walk

$$S_i = \frac{1}{\sqrt{K}} \sum_{k=1}^{i} N_k, \quad N_k \sim iid \ N(0,1), \quad 1 \le k \le K.$$

Figure 1.4 shows a trajectory of the random walk which can be viewed as a function defined on interval $[0, K]$ by $X(t_i) = S_i$, $t_i = i$. Notice some similarity to the price data in Figure 1.2. In the code below, the function smooth.basis is used to create a functional data object Wiener.fd which contains the expansion coefficients c_{nm} as well as information about the basis functions used to expand the random walk function. The following code was used to generate Figure 1.4:

```
Wiener=cumsum(rnorm(10000))/100 # random walk on [0,K], K=10^4
plot.ts(Wiener, xlab="", ylab="")
B25.basis=create.bspline.basis(rangeval=c(0,10000), nbasis=25)
Wiener.fd=smooth.basis(y=Wiener, fdParobj=B25.basis)
lines(Wiener.fd, lwd=3)
```

□

Chapter 5 of Ramsay *et al.* (2009), and several chapters in Ramsay and Silverman (2005), describe more sophisticated methods of constructing functional data objects from raw data. These include smoothing with a roughness penalty and smoothing which preserves monotonicity or positiveness of the data. The last two types of smoothing are easy to understand conceptually. In many studies, data are measurements of the size of an individual or an organ which cannot decrease with time. Similarly, some quantities can only be nonnegative. If one applies a basis expansion (1.3) to such data, one can sometimes obtain curves which are decreasing or negative on small intervals.

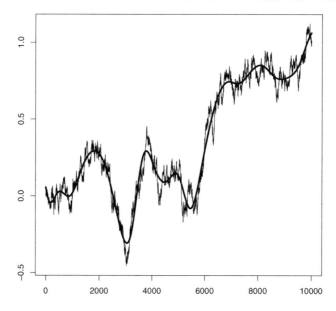

FIGURE 1.4: Random walk and its expansion (1.3) using 25 B-spline basis functions.

Customized R functions can be used to ensure that the output functions are monotone or nonnegative.

1.2 Sample mean and covariance

We now assume that the raw data have been converted to functional objects by a suitable basis expansion, possibly with additional smoothing, and we can work with functional data of the form (1.2). The simplest summary statistics are the pointwise mean and the pointwise standard deviation:

$$\bar{X}_N(t) = \frac{1}{N} \sum_{n=1}^{N} X_n(t), \quad \mathrm{SD}_X(t) = \left\{ \frac{1}{N-1} \sum_{n=1}^{N} (X_n(t) - \bar{X}_N(t))^2 \right\}^{1/2}.$$

In Example 1.2.1 we extend Example 1.1.1 to illustrate these concepts.

EXAMPLE 1.2.1 [Pointwise mean and SD] We simulate a sample of $N = 50$ random walks and convert them to functional objects, as explained in Example 1.1.1. We then plot them, calculate the pointwise mean and SD, and

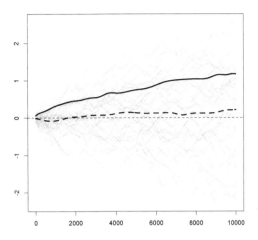

FIGURE 1.5: Random walks converted to functional objects together with the pointwise sample SD (thick continuous line) and the pointwise mean (thick dashed line).

add those to the plot, see Figure 1.5. The following code was used to produce Figure 1.5. The output of the function `smooth.basis` is a list. The values of the functions are extracted using `$fd`.

```
N=50
W.mat=matrix(0, ncol=N, nrow=10000)
for(n in 1:N){W.mat[, n]=cumsum(rnorm(10000))/100}
B25.basis=create.bspline.basis(rangeval=c(0,10000), nbasis=25)
W.fd=smooth.basis(y=W.mat, fdParobj=B25.basis)
plot(W.fd, ylab="", xlab="",col="gray",lty=1)
W.mean=mean(W.fd$fd)
W.sd=std.fd(W.fd$fd)
lines(W.sd, lwd=3); lines(W.mean, lty=2, lwd=3)
```

□

The pointwise sample standard deviation gives us an idea about the typical variability of curves at any point t, but it gives no information on how the values of the curves at point t relate to those at point s. An object which is extensively used in FDA is the *sample covariance function* defined as

$$\hat{c}(t, s) = \frac{1}{N-1} \sum_{n=1}^{N} \left(X_n(t) - \bar{X}_N(t)\right) \left(X_n(s) - \bar{X}_N(s)\right).$$

The interpretation of the values of $\hat{c}(t, s)$ is the same as for the usual variance-covariance matrix. For example, large values indicate that $X_n(t)$ and $X_n(s)$

tend to be simultaneously above or below the average values at these points. Example 1.2.2 shows how to calculate the bivariate function $\hat{c}(t,s)$, which is a bivariate functional data object, `bifd`.

EXAMPLE 1.2.2 [Sample covariance function] We consider the 50 random walks converted to functional objects in Example 1.2.1. The following code computes the sample covariance functions and generates Figure 1.6. In this case, the contour plot is particularly interesting. We will see in Section 11.4, Proposition 11.4.1, that the functional population parameter which the function $\hat{c}(t,s)$ estimates is given by $c(t,s) = \min(t,s)$.

```
#Use the object W.fd generated in the previous example.
W.cov=var.fd(W.fd$fd)  # $fd extracts function values
grid=(1:100)*100
W.cov.mat=eval.bifd(grid, grid, W.cov)
persp(grid, grid, W.cov.mat, xlab="s",
        ylab="t", zlab="c(s,t)")
contour(grid, grid, W.cov.mat, lwd=2)
```

□

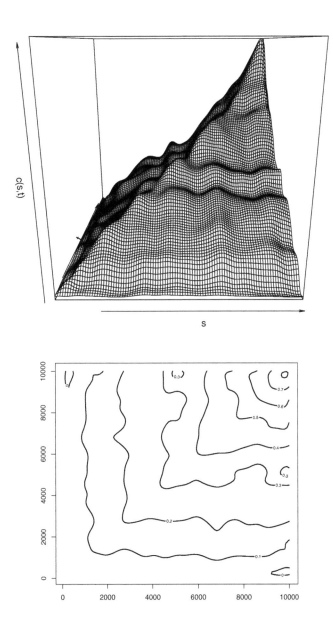

FIGURE 1.6: A perspective plot (top) and contour plot (bottom) of the covariance function of the sample of 50 random walks from Example 1.2.2.

1.3   Principal component functions

One of the most useful and often used tools in functional data analysis is the principal component analysis. Estimated functional principal components, EFPC's, are related to the sample covariance function $\hat{c}(t, s)$, as will be explained in Sections 11.4 and 12.2. Here we only explain how to compute and interpret them.

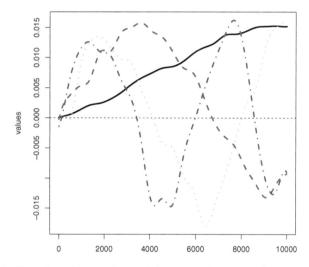

FIGURE 1.7: The first four estimated functional principal components of the 50 random walks from Example 1.2.2.

In expansion (1.3), the basis functions B_m are fixed. The idea of the functional principal component expansion is to find functions $\hat{v}_j$ such that the centered functions $X_n - \bar{X}_N$ are represented as

$$X_n(t) - \bar{X}_N(t) \approx \sum_{j=1}^{p} \hat{\xi}_{nj} \hat{v}_j(t)$$

with p much smaller than M in (1.3). The EFPC's $\hat{v}_j$ are computed from the observed functions $X_1, X_2, \ldots, X_N$ after converting them to functional objects. Figure 1.7 shows the EFPC's $\hat{v}_j, j = 1, 2, 3, 4$, computed for the 50 smoothed trajectories of the random walk. The $\hat{v}_j$ resemble trigonometric functions, and we will see in Sections 11.4 and 12.2 that they indeed are estimates of specific trigonometric functions whose closed form can be found analytically. The first EFPC, $\hat{v}_1$, plotted as the continuous black line, shows the most pronounced pattern of the deviation from the mean function of a randomly selected trajectory. A cursory examination of Figure 1.5 confirms that the shape of $\hat{v}_1$ indeed summarizes the main pattern of variability around the mean function well. For each curve X_n, the coefficient $\hat{\xi}_{n1}$ quantifies the contribution of $\hat{v}_1$ to its shape. The coefficient $\hat{\xi}_{nj}$ is called the *score* of X_n with respect to $\hat{v}_j$. The second EFPC, plotted as the dashed red line, shows the second most important mode of departure from the mean functions of the 50 random walk curves. The EFPC's $\hat{v}_j$ are *orthonormal*, in the sense that

$$\int \hat{v}_j(t)\hat{v}_i(t)dt = \begin{cases} 0 & \text{if } j \neq i, \\ 1 & \text{if } j = i. \end{cases}$$

This is a universal property of the EFPC's which restricts their interpretability. For example, $\hat{v}_2$ is the second most important mode of variability which is *orthogonal* to $\hat{v}_1$.

We will see in Chapter 11 that the total variability of a sample of curves about the sample mean function, can be decomposed into the sum of variabilities explained by each EFPC. For the sample of the 50 random walks, the first EFPC $\hat{v}_1$ explains about 81% of variability, the second about 10%, the third about 4% and the fourth about 2%. Together the first four EFPC's explain over 96% of variability. This justifies the expansion using $p = 4$, or even $p = 2$, as the contribution of the remaining components to the shape of the curves is small. The percentage of the variability explained by $\hat{v}_j$ is related to the size of the scores $\hat{\xi}_{nj}$; the smaller the percentage, the smaller the scores. The following code was used to produce Figure 1.7 and compute the percentage of variability explained by each EFPC.

```
W.pca = pca.fd(W.fd$fd, nharm=4)
plot(W.pca$harmonics, lwd=3)
W.pca$varprop
[1] 0.80513155 0.09509154 0.04099496 0.02119239
```

In the remaining two sections of this chapter we further illustrate the concepts introduced so far by examining specific data applications.

1.4 Analysis of BOA stock returns

Bank of America, ticker symbol BOA, is one of the largest financial institutions in the world and has a history going back over one hundred years. We consider stock values, recorded every minute, from April 9th, 1997 to April 2nd, 2007. Each trading day begins at 9:30 AM (EST) and ends at 4:00 PM, resulting in 6.5 hours of trading time. Thus we can take $t \in (0, 6.5)$. This results in 2511 days worth of data, with each day consisting of 390 measurements. The functional observation we consider is the *cumulative log-return*. If $P_n(t)$ is the

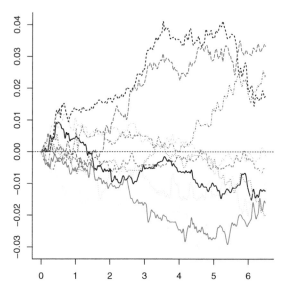

FIGURE 1.8: Plot of first ten cumulative log returns for BOA.

value of the stock on day n at time t, then the cumulative log-return is given by

$$R_n(t) := \log(P_n(t)) - \log(P_n(0)) \approx \frac{P_n(t) - P_n(0)}{P_n(0)}.$$

The function $R_n(t)$ depicts how an investment, made at opening, evolves over the course of the day. The first 10 such days are given in Figure 1.8. There is a substantial outlier that occurs on August 26th, 2004, which is due to a stock split and is thus discarded from further analysis. The plot is generated using the R commands:

```
BOA<-read.table("BOA.txt",header=TRUE)
Dates<-dimnames(BOA)[[1]]
BOA<-data.matrix(BOA)
Outlier<-which(Dates=="08/26/2004")
BOA<-BOA[-Outlier,]
N<-dim(BOA)[1]
M<-dim(BOA)[2]
Times<-seq(0,6.5,length=M)
log_BOA<-log(BOA) - matrix(log(BOA)[,1],nrow=N,ncol=M)
bspline_basis<-create.bspline.basis(rangeval=c(0,6.5),norder
    =4,nbasis=200)
log_BOA_f<-Data2fd(Times,t(log_BOA),basisobj = bspline_basis)
plot(log_BOA_f[1:10],xlab="",ylab="",lwd=1.5)
```

In Figure 1.9 we plot the mean function and, using the standard deviation function, we include point-wise 95% confidence intervals. The returns have a small positive mean function which increases rapidly early in the day and

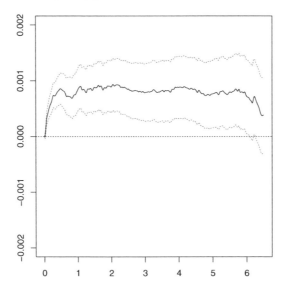

FIGURE 1.9: Plot of mean function for BOA cumulative returns. Point-wise 95% confidence intervals are included in red.

then levels off. Given that stock holders expect a positive return on their investment, but that each day resets at 0, this seems reasonable. The plot was generated using the following R commands.

```
muhat<-mean.fd(log_BOA_f)
sdhat<-sd.fd(log_BOA_f)
SE_hat_U<-fd(basisobj=bspline_basis) # create upper CI bound
SE_hat_L<-fd(basisobj=bspline_basis) # create lower CI bound
SE_hat_U$coefs<-2*sdhat$coefs/sqrt(N) + muhat$coefs
SE_hat_L$coefs<- -2*sdhat$coefs/sqrt(N) + muhat$coefs
plot.fd(SE_hat_U,ylim=c(-0.002,0.002),col='red',lty=2,xlab="",
    ylab="")
plot.fd(SE_hat_L,add=TRUE,col='red',lty=2)
plot.fd(muhat,add=TRUE)
```

The confidence bounds are obtained by adding and subtracting $2N^{-1/2}SD(t)$ from the mean function. The code accomplishes this by using the coordinates of the B-splines basis expansion. Many FDA methods can be applied in the same way; one carries out various FDA manipulations on the basis coefficients.

A plot of the first four EFPCs is given in the left panel of 1.10. In the right panel we also include the first four FPCs of the covariance operator of standard Brownian motion, discussed in the previous section. We see that the shapes are nearly the same. One of the fundamental assumptions in many financial tools is that the log returns can be modeled using Brownian motion (or some variation of it), which agrees with these results.

We finish by offering a third tool for visualizing the covariance surface.

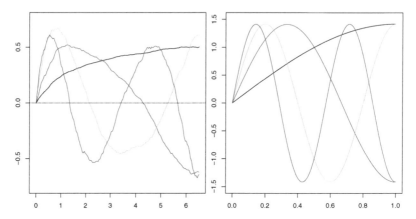

FIGURE 1.10: Plot of first four PCs of BOA cumulative returns (left) and first four eigenfunctions from Brownian motion (right) .

In Figure 1.11 we have added a heat map onto the perspective plot, with red indicating high covariance and blue indicating low covariance. This was constructed using the *persp3D* function in the *plot3D* package in R. We see that the plot is similar to Figure 1.6, which reflects what we found when examining the PCs: cumulative returns show behavior similar to Brownian motion.

1.5 Diffusion tensor imaging

Diffusion tensor imaging, DTI, is a magnetic resonance imaging methodology which is used to measure the diffusion of water in the brain. Water diffuses isotropically (i.e. the same in all directions) in the brain except in white mater where it diffuses anisotropically (i.e. differently in different directions). This allows researchers to utilize DTI to generate images of white matter in the brain. Understanding the structure of the brain is important for a wide range of neurological conditions and diseases including Multiple Sclerosis. The data we examine here were collected at Johns Hopkins University and the Kennedy-Krieger Institute (Goldsmith *et al.*, 2012a; Goldsmith *et al.*, 2012b).

We consider fractional anisotropy tract profiles of the corpus callosum, a portion of the data DTI data set in the refund R package. The corpus collosum, the largest white matter structure in the brain, is a bundle of nerve fibers connecting the two hemispheres of the brain. Fractional anisotropy is a value between 0 and 1 which measures the level of anisotropy, and therefore the quantity of white matter, at a particular location. A total of 376 patients are considered, with each tract measured at 93 equally spaced locations. The

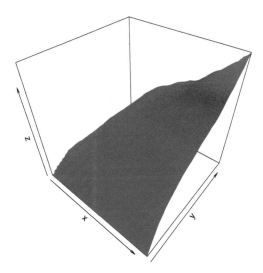

FIGURE 1.11: A perspective plot with a heat map of the covariance surface estimated from the BOA data.

arguments $t_{j,n}$ thus denote a spatial rather than temporal location. The resulting curves are shown in Figure 1.12. To illustrate the spread of the data, we plot green, yellow, and red dashed lines representing one, two, and three standard deviations away from the mean (point-wise) respectively. We will return to this data set further on for deeper analyses, but for now we mention that due to the high within subject correction, the data follows an analog of the classic three standard deviation rule surprisingly well with nearly all curves falling between the red lines.

The following code was used to produce Figure (1.12).

```
library(refund)
data(DTI)
Y<-DTI$cca
Y<-Y[-c(126,130,131,125,319,321),] # missing values
N<-dim(Y)[1]; M<-dim(Y)[2]
argvals<-seq(0,1,length=M)
data_basis<-create.bspline.basis(c(0,1),nbasis=10)
Y.f<-Data2fd(argvals,t(Y),data_basis)

dev.new(width=8,height=6)
plot(Y.f,lty=1,col="gray",xlab="",ylab="",ylim=c(0.1,.9))
lines(mean(Y.f),lwd=2)
lines(mean(Y.f)+std.fd(Y.f),lwd=2,lty=2,col="green")
lines(mean(Y.f)+2*std.fd(Y.f),lwd=2,lty=2,col="yellow")
lines(mean(Y.f)+3*std.fd(Y.f),lwd=2,lty=2,col="red")
lines(mean(Y.f)-std.fd(Y.f),lwd=2,lty=2,col="green")
```

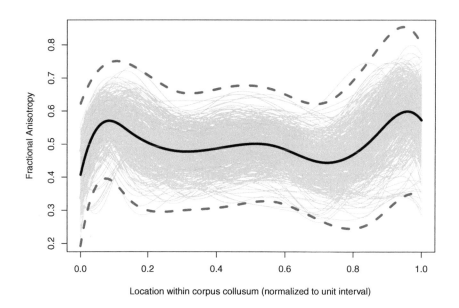

FIGURE 1.12: Plot of fractional anisotropy tract profiles of the corpus collosum. Individual tracts are plotted in gray, while the mean function is plotted in black. Green, yellow, and red dashed lines indicate, respectively, one, two, and three standard deviaitions (point-wise) from the mean.

```
lines(mean(Y.f)-2*std.fd(Y.f),lwd=2,lty=2,col="yellow")
lines(mean(Y.f)-3*std.fd(Y.f),lwd=2,lty=2,col="red")
```

1.6 Chapter 1 problems

1.1 The `pinch` is a dataset included in the `fda` package. It consists of 151 measurements of pinch force for 20 replications (curves).

(a) Convert the pinch data to functional objects using 15 B-splines of order four (cubic splines) and plot the 20 smoothed curves on one graph.

(b) Calculate the pointwise mean and SD and add them to the plot.

(c) Graph the perspective and contour plots of the sample covariance function $\hat{c}(t, s)$ of the pinch curves.

(d) Graph the first four EFPC's of the pinch data. How many components do you need to explain 90% of variation?

1.2 For this problem, download the R package `fds` and use the data set of the United States Federal Reserve interest rates, `FedYieldcurve`, which contains the monthly interest rates from January 1982 to June 2009. The x–values are the maturity terms of 3, 6 , 12, 60, 84 and 120 months which can be identified with the t_j in this chapter. The y–values are the interest rates of the United States Treasury obligations due in x months which can be identified with the $x_n(t_j)$, where n is a month in the range January 1982 to June 2009.

(a) On one graph, plot the interest rates $x(t_j)$ for January 1982 and for June 2009 against the maturity terms t_j. How do the interest rates in these two months compare? Use the following code:

```
library(fds);   library(fda)
yield = FedYieldcurve;   terms = yield$x
plot(terms, yield$y[,1], pch=15, ylab="Yield", ylim=c(0,16))
points(terms, yield$y[,330], pch=16)
```

(b) Convert the yield data to functional objects using bspline basis with four basis functions. Calculate and plot the the mean yield function. What is the average behavior of interest rates as a function of the maturity?

(c) Plot the first principal component of the interest rate curves. What percentage of variance does this component explain? Interpret the plot and the percentage of variance.

1.3 El Nino and La Nina are disruptions in the ocean-atmosphere system in the Tropical Pacific that cause changes in sea surface temperatures (SST's) and have important consequences for global weather and climate. El Nino is characterized by unusually warm SST's while La Nina has unusually cool SST's. Data for monthly SST (in degrees C) in various regions of the south Pacific ocean from 1950 to 2013 are available in the file `ninoSST.csv`. Using the `NINO3` column, treat data for each calendar year as a functional observation. The following code creates the data matrix with 64 functional observations (years):

```
nino = read.csv("S:/G/TH/NB/ninoSST.csv", header = T)
```

```
library(fda)
year = 1950:2013
month = 1:12
SSTmat = matrix(data = NA, nrow = 12, ncol = 64)
count = 1
for(yr in year)
{
        tmp = nino[which(nino$YEAR == yr),]$NINO3
        SSTmat[,count] = tmp
        count = count+1
}
```

Perform the following steps using the package `fda`

(a) Using five Fourier basis functions, convert the data into a functional object `ninofd` containing 64 annual curves and plot these smoothed curves. Add the mean curve in thick black using the code

```
lines(mean(ninofd), col = "black", lwd = 3)
```

(b) The years of 1965, 1972, 1982, and 1997 were pronounced El Nino years. Using the graph you have created in part (a), emphasize these curves in the plot by making them thick red. Emphasize the pronounced La Nina years of 1955, 1973, 1975, 1988, and 1999 with thick blue. Start with the following code:

```
#Pronounced ElNino Years - characterized by unusually warm SST
eln = c(1965, 1972, 1982, 1997)
#Pronounced La Nina Years - characterized by unusually cold
    SST
lan = c(1955, 1973, 1975, 1988, 1999)
elnn = match(eln,year)
lann = match(lan,year)
ELNmat = SSTmat[,elnn]
LANmat = SSTmat[,lann]
```

1.4 This problem illustrates the concept of a *functional boxplot* using the curves of cumulative returns of 30 companies comprising the Dow Jones Industrial Average (DJIA) stock market index in 2013.

(a) Download the data:

```
library("fda")
d = read.csv("Dow_companies_data.csv")
data = d[,2:31]
```

The first column of the object `d` is the date; there were 252 trading days in 2013. The 252 by 30 matrix `data` contains the daily closing prices of the 30 stocks.

Plot the stock value of Exxon–Mobil, ticker symbol XOM. For the same stock, calculate and plot the cumulative return function defined as

$$R_n = 100 \frac{P_n - P_1}{P_1}, \quad 1 \le n \le 252,$$

where P_n is the closing price on day n, starting from the beginning of the year. By how much, in percent, did the Exxon–Mobil stock price increase in 2013?

(b) Create a 252 by 30 matrix `cr` that contains the cumulative returns on all stocks in DJIA. Plot all the cumulative return functions in one plot. Add the mean and median functions in color.

(c) A useful way to visualize the behavior of cumulatve returns is to use a functional boxplot, which can be obtained using the function `fbplot` in the `fda` package. Functional boxplots are analogous to standard boxplots; however, they treat each function as a single data point. Use the following code to display a functional boxplot for the cumulative return data:

```
fbplot(fit = cr, ylim = c(-20,90), xlab="Trading day",
            ylab = "Cumulative return")
```

Functional boxplots are not obtained from pointwise boxplots for each day; a measure of centrality for each function compared to the other functions is used, see Sun and Genton (2011). In the boxplot produced by the above code, the black line is the median curve. The magenta region represents the middle 50% of curves, and the dashed red curves are the outlying functions. Note that functional boxplots produce outlying curves rather than individual outlying returns for each day.

After consulting the help file, produce a functional boxplot which shows three regions corresponding to probability levels of 30, 60 and 90 percent, using a dark color for the central 30 percent and fading colors for the 60 and 90 percent of the curves.

1.5 The Matérn covariance function leads to a very general family of stationary Gaussian processes. The form of the covariance function is

$$C(t,s) = \frac{\sigma^2}{\Gamma(\nu)2^{\nu-1}} \left(\frac{\sqrt{2\nu}|t-s|}{\rho} \right)^\nu K_\nu \left(\frac{\sqrt{2\nu}|t-s|}{\rho} \right), \qquad \nu > 0, \qquad (1.4)$$

where σ^2 is the variance parameter, ν the smoothness parameter, and ρ the range parameter. $K_\nu(\cdot)$ is the modified Bessel function of the second kind. The paths of the Matérn process are k times continuously differentiable for any $\nu > k$, with probability one.

1. Simulate and plot 100 iid mean zero Matérn processes with $\nu = 1/2, \nu = 2$, and $\nu = 4$. Set $\sigma^2 = 1$ and $\rho = 1$. Use a temporal grid with 50 evenly spaced points. Do not use an R package to do this directly. Program it using the *besselK()* function and using it to define the covariance matrix on the grid of time points.

2. Plot the first four EFPCs for each value of ν, comment on any similarities/differences.

3. Plot the explained variance for each ν, comment on any similarities/differences.

4. Using your preferred method, create a plot of the covariance surface for each ν, comment on any similarities/differences.

1.6 In this section we expressed $x_n(t) = \sum_{m=1}^{M} c_{nm} B_m(t)$. One can show that

$$\bar{x}_N(t) = \sum_{m=1}^{M} a_m B_m(t),$$

and

$$\hat{c}(t, s) = \sum_{m=1}^{M} \sum_{k=1}^{M} b_{mk} B_m(t) B_k(s)$$

Find expressions for the coefficients a_m and b_{mk} assuming that the $B_m(t)$ are an orthonormal basis:

$$\int B_m(t) B_k(t) = 1_{m=k}.$$

2

Further topics in exploratory analysis of functional data

In Chapter 1 we discussed motivating examples and some fundamental concepts in FDA. Here we present further topics which, while still fundamental for FDA, are more complex than those introduced in Chapter 1. The concepts discussed here, which do not have direct scalar analogues, include derivatives, penalized smoothing, and curve alignment. We aim only to introduce the reader to the fundamental ideas behind each of these concepts. Additional details can be found in Ramsay *et al.* (2009).

2.1   Derivatives

Incorporating information about derivatives is a distinctive feature of FDA, as compared to traditional multivariate analysis. Once an observation, say $x_n(t)$, has been expressed using a particular basis, say $B_m(t)$, often an easy approximation to the kth order derivative function can be obtained as

$$x_n^{(k)}(t) \approx \sum_{m=1}^{M} c_{nm} B^{(k)}(t).$$

However, there are at least two caveats to the above. The first is that for such a series approximation to hold, one requires some assumptions about the path $x_n(t)$, for example that it is k times continuously differentiable. The second is that the chosen basis functions must have at least k derivatives. For series such as Fourier this is not an issue as they are infinitely differentiable. However, a B–spline basis only posses a number of derivatives equal to its order (and the last derivative is not continuous). One should employ B–splines of a higher order than cubic if going beyond one derivative. After taking a derivative, the function could be re-expressed in terms of the original basis, and this is the approach used in the `deriv.fd` function in the `fda` package.

Derivatives of Fourier series and B–splines take a simple form. The derivative of a cosine (sine) function is simply a constant times a sine (cosine) function. Likewise, the derivative of a B–spline basis function is a constant times

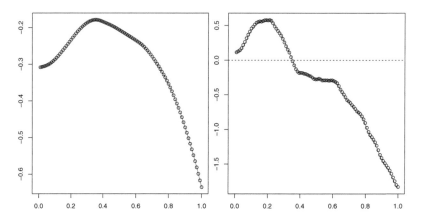

FIGURE 2.1: Circles indicate of one realization of a Matérn process (left) and its numeric derivative (right). The red line in left panel indicates a b–splines curve with 50 basis functions, and the red line in the right panel is its derivative.

another B–spline basis function of a lower order (for example, the derivative of a cubic B–spline function is a quadratic B–spline).

   The following example presents the construction of derivatives for sim-ulated data. To ensure the data possess a derivative we simulate using the Matérn process instead of Brownian motion. An example considering growth curves is given in Section 2.3.

EXAMPLE 2.1.1 The Matérn process is a Gaussian process defined in (1.4). When the smoothness parameter, ν, is an integer plus $1/2$, its covariance takes on a simpler form. For example, when $\nu = 5/2$ and taking all other parameters equal to one, we get that

$$c(t, s) = \left(1 + \sqrt{5}|t - s| + (5/3)|t - s|^2\right) \exp\left(-\sqrt{5}|t - s|\right).$$

In Figure 2.1 we simulate the above Matérn process on 100 evenly spaced time points on $[0, 1]$. The right panel shows the derivative computed using B–splines as well as a traditional numeric derivative (i.e. scaled using differences). We see that the two derivatives are very closely aligned.

□

2.2   Penalized smoothing

In Section 1.1, we explained how to represent functional objects using the basis expansion (1.3). This section is concerned with a very popular approach to smoothing which imposes a penalty on functions that are too "wiggly." Smoothing using expansion (1.3) is recommended in situations when the original data curves are already relatively smooth. Using sufficiently large M changes the curves somewhat, but they essentially look the same as the original data curves. The practical effect is that discrete data points have been converted to functional data objects in R, which can be conveniently manipulated using R functions that take functional objects as arguments. However, if the raw data curves exhibit a substantial level of noise, the functional objects constructed in this way (with large M) will inherit this variability, resulting in very wiggly curves. Based on our knowledge of the data generating mechanism, we often choose to treat the wiggles as the reflection of unimportant features, e.g. measurement error or random variability due to other sources. For example, in climate studies it is assumed that the precipitation at every location follows an annual pattern specific to that location. Northern hemisphere locations far away from oceans typically receive more precipitation in spring and summer, and less in winter months; many coastal locations have almost the same average precipitation in every month, other coastal locations receive most rain in winter. Daily precipitation records extend many decades back, but even when averaged over many decades, there remains a great deal of variability from day to day. For example, the average precipitation on June 30 may be visibly higher than on July 1, but it is unreasonable to believe that there is a fundamental difference between these two days; the observed difference is due to random (sample) variability. A simple approach to smoothing is to use a relatively small M in (1.3). A disadvantage of this approach is that the resulting smooth functions are always linear combinations of the M basis functions, which restrict their shape. To avoid this problem, penalized smoothing uses a large number of basis functions, one can use more basis functions than the number of time points $\{t_j\}$ at which the functions are observed, and then smooths the curves based on criteria which are suitable for the problem at hand. We now explain the basics of this approach. Curve smoothing, including penalized smoothing, has been a subject of extensive research; an interested reader is referred to Ruppert *et al.* (2003).

Suppose we observe values y_j at points $t_j \in [T_1, T_2]$. The values y_j are those denoted $x_n(t_{j,n})$ in (1.1), we suppress the curve index n to focus attention on a single observed noisy curve. The observed data points for this curve are thus (t_j, y_j). We assume that there is a smooth curve $x(t), t \in [T_1, T_2]$ and that $y_j = x(t_j) + \varepsilon_j$, where ε_j is a random error whose expected value is zero. The goal of smoothing is to eliminate the contribution of the ε_j, and obtain an estimate of $x(t)$. Once again, we aim to approximate $x(t)$ using

$$x(t) \approx x_K(t) = \sum_{k=1}^{K} c_k B_k(t),$$

with K often much larger than the M in (1.3); K can be equal to or greater than the number of the points t_j. We introduce a linear differential operator

L such that

$$L(x)(t) = \alpha_0(t)x(t) + \alpha_1(t)x^{(1)}(t) + \cdots + \alpha_m(t)x^{(m)}(t).$$

In other words $L(x)$ is a linear combination of m derivatives of x. For example, L may be the second derivative, so that $L(x)(t) = x^{(2)}(t)$. We want to find a curve x, equivalently the coefficients c_k, which minimize the penalized sum of squares defined as

$$\text{PSS}_\lambda(c_1, c_2, \ldots, c_K) = \sum_j (y_j - x_K(t_j))^2 + \lambda \int_{T_1}^{T_2} [L(x_K)(t)]^2 \, dt. \qquad (2.1)$$

The parameter λ is called the *smoothing parameter*. If $\lambda = 0$, the penalty term does not contribute to the sum of squares, and one ends up with the same expansion as in (1.3). The procedure will then try to minimize all distances $y_j - x_K(t_j)$, resulting in a very rough approximation to x; if K is sufficiently large, x can pass through all data points (t_j, y_j). As λ increases, the second term becomes more and more dominant, and wiggly curves cannot minimize PSS_λ. If λ is very large, the second term in (2.1) dominates, and the estimated curve x will reflect mostly the penalty operator L. For example, if L is the second derivative, then it will be a curve with $x''(t) = 0$, i.e. a straight line. A main difficulty with penalized smoothing is to find a suitable λ which attains a good balance between overfitting (wiggliness) and oversmoothing. A great deal of research has been done in this direction, we discuss merely a single approach, generalized cross-validation, which gives satisfactory results in many cases, but not always. Whenever an automated method of selecting λ is used, a visual inspection of the resulting curves is recommended to check if they reflect the structure of the data well.

To choose the tuning parameter λ, generalized cross-validation (GCV) is commonly employed. This is mainly due to two attributes: (1) GCV gives a close approximation to cross-validation and (2) it is computationally very efficient. Recall that ordinary cross-validation is the process of leaving out an observation, fitting the model, using the model to predict the omitted observation, and then calculating the difference between the observed and predicted values. This process is done to all data points and the squared errors are averaged to get an unbiased estimated of model fit. However, this can present a serious computational burden and thus GCV is often preferred. GCV attempts to approximate the cross-validation calculations without having to recompute parameters as the left out subject varies. It is therefore very fast, though the results should always be examined to make sure the estimates are reasonable. GCV can be implemented in the same fashion for any linear estimator. That is, suppose that for data $Y \in \mathbb{R}^N$ a model is fit with a particular tuning parameter λ. If the predicted values can be expressed as

$$\hat{Y} = \mathbf{H}_\lambda Y,$$

where $\mathbf{H}_\lambda$ is an $N \times N$ matrix, then the GCV score can be calculated as

$$\frac{N^{-1} \sum_{n=1}^{N} (Y_n - \hat{Y}_n)^2}{(1 - N^{-1} \operatorname{trace}(\mathbf{H}_\lambda))^2}.$$

We will return to this formulation often as GCV is used quite frequently for choosing tuning parameters.

EXAMPLE 2.2.1 We return to Example 2.1.1, but now with an additional iid error added to each observed value. We take the error to be $N(0, 1/16)$. In Figure 2.2 we plot the observed values in the left panel (blue) and the values without the additional noise (red). In the right panel we plot the b–splines expansion on the data without noise (red), and the data with noise (blue). The purple line represents a smoothed b–splines approximation with linear differential operator L taken to be the second derivative and λ chosen via GCV. We see that while the smoother is not exactly the same as the underlying curve, it has still eliminated most of the variability added by the noise. □

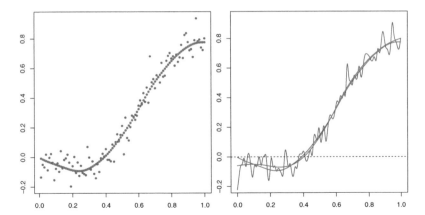

FIGURE 2.2: Left panel plots observations with (blue) and without (red) noise. Right panel plots b–splines expansion on observations with noise (blue) and without (red). The purple line in the right plot a b–splines smoother with penalty on second derivative.

In addition to the second derivative, a penalty operator called the *harmonic acceleration operator* is often used for periodic data. Suppose $T_2 - T_1 = T$, and assume for simplicity that $T_1 = 0$. Suppose that the Fourier basis is used and let $K = 2J + 1$, then $x_J(t)$ takes the form

$$x_J(t) = c_0 + \sum_{j=1}^{J} [a_j \sin(\omega j t) + b_j \cos(\omega j t)], \qquad \omega = \frac{2\pi}{T}.$$

The harmonic acceleration operator is defined by

$$L(x)(t) = \omega^2 x^{(1)}(t) + x^{(3)}(t), \quad \omega = \frac{2\pi}{T}.$$

It is suitable for functions defined on $[0, T]$ such that $x(0) = x(T)$, so that the Fourier expansion displayed above can be used. A calculation using the properties of trigonometric functions shows that

$$\int_0^T [L(x)(t)]^2 \, dt = \pi \omega^5 \sum_{j=2}^J j^2 (j^2 - 1)^2 (a_j^2 + b_j^2). \tag{2.2}$$

Notice that in (2.2) no penalty is imposed on the term $a_1 \sin(\omega t) + b_1 \cos(\omega t)$. As j increases, the squared amplitude of the higher frequency terms, $a_j^2 + b_j^2$, is multiplied by rapidly increasing coefficients $j^2(j^2 - 1)^2$. This forces the coefficients a_j and b_j to be small for large j, thus eliminating the high frequency behavior.

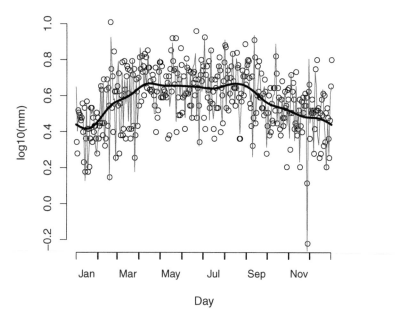

St. Johns

FIGURE 2.3: Average log–precipitation at St. Johns, Canada. The red line is fitted without smoothing. The thick line shows a smooth with the harmonic acceleration penalty and the smoothing parameter λ minimizing the GCV.

EXAMPLE 2.2.2 We illustrate the entire discussed procedure using log–precipitation data. This data was collected across several stations in Canada and is commonly used in the FDA literature to illustrate new methodologies. The code below produces graphs, without any smoothing, for six locations (first 5 out of 35 in the data set). The red line in Figure 2.3 shows the result for the first location in the data set, St. Johns.

```
# set up a saturated basis: as many basis
# functions as observations
nbasis = 365
yearRng = c(0,365)
daybasis = create.fourier.basis(yearRng, nbasis)
logprecav = CanadianWeather$dailyAv[, , 'log10precip']
dayprecfd <- with(CanadianWeather, smooth.basis(day.5,
logprecav, daybasis,
fdnames=list("Day", "Station", "log10(mm)"))$fd )
for(i in 1:5){
plot(logprecav[,i], axes=FALSE, xlab="Day", ylab="log10(mm)",
main=CanadianWeather$place[i])
lines(dayprecfd[i], col=2)
axisIntervals(1)
axis(2)
readline("Press <return to continue")
}
```

We now turn to smoothing using penalty imposed by the harmonic acceleration operator. The following code sets up this operator.

```
Lcoef = c(0, (2*pi/diff(yearRng))^2,0,1)
harmaccelLfd = vec2Lfd(Lcoef, yearRng)
```

Next, we find λ which minimizes the GCV. The initial range of the values of λ can be established through trial and error. In the code below, six values $\lambda = 10^i, 4 \le i \le 9$, are used, and the value which minimizes the GCV is found by plotting it against i. The plot shows that $\lambda = 10^6$ minimizes the GCV.

```
loglam = 4:9
nlam = length(loglam)
dfsave = rep(NA,nlam)
names(dfsave) = loglam
gcvsave = dfsave
for (ilam in 1:nlam) {
cat(paste('log10 lambda =',loglam[ilam],'\n'))
lambda = 10^loglam[ilam]
fdParobj = fdPar(daybasis, harmaccelLfd, lambda)
smoothlist = smooth.basis(day.5, logprecav,
fdParobj)
dfsave[ilam] = smoothlist$df
gcvsave[ilam] = sum(smoothlist$gcv)
}
plot(loglam, gcvsave, type='b', lwd=2)
```

Using the smoothing parameter $\lambda = 10^6$, we can now obtain smoothed log–precipitation curves. For the first location, such a curve is shown as the thick black line in Figure 2.3.

```
lambda = 1e6
fdParobj = fdPar(daybasis, harmaccelLfd, lambda)
logprec.fit = smooth.basis(day.5, logprecav, fdParobj)
logprec.fd = logprec.fit$fd
fdnames = list("Day (July 1 to June 30)",
"Weather Station" = CanadianWeather$place, "Log 10
    Precipitation (mm)")
logprec.fd$fdnames = fdnames
# plot smoothed curves for the first five locations
for(i in 1:5){
plot(logprecav[,i],axes=FALSE,xlab="Day",ylab="log10(mm)",
main=CanadianWeather$place[i])
lines(dayprecfd[i],col=2)
axisIntervals(1)
axis(2)
lines(logprec.fd[i], lwd=3)
readline("Press <return to continue")
}
```

□

2.3   Curve alignment

In many situations, the observed curves have similar shapes, but they are in some way shifted. Figure 2.4 shows a simple example of five sine curves shifted by the same amount. Statistics computed from such samples may be misleading. For example, the sample mean function in Figure 2.4 does not represent the structure of the sample well. The amplitude of the mean is smaller than that of any of the five functions, while its support is longer. In such situations it is advisable to align the curves in the sample in some way and then compute statistics like the mean, the covariance function and the functional principal components. Before aligning, three functional principal components are needed to explain the variability of the five curves in Figure 2.4, respectively, 55, 39 and 5 percent. After aligning, the five curves are identical, and one component trivially explains 100 percent of variance. This is what often happens for real data, after an alignment the variability of the curves can be explained using a smaller number of EFPC's.

As we will discuss further at the end of this section, the variability in a sample curves can be decomposed into two sources. The first is the random curve-to-curve variation we are familiar with; this is called *amplitude variation*.

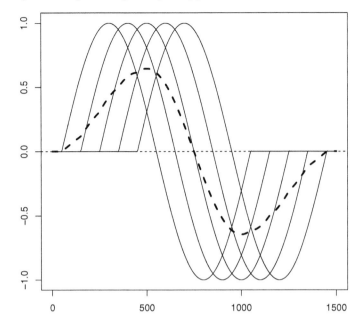

FIGURE 2.4: Five phase shifted sine curves and their mean.

The second, as just discussed, constitutes possible shifting of the curves with respect to the domain; this is called *phase variation.*

A well–known example of a sample of this type are human growth curves. Various phases of growth occur at different ages for different individuals. Every individual can, however, be thought of as having a biological clock which governs various phases of growth, and these phases appear at the same time according to this internal clock. In most situations, phase variability in the sample goes together with amplitude variability. For example, the water level in a mountain stream measured over the spring months provides one curve per year. Depending on the total winter snowfall and the temperature pattern in the spring, such curves will exhibit both amplitude and phase variability. A problem that has received attention is the separation of these two sources of variability. In this section, we address these issues in some detail.

Figure 2.5 shows growth curves of 54 girls. These data are part of the object growth in the fda package. The following code was used to generate Figure 2.5. The code involves penalized smoothing, see Section 2.2

```
age=growth$age
heightBasis=create.bspline.basis( c(1,18), 35, 6, age)
heightPar=fdPar(heightBasis, 3, 10^(-0.5))
heightSmooth=smooth.basis(age, growth$hgtf, heightPar)
```

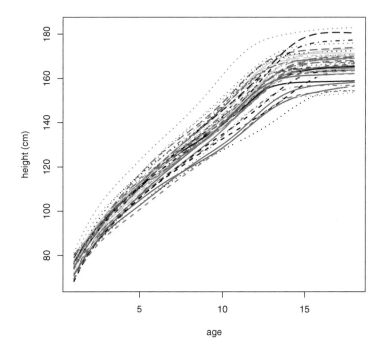

FIGURE 2.5: Smoothed growth curves of 54 girls age 1–18.

```
plot(heightSmooth, lwd=2, xlab="age", ylab="height (cm)")
```

Figure 2.5 shows that individual growth curves differ by level (amplitude), some girls are tall and some are short, and also by the timing of certain landmark events. For example, some girls practically stop growing around the age of 13, while for others this happens past the age of 15. However, it is generally difficult to glean a lot of information directly from the growth curves. It is useful to consider *acceleration curves* defined as second derivatives of the growth curves. These are shown in Figure 2.6, which was obtained with the following code:

```
accelUnreg = deriv.fd(heightSmooth$fd, 2)
plot(accelUnreg[,1], lwd=2, xlab="Age", ylab="Acceleration",
  ylim=c(-4,2))
mean.accelUnreg=mean(accelUnreg)
lines(mean.accelUnreg, lwd=4, col="black")
```

The curves in Figure 2.6 exhibit both amplitude and phase variation. Consider the time the curves cross the zero line for the last time. This is the age when the acceleration drops to zero, so this is the age of the fastest (puber-

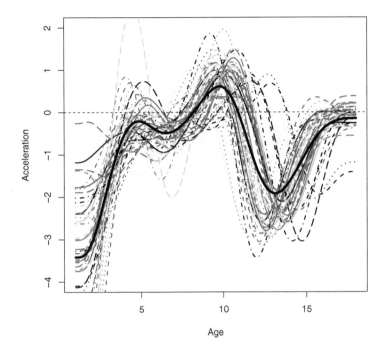

FIGURE 2.6: Acceleration curves of 54 girls with the mean function in bold.

tal) growth before reaching the final height. After that age, the acceleration becomes negative, i.e. the growth slows down. This time corresponds to ages between 10 and 15 years for individuals, with the average of about 11.7 years, roughly when the mean curve crosses zero for the last time. As in the toy example shown in Figure 2.4, the mean curve is not as good a representation of individual acceleration curves, which generally show more amplitude variation over shorter periods of time. We may therefore wish to align the acceleration curves in some way. If our interest is on the time of the fastest pubertal growth, we may use the time when a curve crosses zero for the last time as a *landmark*. The resulting procedure is therefore known as *landmark registration*. We now describe how it works.

A common approach to aligning curves in a sample is to use time warping functions. For each individual n, we assume that there exists a *time warping function* $h_n(t)$ which stretches or shrinks time. Here t denotes physical time while $h_n(t)$ denotes the subject specific time. For example, a one girl may hit her pubertal growth spurt at t_a while a second girl might hit hers later at $t_b > t_a$. So, while $t_a \neq t_b$ it would still be the case that $h_n(t_a) = h_n(t_b)$.

We assume that the observed curves $X_n(t)$ are given by

$$X_n(t) = X_n^*(h_n(t)),$$

where the $X_n^*(t)$ are properly aligned curves. The sample we would like to analyze is therefore $\{X_n^*(t) = X_n(h_n^{-1}(t))\}$. To estimate the $h_n(t)$ using landmark registration, one manually chooses landmarks/features in the data and, for each individual, finds which individual times correspond to those landmarks. One then uses those times to construct the warping functions. As an illustration, consider the growth curve example. Let t_n denote the (physical) time when individual n grows the fastest and t_a is the sample average of these times, i.e. t_a is the average time of the fastest growth. If the timing of the fastest growth is an important event in a study, we want to ensure that $h_n(t_n) = t_a$, i.e. all individual times t_n get mapped to the same value t_a. If the functions are defined on a common interval, say $[0, T]$, we also require that $h_n(0) = 0$, $h_n(T) = T$. and that h_n is a smooth monotonically increasing function. We thus, for each individual, have three points in the plane, $(0, 0), (t_n, t_a), (T, T)$, and there is a unique parabola $t \mapsto h_n(t)$ passing through these points. If X_n is the original curve, the registered curve is defined as $X_n^{\text{reg}}(t) = X_n(h_n^{-1}(t))$. If the X_n are the acceleration curves satisfying $X_n(t_n) = 0$, then $X_n^{\text{reg}}(t_a) = X_n(h_n^{-1}(t_a)) = X_n(t_n) = 0$. While this method of curve registration is intuitively appealing, it is tiresome if the number of curves to be registered is large, as the landmarks have to be identified separately for each curve using the R function `locator()`. Alternatively, a code of some complexity must be prepared to locate the position of the landmarks. In some cases, it is not clear how to choose the landmarks. A more automated allignment procedure is implemented with the function `register.fd`. By default, this functions calculates increasing warping functions, h_n, so that the registered curves become close to their mean. A target curve different than the mean can be specified. The following code uses default settings:

```
regList=register.fd(yfd=accelUnreg)
accelReg=regList$regfd
# $regfd extracts the registered functions
plot(accelReg, xlab="Age", ylab="Acceleration",
         ylim=c(-4,3))
```

This type of alignment is known as *continuous registration*, as no discrete landmarks are used. The resulting curves are shown in Figure 2.7. The continuously registered curves show less *phase variation* than the unregistred curves in Figure 2.6.

Kneip and Ramsay (2008) developed a theory that decomposes the total sample variance of unregistered curves X_i,

$$\text{MSE}_{total} = \frac{1}{N} \sum_{n=1}^{N} \int [X_n(t) - \bar{X}(t)]^2 dt,$$

into two components, MSE_{amp} and MSE_{pha}, which, respectively, quantify

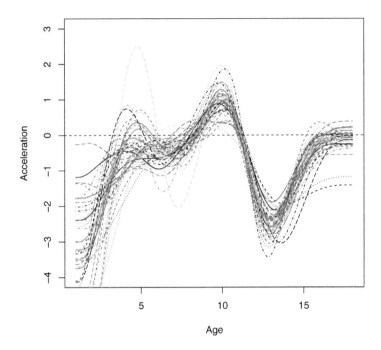

FIGURE 2.7: Continuously registered acceleration curves of the 54 girls.

the variance due to amplitude and phase variability. The decomposition $\text{MSE}_{total} = \text{MSE}_{amp} + \text{MSE}_{pha}$ is not only the property of the sample, but also of the method used to register the curves. Broadly speaking, MSE_{pha} is the part of variance removed by the registration process. For the growth curves registered as above, $\text{MSE}_{amp} = 5.92$ and $\text{MSE}_{pha} = 3.78$. A useful summary is the proportion of the variance due to phase variation which is defined as the ratio $\text{MSE}_{pha}/\text{MSE}_{total}$. The following code shows how this ratio can be computed for the growth curves registered with the funtion `register.fd`, as explained above:

```
warpFunctions=regList$warpfd
# $warpfd extracts the warping functions
APList=AmpPhaseDecomp(xfd=accelUnreg, yfd=accelReg, hfd=
    warpFunction)
APList$RSQR       [1] 0.3900585
# $RSQR extract proportion of the total variation due to phase
      variation
APList$MS.amp [1] 5.917859;    APList$MS.pha  [1] 3.784479
```

2.4   Further reading

Curve alignment and registration have been subject of continued research. Marron *et al.* (2015) provide a thorough review of recent developments; Earls and Hooker (2016) develop a Bayesian perspective. An important research area which has some relation to FDA has been the application of smoothing methodology to differential equations; Ramsay *et al.* (2007) is a frequently cited work in this area.

As noted in the Preface, this textbook does not cover all directions of FDA research. Classification and clustering of functional data are important in many applications. A classification problem arises when two or more groups of functions are defined, and we want to classify a newly observed curve as belonging to one of these predefined groups. A clustering problem arises when we have a sample of curves, and we suspect that these curves may belong to several distinct groups. We want to devise an algorithm that identifies the clusters, including their number, and assigns each curve to one of them. In the field of machine learning, classification is called supervised learning and clustering is called unsupervised learning. The review papers of Wang *et al.* (2016) and Cuevas (2014) include sections that introduce these concepts in greater detail. (Cuevas (2014) also discusses concepts of the median and quantiles of a sample of functions and bootstrap techniques for functional data.) The number of research papers on classification and clustering of functional data is very large. To give some initial guidance, we refer the readers to Górecki *et al.* (2015) (classification) and Jacques and Preda (2014) (clustering). Delaigle and Hall (2012) develop profound theoretical insights into the classification of functional data and emphasize the differences to the classification of multivariate data.

Another direction of FDA research is the analysis of functional data defined on manifolds. Such manifolds may arise as result of a meaningful time transformation (time warping), as explained in Wang *et al.* (2016), or in a specific science problem. For example, the domain of the data may be the brain or a blood vessel. Ettinger *et al.* (2016) is recent contribution in this direction. Throughout this book, the functions are defined on the same domain. In some applications, the functions may be defined on different domains, see e.g Liebl (2013).

2.5   Chapter 2 problems

2.1 Verify equality (2.2).

2.2 Download the R package `fds` and use the data set `FedYieldcurve`, which contains the monthly Federal Reserve interest rates, cf. Problem 1.2.

(a) Smooth the interest rates (yields) in January 1982 using a B–spline basis with four basis functions. Plot the raw and smoothed interest rates on one graph.

(b) Re–fit the January 1982 yields using a penalized smoothing based on six basis functions (as many as data points) with with the smoothing parameter $\lambda = 1$, and the second derivative as the penalty operator. Add the smooth in red to the graph you obtained in part (a) and comment on the result.

(c) Repeat part (b) with several other smoothing parameters λ. Which λ gives the most informative smooth curve?

2.3 Suppose that the coefficient functions in the linear differential operator are constants, i.e.,

$$L(x)(t) = \sum_{m=0}^{m} \alpha_m x^{(m)}(t).$$

Derive an expression for $\hat{c}^\top = (\hat{c}_1, \ldots, c_K)$ defined in 2.1.

2.4 Consider the DTI data in Section 1.5.

(a) Use penalized smoothing with 100 basis functions and GCV to convert the data to functional objects. Plot the data and the mean function. Comment on any differences with the direct splines expansion plotted in Figure 1.12.

(b) Use continuous registration to align the curves. Plot the resulting curves and mean function. Comment on any differences from the plot in (a).

(c) Carry out a PCA for both (a) and (b). Comment on any differences between FPCs and explained variance.

(d) In your opinion, was curve alignment necessary for this data? Explain.

2.5 In this problem we will explore some potential dangers of curve alignment. A *bump* function centered at a value c_0, with radius r_0, and amplitude a_0 is defined as

$$f(x) = \begin{cases} a_0 \exp\left\{ -\left(1 - \left(\frac{x-c_0}{r_0}\right)^2\right)^{-1} \right\} & \text{if } |x - c_0| < r_0 \\ 0 & \text{otherwise} \end{cases}.$$

(a) Simulate a functional sample over the unit interval each with a sample size of 50 from the Matérn process. For the first half of the sample, set the mean function equal to the the bump function with parameters $(c_0, r_0, a_0) = (3/8, 1/4, 5)$. For the second half use $(c_0, r_0, a_0) = (5/8, 1/4, 5)$. You may choose the values for the Matérn covariance function as well as the number of points sampled per curve. Plot all of the curves and include a curve for the overall mean function.

(b) Align the curves using continuous registration. Plot the resulting curves and include a mean function. Comment on any differences with (a) and if the registered curves exhibit any odd patterns.

(c) Carry out an FPCA with one PC on the unaligned and aligned curves separately. For each, do a simple linear regression of the score onto a dummy variable (coded 0/1) indicating which type of mean the function had (i.e. is it from the first or second half of the sample). Calculate a p-value to determine if the estimated slope parameters you get are significant. Compare with the aligned and unaligned curves. What did aligning do to the p-value? You may want to rerun your simulations a few times to see how the p-values change.

(d) Come up with one potential setting/application where you might lose something if you align. Make up whatever scenario you like, but think it through.

3

Mathematical framework for functional data

In previous chapters, we have seen examples of functional data and learned how to perform certain exploratory analyses. In order to perform inference, e.g. hypothesis tests and estimation with associated measures of uncertainty, we must treat the observed functions as elements of some space. In elementary statistics, observations are numbers (scalars), and are elements of the real line. In multivariate statistics, observations are vectors, and so are elements of the Euclidean space $\mathbb{R}^d$, where d is the dimension of the observed vectors. Even though functions are observed at discrete points t_j, in FDA they are treated as infinitely dimensional objects. We have seen that the first step in analysis of functional data is to convert functions observed at discrete time points to functional objects using basis expansions. For example, if the basis consists of trigonometric functions, each functional object is a function defined on the whole interval because each sine and cosine function is defined on the whole interval. So the question we must address is what kind of space these functional objects, which we simply call functions, live in. Based on the examples we have seen, we might require that the functions we work with are continuous and so are elements of the space of continuous function on the interval $[0, 1]$. If we are concerned with the estimation of derivatives, we may require that the functions are not only continuous, but also have sufficiently many derivatives, which are also continuous functions themselves. It turns out, that the simplest setting arises if the functions are treated as elements of a much larger space, the space of square integrable functions. Our objective in this chapter is to outline basic ideas related to data living in such a space. Our exposition focuses on introducing suitable notation and explaining the fundamental concepts so that the remaining chapters can be understood without searching for further references. The concepts we discuss in this chapter are revisited in Chapters 10 and 11, which contain a more rigorous exposition and a large number of problems which help to absorb these ideas more fully. An excellent and very extensive account of mathematical foundations of FDA is given in the monograph of Hsing and Eubank (2015).

3.1   Square integrable functions

To simplify the notation, we assume that all functions are defined on the unit interval $[0, 1]$. (If the argument t of a function f is in the interval $[a, b]$, then we simply consider the function $f^\star$ defined by $f^\star(u) = f(t)$, where $u = (t - a)/(b - a)$.) A function f is said to be square integrable if

$$\int f^2(t)dt = \int_0^1 \{f(t)\}^2 dt < \infty.$$

In the following, if the limits of integration are not specified, it is understood that the integration is over the interval $[0, 1]$. We denote the set of all square integrable functions by L^2.

Square integrable functions form a vector space. This means that if $f, g \in L^2$, then $af + bg \in L^2$ for any scalars a and b. Multiplication by scalars and addition are defined pointwise, i.e.

$$(af + bg)(t) = af(t) + bg(t), \quad t \in [0, 1].$$

Readers aware of measure theoretic foundations of integration, will recall that the elements of L^2 and operations on them do not need to be defined for every point t, they are defined for *almost all* points t. In this chapter, we will ignore measure–theoretic considerations.

What makes the space L^2 so convenient is that we can define the inner product of two functions by

$$\langle f, g \rangle = \int f(t)g(t)dt.$$

This inner product is analogous to the inner (or scalar) product of two vectors. Recall that if $\mathbf{x} = [x_1, x_2, \dots x_d]^T$, $\mathbf{y} = [y_1, y_2, \dots y_d]^T$, then $\langle \mathbf{x}, \mathbf{y} \rangle = \sum_{j=1}^d x_j y_j$. The inner product allows us to entertain the notion of orthogonality, which makes the geometric structure of the space L^2 very similar to that of the usual finite dimensional Euclidean space, at least at an intuitive level. We say that functions f and g are orthogonal (or perpendicular) if $\langle f, g \rangle = 0$. The inner product also allows us to introduce the notion of distance between functions. This is done by defining the *norm*

$$\|f\| = \sqrt{\langle f, f \rangle} = \left\{ \int f^2(t)dt \right\}^{1/2}.$$

The reader will notice that the norm is analogous to the length of a finite dimensional vector. The distance between two functions in L^2 is then the norm of their difference, i.e. $d(f, g) = \|f - g\|$. There is a very useful relation between the inner product of two functions and their norms:

$$\left| \int f(t)g(t)dt \right| = |\langle f, g \rangle| \leq \|f\| \|g\| = \left\{ \int f^2(t)dt \right\}^{1/2} \left\{ \int g^2(t)dt \right\}^{1/2}.$$
(3.1)

Relation (3.1) is known as the *Cauchy–Schwarz inequality* . Other properties of the norm and the inner product are fully analogous to the properties of the length of a vector and of the inner product in a finite dimensional space. For example, $\|f + g\| \leq \|f\| + \|f\|$ (triangle inequality) and $\langle af + bg, h \rangle = a \langle f, h \rangle + b \langle g, h \rangle$, which is actually a defining property of any inner product. We will use such relations in the following chapters; they are listed in a systematic way in Chapter 10.

As we have already seen, basis expansions play an important role in methodology for functional data. We say that a set of functions $\{e_1, e_2, e_3, \dots\}$ is a basis in L^2 if every function $f \in L^2$ admits a unique expansion

$$f(t) = \sum_{j=1}^{\infty} a_j e_j(t).$$

We say that $\{e_1, e_2, e_3, \dots\}$ is an *orthonormal basis*, if, in addition, $\langle e_j, e_{j'} \rangle = 0$ whenever $j \neq j'$ and $\|e_j\| = 1$. Trigonometric functions form an orthonormal basis; B–splines form a basis which is not orthonormal (spline functions are not orthogonal). For an orthonormal basis, $a_j = \langle f, e_j \rangle$, and we have the following *Parseval's equality*:

$$\int f^2(t)dt = \|f\|^2 = \sum_{j=1}^{\infty} \langle f, e_j \rangle^2 = \sum_{j=1}^{\infty} \left\{ \int f(t)e_j(t)dt \right\}^2. \qquad (3.2)$$

3.2   Random functions

The starting point of a statistical analysis is a sample. For example, $x_1, x_2, \dots, x_N$ can be heights of females age 20 from a certain population. To perform inference, these numbers are treated as realizations of random variables $X_1, X_2, \dots, X_N$. A setting which is often considered is that the X_i are independent and share a common distribution, for example each X_i is normal with mean μ and variance σ^2. Statistical inference is then concerned with the estimation, construction of confidence intervals and testing hypotheses related to these parameters. The reader will be aware of many other inferential procedures which are not related to the normal parameters. The central point is that they all involve measures of statistical uncertainty like significance levels and P–values. To compute such measures, the observations x_i must be treated as realizations of random variables X_i which have a certain distribution, e.g. a density or a probability mass function.

In FDA, the observations are functions. For example, x_i can be the growth curve of female i, cf. Section 2.3. The 54 females considered in Section 2.3 form a sample of size $N = 54$ from a large population. Since these females were randomly selected, we treat the growth curve x_i as a realizations of a *random function* X_i. We can think of X_i as the growth curve of the ith female before she has been selected. The shape and other attributes of X_i are thus random. In this section, we describe how they can be quantified. We will focus on attributes similar to the mean and variance of scalar observations, and we will explain what it means that a random function is Gaussian. These topics are revisited in much greater detail in Chapter 11.

In the remainder of this section X denotes a random function. Just as a random variable, a random function is defined on a probability space, say Ω, so for each $\omega \in \Omega$, $X(\omega)$ is a deterministic function. We assume that all realizations $X(\omega), \omega \in \Omega$, are elements of the space L^2 of square integrable functions. This means that of each $\omega \in \Omega$,

$$\|X(\omega)\|^2 = \int \{X(\omega)(t)\}^2 \, dt < \infty.$$

The function $\omega \mapsto \|X(\omega)\|$ is thus a random variable (it takes nonnegative values), and we can ask if it has a finite second moment or not. If it does, i.e. if $E\|X\|^2 < \infty$, we say that the random function X is *square integrable*. It is important to see the distinction between a square integrable deterministic function (integration refers to the interval $[0,1]$) and a square integrable random function (integration refers to the probability space Ω). It might be more pedagogical to say that a random function has a finite second moment, rather that to say that it is square integrable, but we will stick to this more usual terminology. In the following, we assume that the random function X is square integrable; this ensures that all objects we study exist. In particular, the analog of the variance, the covariance function, exists in a suitable space.

In Chapter 1, we considered the *sample* mean and covariance functions. Recall that if we observe curves $x_1, x_2, \ldots, x_N$, then the sample mean function is defined by $\hat{\mu}(t) = N^{-1} \sum_{i=1}^{N} x_i(t)$, and the sample covariance function by

$$\hat{c}(t, s) = \frac{1}{N} \sum_{i=1}^{N} (x_i(t) - \hat{\mu}(t))(x_i(s) - \hat{\mu}(s)).$$

In inferential procedures, we often treat the curves x_i as realizations of a random function X, for which we define the mean and covariance functions by

$$\mu(t) = EX(t), \quad c(t, s) = E[(X(t) - \mu(t))(X(s) - \mu(s))]. \tag{3.3}$$

The sample functions $\hat{\mu}$ and $\hat{c}$ are viewed as estimators of the population parameters μ and c.

The chief importance of the covariance function in FDA is that it leads to Functional Principal Component Analysis. We explain here the idea at an intuitive level. Every square integrable function X can be represented as

$$X(t) = \mu(t) + \sum_{j=1}^{\infty} \xi_j v_j(t). \tag{3.4}$$

The v_j depend on the covariance function c of X. Specifically, they are defined as the eigenfunctions of c, i.e. they are solutions to the equation

$$\int c(t, s)v(s)ds = \lambda v(t).$$

There are countably many solutions, $(\lambda_1, v_1), (\lambda_2, v_2), \ldots$, which we call the *eigenelements* of X (or of the covariance function c). The λ_j are called the eigenvalues of X (or c). The eigenvalues, and so the corresponding eigenfunctions v_j are arranged in nonincreasing order: $\lambda_1 \geq \lambda_2 \geq \ldots$. The random variables ξ_i, called the *scores*, are given by

$$\xi_j = \langle X - \mu, v_j \rangle = \int (X(t) - \mu(t))v_j(t)dt.$$

It can be shown that

$$E\xi_j = 0, \quad E\xi_j^2 = \lambda_j, \quad \mathrm{Cov}(\xi_j, \xi_k) = 0, \text{ if } j \neq k.$$

and

$$E \int (X(t) - \mu(t))^2 dt = E \|X - \mu\|^2 = \sum_{j=1}^{\infty} \lambda_j. \tag{3.5}$$

The interpretation of the above formula is that λ_j is the variance of the random function X in the principal direction v_j. The sum of these variances is the total variance of X. Relation (3.5) is thus a decomposition, or analysis, of variance.

Decomposition (3.4) is optimal in a sense that X can be well approximated using only very few initial v_j, in many applications 2–3 of them. We will refer to (3.4) as the *Karhunen–Loéve expansion* of X. The functions v_j form an orthonormal basis and are called the *Functional Principal Components* (FPC's) of X.

EXAMPLE 3.2.1 [Brownian motion] In Chapter 1 we used 50 realizations (a sample of size $N = 50$) of the random walk to illustrate the computation of the sample mean and covariance functions, as well as of the estimated FPC's. We generated the random walk on a grid of ten thousand points. As the grid becomes finer and finer, a realization of random walk approaches a realization of a random function known as the *Brownian motion* or the *Wiener process*. This random function can be defined on the whole positive half–line, but to stick to the setting of this section, we consider its restriction to the interval $[0, 1]$. The Wiener process we consider is thus the random function $W = \{W(t), t \in [0, 1]\}$. It can be shown that

$$\mu(t) = EW(t) = 0, \quad c(t, s) = \text{Cov}(W(t), W(s)) = \min(t, s),$$

$$\lambda_j = \frac{1}{(j - \frac{1}{2})^2 \pi^2}, \quad v_j(t) = \sqrt{2} \sin\left(\left(j - \frac{1}{2}\right)\pi t\right).$$

The Karunen–Loéve expansion (3.4) becomes

$$W(t) = \sum_{j=1}^{\infty} \frac{Z_j}{(j - \frac{1}{2})\pi} \sqrt{2} \sin\left(\left(j - \frac{1}{2}\right)\pi t\right), \tag{3.6}$$

where the Z_j are independent standard normal. The scores ξ_j are thus given by $\xi_j = ((j - 1/2)\pi)^{-1} Z_j$. The decomposition of variance (3.5) becomes

$$E \int_0^1 W^2(t)dt = \int_0^1 t\,dt = \frac{1}{2} = \sum_{j=1}^{\infty} \frac{1}{(j - \frac{1}{2})^2 \pi^2}.$$

Note that expansion (3.6) offers a way of generating smooth approximations to the Brownian motion; the sum must be truncated at some finite j. This is an alternative to a step–wise approximation by a random walk. The latter approach is often preferred on "visual grounds" because it produces jugged realizations which reflect the fact that realizations of W are not differentiable functions. However, if this aspect is not important, a truncated version of (3.6) can be used as well, and leads to a computationally faster way of generating approximate realizations of the Wiener process.   □

We conclude this section with the definition of a *Gaussian random function*. Expansion (3.6) shows that the Wiener process is a linear combination of orthonormal functions v_j with scores which are normal random variables. This observation leads to a general definition. *A random function X is Gaussian if the scores ξ_j in (3.4) are Gaussian random variables.* An equivalent definition is that X is Gaussian if for any function $u \in L^2$, the random variable $\langle X, u \rangle$ is Gaussian.

3.3   Linear transformations

Linear transformation play a very important role in statistics. In the straight line regression $y_i - \mu_Y = \beta(x_i - \mu_X) + \varepsilon_i$, the centered explanatory observations are multiplied by the slope β to yield centered responses observed with an error. In FDA, many types of regressions are considered; they involve linear transformations of functions to functions, functions to scalars and vectors to functions. They will be discussed in subsequent chapters. In this section, we introduce some basic concepts related to general linear transformations to help the reader follow the chapters on functional regressions. A more detailed treatment is given in Chapter 10.

Linear transformations, often called operators, are defined on vector spaces. If $\mathcal{V}_1, \mathcal{V}_2$ are two vector spaces, a function $L : \mathcal{V}_1 \to \mathcal{V}_2$ is called linear if $L(ax + by) = aL(x) + bL(y)$, for any $x, y \in \mathcal{V}_1$ and any scalars a, b.

EXAMPLE 3.3.1 Suppose $\mathcal{V}_1 = \mathbb{R}^2, \mathcal{V}_2 = \mathbb{R}$, and consider the following two transformations:

$$L((x_1, x_2)) = 2x_1 + 3x_2, \quad G((x_1, x_2)) = x_1 x_2.$$

It is easy to check that the L is linear and G is not. □

Two situations commonly encountered in FDA are those with $\mathcal{V}_1 = L^2, \mathcal{V}_2 = \mathbb{R}$ and $\mathcal{V}_1 = L^2, \mathcal{V}_2 = L^2$. In both cases, we use *integral operators* defined, respectively, as follows

$$L_1(x) = \int \psi(t)x(t)dt, \quad L_2(x)(t) = \int \psi(t, s)x(s)ds.$$

The operator L_1 transforms a function x to a real number $L_1(x)$; L_2 transforms a function x to another function, $L_2(x)$. The reader is encouraged to check that both transformations are indeed linear. To ensure that the number $L_1(x)$ is finite, we must assume that $\int \psi^2(t)dt < \infty$; to ensure that the function $L_2(x)$ is square integrable whenever x is so, we must assume that

$$\iint \psi^2(t, s)dtds < \infty. \tag{3.7}$$

These assumptions are stronger than what is actually needed, but they are common and simplify many arguments. We emphasize that no similar assumptions are needed in the theory of scalar or vector regressions because such regressions involve matrices consisting of a finite number of entries, and all sums are consequently finite. The functions ψ are called kernels of the respective operators. An integral operator with the kernel satisfying (3.7) is called a *Hilbert–Schmidt* operator.

An operator that is used very often in FDA is the *covariance operator*. The covariance function c defined in (3.3) is its kernel, so its action on a function $x \in L^2$ is defined by $C(x)(t) = \int c(t, s)x(s)ds$. The FPC's v_j introduced in Section 3.2 are thus the eigenfunctions of C: $C(v_j) = \lambda_j v_j$. The operator C has a very convenient representation

$$C(x) = \sum_{j=1}^{\infty} \lambda_j \langle x, v_j \rangle v_j, \tag{3.8}$$

which is known as its *spectral decomposition*. Representation (3.8) is analogous to the diagonal representation of the covariance matrix by means of the Cholesky decomposition.

4

Scalar–on–function regression

Linear regression is one of the most fundamental tools in statistics. In the usual linear model

$$y_i = x_{i1}\beta_1 + x_{i2}\beta_2 + \ldots x_{ip}\beta_p + \varepsilon_i, \quad i = 1, 2, \ldots, N, \tag{4.1}$$

all variables and parameters are scalars, and the regressors x_{ik} are typically assumed to be known scalars. In a functional linear model, some of these quantities are curves, and analogs of the coefficients β_k must be then appropriately defined.

Functional regression models are subdivided into three broad categories, depending on whether the responses or the regressors, or both, are curves. We assume for simplicity that the responses and the regressors have mean zero.

Scalar–on–function regression:

$$Y_i = \int \beta(s)X_i(s)ds + \varepsilon_i. \tag{4.2}$$

The regressors are curves, but the responses are scalars. The parameter is the regression function β.

Function-on-scalar regression:

$$Y_i(t) = \sum_{k=1}^{p} x_{ik}\beta_k(t) + \varepsilon_i(t). \tag{4.3}$$

The responses are curves, but the regressors are scalars. The parameters are the coefficient functions β_k.

The function–on–function regression:

$$Y_i(t) = \int \beta(t, s)X_i(s)ds + \varepsilon_i(t). \tag{4.4}$$

The responses Y_i are curves, and so are the regressors X_i. The parameter is the kernel (surface) β.

The above formulations are just prototypes intended to illustrate the general idea. The main point is that the functional parameters are infinite dimensional objects which must be estimated from a finite sample. Without any restrictions on the functional parameters, a perfect fit is possible (all residuals are zero), and the resulting estimates are erratic, noisy functions which do not provide useful insights. Estimation therefore often involves additional restrictions, either by imposing smoothness conditions or restricting the action of the corresponding operators to appropriate subspaces. Several chapters of Ramsay and Silverman (2005) are dedicated to functional linear models and contain the discussion of some issues which are not addressed in this book. In this chapter, we focus on the scalar response model which bears the closest resemblance to the standard model (4.1); the discrete index $k = 1, 2, \ldots, p$ is replaced by the continuous index s. This model also illustrates many issues common to all models with parameters which are functions. Chapter 5 is dedicated to regressions with functional responses.

This chapter proceeds as follows. In Section 4.1, we present several examples of data that can be modeled using scalar–on–function regression. We follow with Section 4.2 which reviews the standard regression theory, and then with Section 4.3 which explains chief differences between the cases of functional and vector regressors. The next three sections explain several approaches to the estimation the regression function. Implementation of these approaches in the R package refund is illustrated in Section 4.7. We conclude the chapter with Section 4.8 which discusses nonlinear approaches to scalar–on–function regression.

4.1 Examples

In this section, we present several data sets that have been analyzed using scalar–on–function regression. The first data sets consists of data for $N = 60$ gasoline samples. It is available as gasoline in the refund package, which is discussed in greater detail in Section 4.7. The responses $Y_i, 1 \leq i \leq N$, shown in Figure 4.1, are octane ratings obtained through a careful chemical analysis of each gasoline sample. The explanatory curves are near infrared spectral curves. The curve for the first gasoline sample is shown in the left panel of Figure 4.2. The differences between the 60 curves are small relative to their range, two representative differences are shown in the right panel of Figure 4.2. The purpose of regression analysis is to determine the form of dependence between the octane rating Y and the spectral curve X. If we can find a functional $g : L^2 \to R$ such that $Y \approx g(X)$, then it is easy to determine an approximate octane rating of a gasoline sample from its spectral curve. The spectral curves are easy to obtain, while chemical analysis is time consuming and expensive.

The second data set consists of data for $N = 215$ meat samples. It is available as tecator in the R package fda.usc. Similarl to the gasoline example, the explanatory curves are spectral curves (defined at shorter wavelengths).

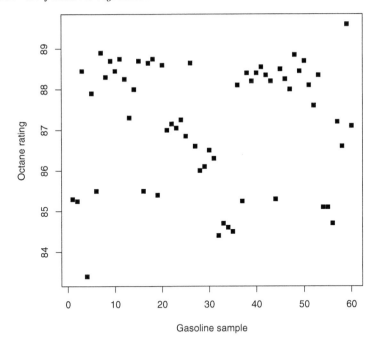

FIGURE 4.1: Octane ratings of 60 gasoline samples.

These curves are shown in the left panel of Figure 4.3. The right panel shows their estimated derivatives which reveal features of the curves not readily seen in the left panel. There are three scalar responses for each meat sample: fat, water and protein content in percent, obtained by chemical analysis. As for the gasoline samples, the objective is to determine the form of dependence between the spectral curves and the response variables to quickly determine the response without performing a chemical analysis which may be impratical in an industrial setting.

Scalar–on–function regression has been applied to Diffusion Tensor Imaging data discussed in Section 1.5. The regressors are curves like those shown in Figure 1.12, or similar curves obtained from a brain imaging. The responses are various scores of patients, for example The PASAT (Paced Auditory Serial Addition Test) scores. These scores measure cognitive function and assess auditory information processing speed and flexibility, as well as calculation ability.

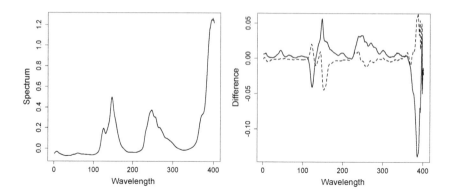

FIGURE 4.2: Left: Near infrared spectrum of a gasoline sample with index 1; Right: differences between the spectrum of the samples with indexes 2 and 1 (continuous) and 5 and 1 (dashed).

4.2 Review of standard regression theory

There are dozens of excellent textbooks on linear regression. For example, Faraway (2009) provides an exposition which carefully addresses practical aspects while Seber and Lee (2003) focus on the theory. In this section, we review only the most fundamental properties to provide some reference points.

In the standard linear model (4.1), the response for the ith subject, y_i, is equal to a linear combination of the observed values, $x_{i1}, x_{i2}, \ldots, x_{ip}$ of the vector $\mathbf{x} = [x_1, x_2, \ldots, x_p]$ of explanatory variables (regressors) plus a mean zero random error ε_i. It is often assumed that

$$y_i, \quad \mathbf{x}_i = [x_{i1}, x_{i2}, \ldots, x_{ip}]^T, \quad \varepsilon_i$$

are iid realizations of the random variables y, $\mathbf{x} = [x_1, x_2, \ldots, x_p]^T$, ε. The population model for (4.1) thus is $y = \mathbf{x}^T \boldsymbol{\beta} + \varepsilon$. If the regressors $\mathbf{x}_i$ are treated as deterministic, then it is enough to assume that the errors ε_i are iid. If, in addition, the errors are assumed to be normal, then the least square estimates of the parameters $\beta_1, \beta_2, \ldots, \beta_p$ are jointly normal with standard errors which can be readily estimated.

The least squares estimation of the parameter vector $\boldsymbol{\beta} = [\beta_1, \beta_2, \ldots, \beta_p]^T$

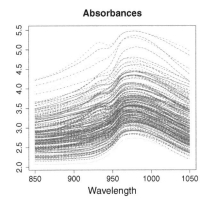 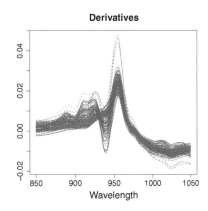

FIGURE 4.3: Left: Absorbance curves of 100 meat samples; Right: Derivatives of these curves. Curves of samples with fat content of less than 20% are in red.

proceeds as follows. Introduce the following vectors and matrices:

$$\mathbf{Y} = \begin{bmatrix} y_1 \\ y_2 \\ \vdots \\ y_N \end{bmatrix}, \quad \mathbf{X} = \begin{bmatrix} x_{11} & x_{12} & \cdots & x_{1p} \\ x_{21} & x_{22} & \cdots & x_{2p} \\ \vdots & \vdots & \vdots & \vdots \\ x_{N1} & x_{N2} & \cdots & x_{Np} \end{bmatrix}, \quad \boldsymbol{\beta} = \begin{bmatrix} \beta_1 \\ \beta_2 \\ \vdots \\ \beta_p \end{bmatrix}, \quad \boldsymbol{\varepsilon} = \begin{bmatrix} \varepsilon_1 \\ \varepsilon_2 \\ \vdots \\ \varepsilon_N \end{bmatrix}. \tag{4.5}$$

In this notation, relations (4.1) can be succinctly written as

$$\mathbf{Y} = \mathbf{X}\boldsymbol{\beta} + \boldsymbol{\varepsilon}. \tag{4.6}$$

The least squares estimator of $\boldsymbol{\beta}$ is designed to minimize the Euclidean norm of the difference $\mathbf{Y} - \mathbf{X}\boldsymbol{\beta}$, i.e. the distance between $\mathbf{Y}$ and $\mathbf{X}\boldsymbol{\beta}$. Denote by L_X the subspace of $\mathbb{R}^N$ spanned by the columns of $\mathbf{X}$ and denote by $\boldsymbol{\theta} = \mathbf{X}\boldsymbol{\beta}$ a generic element of L_X. By the projection theorem, cf. Theorems 10.2.1 and 10.2.2, $\mathbf{X}\hat{\boldsymbol{\beta}}$ must be the projection $\hat{\boldsymbol{\theta}}$ of $\mathbf{Y}$ onto L_X. Thus $\hat{\boldsymbol{\theta}}$ is the unique vector minimizing the length of $\mathbf{Y} - \boldsymbol{\theta}$ over $\boldsymbol{\theta} \in L_X$. The least square estimator $\hat{\boldsymbol{\beta}}$ must thus satisfy $\hat{\boldsymbol{\theta}} = \mathbf{X}\hat{\boldsymbol{\beta}}$. While $\hat{\boldsymbol{\theta}}$ is always unique (as a projection), $\hat{\boldsymbol{\beta}}$ can be uniquely determined only if the columns of $\mathbf{X}$ are linearly independent, i.e. if $\mathbf{X}$ has rank p. Assuming this is the case, we now explain how $\hat{\boldsymbol{\beta}}$ is computed. The vector $\mathbf{Y} - \hat{\boldsymbol{\theta}}$ is orthogonal to L_X, so $\mathbf{X}^T(\mathbf{Y} - \hat{\boldsymbol{\theta}}) = 0$, i.e. $\mathbf{X}^T\hat{\boldsymbol{\theta}} = \mathbf{X}^T\mathbf{Y}$. Therefore $\hat{\boldsymbol{\beta}}$ satisfies the *normal equations*

$$\mathbf{X}^T\mathbf{X}\hat{\boldsymbol{\beta}} = \mathbf{X}^T\mathbf{Y}.$$

It can be shown, Problem 4.1, that if $\mathbf{X}$ is of rank p, then $\mathbf{X}^T\mathbf{X}$ is nonsingular, and so the least squares estimator of $\boldsymbol{\beta}$ is given by

$$\hat{\boldsymbol{\beta}} = (\mathbf{X}^T\mathbf{X})^{-1}\mathbf{X}^T\mathbf{Y}. \tag{4.7}$$

Combining relations (4.6) and (4.7), it can be shown that the estimator $\hat{\boldsymbol{\beta}}$ is unbiased. Under the assumption that the errors ε_i are iid normal, the standard errors of the components of $\hat{\boldsymbol{\beta}}$ can be estimated, see e.g. Chapter 3 of Seber and Lee (2003).

In fact, the least squares estimator works well under a wide variety of assumptions. Notice that we can write

$$\hat{\boldsymbol{\beta}} = (\mathbf{X}^T \mathbf{X})^{-1} \mathbf{X}^T (\mathbf{X}\boldsymbol{\beta} + \boldsymbol{\varepsilon}) = \boldsymbol{\beta} + (\mathbf{X}^T \mathbf{X})^{-1} \mathbf{X}^T \boldsymbol{\varepsilon}.$$

To ensure that the estimator is consistent, i.e. $\hat{\boldsymbol{\beta}} \xrightarrow{P} \boldsymbol{\beta}$, as $N \to \infty$, it is enough if

$$N^{-1} \mathbf{X}^T \mathbf{X} \to \boldsymbol{\Sigma}_X \quad \text{and} \quad N^{-1} \mathbf{X}^T \boldsymbol{\varepsilon} \xrightarrow{P} 0.$$

If the matrix $\mathbf{X}$ is deterministic, the existence of $\boldsymbol{\Sigma}_X$ is assumed. The second condition follows from the law of large numbers, $N^{-1} \boldsymbol{\varepsilon} \xrightarrow{P} 0$, which holds under very general assumptions on the errors. If the regressors are random, the above conditions can be established under a wide range of assumptions including dependence, heterogeneous errors, and increasing dimensions. Though the least squares estimator may not be optimal in these situations, it is often still consistent and asymptotically normal.

A problem that often arises in regression analysis is that of testing the significance of some coefficients β_ℓ in (4.1). For example, if β_p is not significantly different form zero, then a simpler model with only the first $p-1$ regressors should be used. To streamline the exposition, suppose that the explanatory variables are ordered in such a way that the simpler model has the form $y_i = x_{i1}\beta_1 + \ldots + x_{im}\beta_m + \delta_i$, for some $m < p$. We want to test the null hypothesis

$$H_0 : \quad \beta_{m+1} = \ldots = \beta_p = 0.$$

Assuming that the errors in both models are iid normal, the standard procedure is to use the F-test. To define the test statistic, we introduce the following quantities

$$R_p = \sum_{i=1}^{N} \left(y_i - \sum_{j=1}^{p} x_{ij}\hat{\beta}_j^{(p)} \right)^2, \qquad R_m = \sum_{i=1}^{N} \left(y_i - \sum_{j=1}^{m} x_{ij}\hat{\beta}_j^{(m)} \right)^2.$$

The estimators $\hat{\beta}_j^{(p)}$ and $\hat{\beta}_j^{(m)}$ are the least squares estimators computed, respectively, using the model with p and m regressors. Each of the quantities, R_p and R_m, is called the residual error sum of squares. Since $m < p$, $R_m \geq R_p$. If $\beta_{m+1} = \ldots = \beta_p = 0$. the difference $R_m - R_p$ will be small, otherwise it will be large. The F-statistic is basically a suitably normalized difference $R_m - R_p$ defined by

$$F = \frac{(R_m - R_p)/(p-m)}{R_p/(N-p)}. \tag{4.8}$$

If H_0 is true, then F has the $F_{p-m,N-p}$ distribution (assuming iid normal errors). The null hypothesis is rejected if the observed F statistic exceeds a suitable upper quantile of this distribution.

4.3 Difficulties specific to functional regression

This section contains a discussion of those aspects of functional regression which are peculiar to functional setting, or much more pronounced in this setting than in the usual multiple regression (4.1). An important distinction is that in functional regression, we are interested not only in computing an estimate of the function β in (4.2), but we must also ensure that this estimate has a useful interpretation, and can be used for effective prediction of responses from explanatory functions. Intervals of the s values where $|\beta(s)|$ is large are influential for the responses, and the sign of $\beta(s)$ on those intervals implies either negative or positive association. If an estimate $\hat{\beta}$ jumps chaotically, no such interpretation is possible; predictions tend to center around the mean of the responses and have large variance.

To see differences between the usual regression model (4.1) and the scalar–on–function regression, it is illustrative to begin with an alternative derivation of the estimator (4.7). Consider the population version of (4.1), i.e. assume that

$$Y = \sum_{i=1}^{p} \beta_i X_i + \varepsilon$$

in which $[X_1, X_2, \ldots, X_p, \varepsilon]$ is a random vector. We will first find β which minimizes

$$R(\beta) = E\left(Y - \sum_{i=1}^{p} \beta_i X_i\right)^2.$$

This is done by equating the partial derivatives $\partial R(\beta)/\partial \beta_\ell$ to zero for $\ell = 1, 2, \ldots, p$. In Problem 4.3, the reader is asked to verify that this leads to the solution

$$\beta = \mathbf{C}_X^{-1} \mathbf{C}_{XY}, \tag{4.9}$$

where

$$\mathbf{C}_X = [E[X_i X_j], \ 1 \le i, j \le p], \quad \mathbf{C}_{XY} = [E[X_i Y], \ 1 \le i \le p]^T.$$

We emphasize that to obtain (4.9), we must assume that the $p \times p$ matrix $\mathbf{C}_X$ is invertible. In Problem 4.4, the reader is asked to verify that the invertibility of $\mathbf{C}_X$ is equivalent to the linear independence of the variables X_i, i.e. to the implication

$$\sum_{i=1}^{p} \alpha_i X_i = 0 \quad \Longrightarrow \quad \alpha_1 = \alpha_2 = \ldots = \alpha_p = 0. \tag{4.10}$$

Linear independence of explanatory variables is a common assumption because otherwise some of them could be expressed as linear combinations of others, and would not be needed. To obtain an estimator based on expression (4.9), we replace the matrix $\mathbf{C}_X$ and the vector $\mathbf{C}_{XY}$ by their method of moments estimators whose entries are defined by

$$\widehat{C}_X(i,j) = \frac{1}{N} \sum_{n=1}^{N} X_{ni} X_{nj}, \quad \widehat{C}_{XY}(i) = \frac{1}{N} \sum_{n=1}^{N} X_{ni} Y_n.$$

Using the notation introduced in Section 4.2, these estimators can be represented as

$$\widehat{\mathbf{C}}_X = N^{-1} \mathbf{X}^T \mathbf{X}, \quad \widehat{\mathbf{C}}_{XY} = N^{-1} \mathbf{X}^T \mathbf{Y},$$

leading to the usual least squares estimator (4.7).

We now turn to the functional regression (4.2) for which the population model is

$$Y = \int \beta(t) X(t) dt + \varepsilon.$$

Our objective is to retrace the above steps and obtain an estimator of the function $\beta(\cdot)$. We will see the difficulties such an extension entails. We cannot compute the partial derivatives to find the minimum of $R(\beta) = E(Y - \int \beta(t) X(t) dt)^2$ because there are infinitely many, in practice too many, values $\beta_t = \beta(t)$. We can however attempt to derive an analog of (4.9) is a different way. Define the analogs of the matrix $\mathbf{C}_X$ and the vector $\mathbf{C}_{XY}$ by

$$c_X(t,s) = E[X(t)X(s)], \quad c_{XY}(t) = E[X(t)Y].$$

Assuming that X and ε are independent, we obtain, Problem 4.6,

$$c_{XY}(t) = \int c_X(t,s)\beta(s)ds. \tag{4.11}$$

The above relation can be viewed as a system of infinitely many linear equations, in practice of a large number of equations, which is impossible, or difficult, to solve. Even if it is possible to solve, experience shows that the resulting function $\beta(\cdot)$ evaluated at a large number of points t_j is very noisy and difficult to interpret. This is because solving equation (4.11) without any further restrictions does not utilize any intuition that the function $\beta(\cdot)$ should be smooth. One must postulate some special form of this function to make the problem practically solvable with an interpretable solution. We will describe such approaches in Sections 4.4 and 4.5. There is another justification why, in general, it is not possible to solve equation (4.11). Assuming $EX(t) = 0$, we recognize the kernel $c_X(t,s)$ as the covariance functions of the random function X, cf. (3.3). In Chapter 3, we also observed, cf. (3.8), that the right–hand side of (4.11) can be written as $C_X(\beta)$, where C_X is the integral operator with kernel $c_X(t,s) = \sum_{j=1}^{\infty} \lambda_j v_j(t) v_j(s)$, with nonnegative eigenvalues λ_j and eigenfunctions v_j which form an orthonormal basis in the space L^2. In this operator notation, equation (4.11) can be written as $C_X(\beta) = c_{XY}$. It cannot be always solved because the inverse operator C_X^{-1} does not exist, cf. Problem 4.5. Arguments based on such principal components expansion motivate the approach described in Section 4.6.

Another elucidating perspective on methods presented in this Chapter can be gained by viewing the scalar response model (4.2) as an approximation to the regression

$$Y_i = \sum_{j=1}^{J} \beta(t_j) X_i(t_j) + \varepsilon_i$$

with a large number of points t_j. If t_j is close to $t_{j'}$, then, for every i, the values $X_i(t_j)$ and $X_i(t_{j'})$ will be close. If the curves are approximately periodic, the values $X_i(t_j)$ and $X_i(t_{j'})$ can be close even for t_j not geometrically close to $t_{j'}$. The point is that there will, in general, be many column vectors $\mathbf{X}(t_j) = [X_1(t_j), X_2(t_j), \ldots, X_N(t_j)]^T$ which are strongly correlated, or nearly linearly dependent in some complex way. If the columns of the matrix $\mathbf{X}$ in (4.5) are nearly linearly dependent, we say that the regression exhibits *colinearity*, or *multicolinearity*. In such cases, the variances of some components of $\hat{\boldsymbol{\beta}}$ given by (4.7) become large, so precise estimation is not possible. This is intuitively understandable because if two regressors are linearly related, strongly correlated, then their coefficients can be almost arbitrarily chosen as long as their linear combination has a certain value. To illustrate, consider the regression

$$Y_i = \beta_1 x_{i1} + \beta_2 x_{i2} + \varepsilon_i$$

with just two regressors. One can show that if the correlation coefficient between the vectors $[x_{11}, x_{21}, \ldots, x_{N1}]^T$ and $[x_{12}, x_{22}, \ldots, x_{N2}]^T$ is r, then the variances of the least squares estimators are

$$\mathrm{Var}[\hat{\beta}_1] = \frac{\sigma_\varepsilon^2}{(1 - r^2)s_1^2}, \quad s_1^2 = \sum_{i=1}^{N} x_{i1}^2;$$

$$\mathrm{Var}[\hat{\beta}_2] = \frac{\sigma_\varepsilon^2}{(1 - r^2)s_2^2}, \quad s_2^2 = \sum_{i=1}^{N} x_{i2}^2.$$

The simplest, and often very effective, solution to collinearity is to remove regressors which can be effectively replaced by other regressors, i.e. to reduce the number of explanatory variables. This is not reasonable in the context of functional explanatory variables because it would mean removing some chunks of curves over certain intervals. More sophisticated solutions to collinearity are ridge regression and principal component regression. Ridge regression is similar to the estimation with a roughness penalty, Section 4.5, and the principal component regression is analogous to the method described in Section 4.6. A reader interested in these methods in the context of the standard model (4.1) is referred to Seber and Lee (2003), or any other recent textbook on regression analysis.

4.4   Estimation through a basis expansion

To enhance practical applicability, in this and the following sections, we consider the model (4.2) with an intercept term, i.e.

$$Y_i = \alpha + \int \beta(s)X_i(s)ds + \varepsilon_i, \quad i = 1, 2, \ldots, N. \quad (4.12)$$

The simplest approach is to expand the function β using deterministic basis functions, and so to reduce the problem to the standard model (4.6). Assume then that

$$\beta(t) = \sum_{k=1}^{K} c_k B_k(t). \quad (4.13)$$

The selected basis functions B_k influence the shape of the estimate. The number K of basis functions is generally chosen to be smaller than the number of time points t_j at which the functions X_i are observed. It reflects intuition as to how smooth the resulting estimate should be. There are data driven methods of selecting K, which we discuss below, but in some applications they lead to estimates that are difficult to interpret from the angle of a specific applied problem. It is always useful to experiment with several selections of K.

Using expansion (4.13), we obtain

$$\int \beta(t)X_i(t)dt = \sum_{k=1}^{K} c_k \int B_k(t)X_i(t)dt =: \sum_{k=1}^{K} x_{ik}c_k.$$

Model (4.12) thus reduces to (4.6), and the parameter vector $\mathbf{c} = [\alpha, c_1, c_2, \ldots, c_K]^T$ (corresponding to β) is estimated by $\hat{\mathbf{c}} = (\mathbf{X}^T\mathbf{X})^{-1}\mathbf{X}^T\mathbf{Y}$, where

$$\mathbf{X} = \begin{bmatrix} 1 & x_{11} & x_{12} & \cdots & x_{1K} \\ 1 & x_{21} & x_{22} & \cdots & x_{2K} \\ \vdots & \vdots & \vdots & \vdots & \vdots \\ 1 & x_{N1} & x_{N2} & \cdots & x_{NK} \end{bmatrix} \quad (4.14)$$

A disadvantage of this method is that the resulting estimate

$$\hat{\beta}(t) = \sum_{k=1}^{K} \hat{c}_k B_k(t)$$

depends on the shape of the basis functions B_k and on their number K, both being subjectively chosen. A reasonable approach is to use the same basis functions as in expansions (1.3) of the regressor functions. Furthermore, the theory supporting this estimation procedure, assumes that β is exactly equal to a linear combination of the basis functions. In practice this only holds approximately which must be respected to produce proper inferential procedures.

The reduction to the standard model (4.6) also leads to approximate confidence intervals for the unknown regression function β. If the errors, ε_i, are independent $\mathcal{N}(0, \sigma_\varepsilon^2)$, then the covariance matrix of $\hat{\mathbf{c}}$ is $\mathrm{Var}[\hat{\mathbf{c}}] = \sigma^2(\mathbf{X}^T\mathbf{X})^{-1}$, Problem 4.2. The error variance, σ_ε^2, is estimated by the sample variance, $\hat{\sigma}_\varepsilon^2$, of the residuals

$$\hat{\varepsilon}_i = y_i - \hat{\alpha} - \int \hat{\beta}(t)X_i(t)dt = y_i - \hat{\alpha} - \sum_{k=1}^{K} x_{ik}\hat{c}_k.$$

The variance of $\hat{c}_k$ is thus estimated by the corresponding diagonal entry of the matrix $\hat{\sigma}_\varepsilon^2(\mathbf{X}^T\mathbf{X})^{-1}$. Denoting this estimate by $\hat{\sigma}_k^2$, an approximate 95% confidence interval for the regression function β is

$$\sum_{k=1}^{K} \hat{c}_k B_k(t) \pm 1.96 \sum_{k=1}^{K} \hat{\sigma}_k B_k(t).$$

The above confidence interval can be used only as an exploratory tool. Precise confidence statements can be made only about the coefficients c_k in the expansion (4.13), which itself is only an approximation.

We now provide a discussion of the bias resulting from using expansion (4.13) which involves only a finite number K of basis function. This discussion further illustrates the issues arising in the functional case which are absent in the standard multivariate case. The function β can be expanded, without any approximation error, only using all basis functions, i.e. $\beta(t) = \sum_{k=1}^{\infty} c_k B_k(t)$. Therefore,

$$\beta(t) = \sum_{k=1}^{K} c_k B_k(t) + \sum_{k=K+1}^{\infty} c_k B_k(t) = \sum_{k=1}^{K} c_k B_k(t) + \delta(t),$$

where $\delta(t)$ is the truncation error. In this case, we have

$$\int \beta(t)X_i(t)\, dt = \sum_{k=1}^{K} x_{ik}c_k + \int \delta(t)X_i(t)dt := \sum_{k=1}^{K} x_{ik}c_k + \delta_i.$$

The model is now not the same as the traditional regression model as $\mathbf{Y} = \mathbf{Xc} + \boldsymbol{\delta} + \boldsymbol{\varepsilon}$. The least squares estimate of the coefficients, $\mathbf{c}$, in (4.13), is given by

$$\hat{\mathbf{c}} = \mathbf{c} + (\mathbf{X}^\top\mathbf{X})^{-1}\mathbf{X}^\top\boldsymbol{\delta} + (\mathbf{X}^\top\mathbf{X})^{-1}\mathbf{X}^\top\boldsymbol{\varepsilon}.$$

As opposed to traditional multivariate methods, the LS estimator is now actually biased. The analysis of this bias explains why to obtain an asymptotically consistent estimate in the functional case, it is necessary to assume that K increases to infinity with the sample size N. The discussion that follows should be compared with the discussion of the consistency of $\hat{\beta}$ presented at the end of Section 4.2. We focus on $\mathbf{X}^\top \boldsymbol{\delta}$, but note that inverting $\mathbf{X}^\top \mathbf{X}$ can also cause practical problems. Recall that $\delta(t) = \sum_{j=K+1}^{\infty} c_j B_j(t)$ and $\delta_i = \int \delta(s) X_i(s) ds$. The kth, $1 \leq k \leq K+1$, component of $N^{-1} \mathbf{X}^\top \boldsymbol{\delta}$ is

$$N^{-1}(\mathbf{X}^\top \boldsymbol{\delta})(k) = N^{-1} \sum_{i=1}^{N} x_{ik} \delta_i$$

$$= N^{-1} \sum_{i=1}^{N} \left(\int B_k(t) X_i(t) dt \right) \left(\int \delta(s) X_i(s) ds \right)$$

$$= \iint B_k(t) \hat{c}_X(t, s) \delta(s) dt ds$$

$$= \sum_{j=K+1}^{\infty} c_j \iint B_k(t) \hat{c}_X(t, s) B_j(s) dt ds,$$

where $\hat{c}_X(t, s) = N^{-1} \sum_{i=1}^{N} X_i(t) X_i(s)$. Thus, for a fixed K,

$$\lim_{N \to \infty} N^{-1}(\mathbf{X}^\top \boldsymbol{\delta})(k) = \sum_{j=K+1}^{\infty} c_j \iint B_k(t) c_X(t, s) B_j(s) dt ds,$$

which, in general, is a nonzero number. To make the above limit vanish, we must assume that $K = K(N) \to \infty$, and impose assumptions on the rate of decay of the c_j.

4.5 Estimation with a roughness penalty

The parameter K in the previous section should be viewed as a tuning parameter; adjusting K adjusts the smoothness of the resulting estimator $\hat{\beta}(t)$. From this perspective, it is often more desirable to smooth by using a roughness penalty term. As in the previous approach, the starting point is expansion (4.13), but now K is taken to be some large value so that $\hat{\beta}$ is no longer sensitive to the value of K. It is often recommended to use K equal to the number of points t_j at which curves X_i are observed. The roughness penalty approach shifts the control of smoothness from K to the smoothing parameter λ and a differential operator L :

$$P_\lambda(\alpha, \beta) = \sum_{i=1}^{N} \left\{ Y_i - \alpha - \int \beta(t) X_i(t) dt \right\}^2 + \lambda \int [(L\beta)(t)]^2 dt. \qquad (4.15)$$

In (4.15), L is a differential operator that acts on the function β and λ is a smoothing parameter. The idea is to force smoothness by penalizing functions which are too rough with a penalty term $\lambda \int [(L\beta)(t)]^2 dt$. A common choice of L is the second derivative: $(L\beta)(t) = \beta''(t)$. The most critical aspect of this method is the selection of λ; if it is too large, the estimate of β is too smooth and important aspects of its shape can be suppressed; if it is too small, the estimate of β will reflect random errors. We first explain how the estimate of β is computed for a fixed λ, then we comment on methods of selecting it. Using the notation introduced in Section 4.4, we can rewrite (4.15) as

$$P_\lambda(\alpha, \beta) = P_\lambda(\alpha, c_1, c_2, \ldots, c_K)$$

$$= \sum_{i=1}^{N} \left\{ Y_i - \alpha - \sum_{k=1}^{K} x_{ik} c_k \right\}^2 + \lambda \int \left[\sum_{k=1}^{K} c_k (LB_k)(t) \right]^2 dt$$

$$= \sum_{i=1}^{N} \left\{ Y_i - \alpha - \sum_{k=1}^{K} x_{ik} c_k \right\}^2 + \lambda \sum_{k,k'=1}^{K} c_k c_{k'} R_{kk'},$$

where

$$R_{kk'} = \int (LB_k)(t)(LB_{k'})(t) dt.$$

In Problem 4.7, the reader is asked to show that the parameter vector $\mathbf{c} = [\alpha, c_1, c_2, \ldots, c_K]^T$ minimizing this expression is given by

$$\hat{\mathbf{c}} = (\mathbf{X}^T \mathbf{X} + \lambda \mathbf{R})^{-1} \mathbf{X}^T \mathbf{Y}, \tag{4.16}$$

where $\mathbf{X}$ is the matrix (4.14) and $\mathbf{R}$ is the $(K+1) \times (K+1)$ matrix given by

$$\mathbf{R} = \begin{bmatrix} 0 & 0 & 0 & \cdots & 0 \\ 0 & R_{11} & R_{12} & \cdots & R_{1K} \\ 0 & R_{21} & R_{22} & \cdots & R_{2K} \\ \vdots & \vdots & \vdots & \vdots & \vdots \\ 0 & R_{K1} & R_{K2} & \cdots & R_{KK} \end{bmatrix}.$$

We conclude this section by commenting on how the λ in (4.16) can be selected. Broadly speaking, the choice of λ reflects the researchers belief on how smooth an estimate of β should be. In practice, it is reasonable, to obtain estimates for several values of λ and choose one based on a visual inspection. In statistical theory, the belief of a researcher is quantified by mathematical assumptions on the unknown function β. Mathematical derivations then lead to an optimal λ based on this assumed smoothness of β. One can however never be certain that the postulated smoothness conditions actually hold. The optimal λ will also depend on the measure of distance between the target β and its estimator.

An approach that has gained a fairly broad acceptance is known as *cross–validation* (CV). It uses the quality of predictions of the Y_i as a criterion, rather than a postulated smoothness of β. As a result, CV often produces a $\hat{\beta}(t)$ that appears too noisy, as it attempts to optimize the predictive fit.

The method works as follows. For a fixed λ, compute the estimates $\hat{\alpha}_\lambda^{(-i)}$ and $\hat{\beta}_\lambda^{(-i)}$ using the sample without (Y_i, X_i). Then compute the best prediction of Y_i based on these estimates, i.e.

$$\widehat{Y}_\lambda^{(-i)} = \hat{\alpha}_\lambda^{(-i)} + \int \hat{\beta}_\lambda^{(-i)}(s) X_i(s) ds.$$

The preferred value of λ should make the differences $Y_i - \widehat{Y}_\lambda^{(-i)}$ collectively small. The CV method thus selects λ which minimizes

$$S_N(\lambda) = \frac{1}{N} \sum_{i=1}^{N} \left\{ Y_i - \widehat{Y}_\lambda^{(-i)} \right\}^2.$$

In software implementations, the user can generally specify the range of the values of λ and the resolution of the grid. The idea of cross–validation described above has been extended in many directions. For example, the function S_N can be replaced by a different measure of predictive fit, samples with more than one pair of observations removed can be used.

Other approaches to the selection of smoothing parameters in regression models include information criteria and restricted maximum likelihood (REML). These approaches require more background than can be presented in this textbook; Krivobokova and Kauermann (2007) is a good reference. In the context of scalar–on–function regression of this section, an approach based on a connection to a mixed model formulation has been found particularly useful. Advantage can be taken of existing software to estimate random effect variance using REML, and this variance acts as the tuning parameter that controls the smoothness of the coefficient function $\beta(\cdot)$. The numerical implementation discussed in Section 4.7 is based on this approach. An interested reader is referred to Goldsmith *et al.* (2011) and Ruppert *et al.* (2003), Chapter 5.

4.6   Regression on functional principal components

We have seen in Section 3.2 that every random function X in L^2 admits the expansion

$$X(t) = \mu(t) + \sum_{j=1}^{\infty} \xi_j v_j(t),$$

where the v_j are the functional principal components, i.e. the eigenfunctions of the covariance operator of X. Denote by $\hat{v}_j$ the estimated functional principal components (EFPC's) of the regressor functions X_i. In the FPC regression, reduction to the standard model (4.6) is achieved by using the approximation

$$X_i(t) \approx \hat{\mu}(t) + \sum_{j=1}^{p} \hat{\xi}_{ij} \hat{v}_j(t), \quad \hat{\xi}_{ij} = \int [X_i(t) - \hat{\mu}(t)] \hat{v}_j(t) dt.$$

Replacing this approximation by an equality, informally reduces the scalar response model (4.2) to

$$Y_i = \alpha + \int \beta(t) \left(\hat{\mu}(t) + \sum_{j=1}^{p} \hat{\xi}_{ij} \hat{v}_j(t) \right) dt + \varepsilon_i$$

$$= \beta_0 + \sum_{j=1}^{p} \hat{\xi}_{ij} \beta_j + \varepsilon_i,$$

where

$$\beta_0 = \alpha + \int \beta(t) \hat{\mu}(t) dt \quad \text{and} \quad \beta_j = \int \beta(t) \hat{v}_j(t) dt$$

are treated as unknown parameters. Defining the $N \times (p+1)$ matrix

$$\Xi = \begin{bmatrix} 1 & \hat{\xi}_{11} & \hat{\xi}_{12} & \cdots & \hat{\xi}_{1K} \\ 1 & \hat{\xi}_{21} & \hat{\xi}_{22} & \cdots & \hat{\xi}_{2K} \\ \vdots & \vdots & \vdots & \vdots & \vdots \\ 1 & \hat{\xi}_{N1} & \hat{\xi}_{N2} & \cdots & \hat{\xi}_{NK} \end{bmatrix},$$

we estimate the parameter vector $\boldsymbol{\beta} = [\beta_0, \beta_1, \ldots \beta_p]^T$ by the least squares estimator in the regression $\mathbf{Y} = \Xi\boldsymbol{\beta} + \boldsymbol{\varepsilon}$. Denoting the estimates thus obtained by $\hat{\beta}_0, \hat{\beta}_1, \ldots, \hat{\beta}_p$, the estimates of the parameters in (4.2) are

$$\hat{\beta}(t) = \sum_{j=1}^{p} \hat{\beta}_j \hat{v}_j(t), \quad \hat{\alpha} = \hat{\beta}_0 - \sum_{j=1}^{p} \hat{\beta}_j \int \hat{v}_j(t) \hat{\mu}(t) dt.$$

An important aspect of this method is the selection of the number of EFPC's, p. A simple but effective approach uses the smallest number p such the the first p EFPC's explain 85 or 90 percent of cumulative variance, cf. Section 1.3.

Despite its appealing simplicity, the principal component regression leads to estimates whose large sample properties are far from obvious. In addition to the selections of the number of EFPC's, p, which is analogous to the selection of λ in Section 4.5, it introduces another approximation error by treating the entries of the design matrix Ξ as deterministic, whereas, in fact, these are random quantities with a complex dependence structure; the EFPC's $\hat{v}_j$ are computed using all regression functions X_i. All methods described in this chapter must be approximate because they aim at estimating an infinite dimensional function β using only a finite sample. Their consistency can be established by imposing smoothness conditions on the target function β and allowing the tuning constants like K, λ, p to depend on the sample size N.

4.7   Implementation in the `refund` package

In this section, we present implementation in the `refund` package of the methods discussed in Sections 4.4, 4.5 and 4.6. They are also implemented in the `fda` package, see Ramsay *et al.* (2009). We use several packages than enable graphics functions. The analysis to follow must thus be preceded by the call to the required packages:

```
rm(list=ls())
library(refund);   library(ggplot2)
library(dplyr);  library(reshape2)
set.seed(9000)
```

When using many packages, it is advisable to clear the work space with the `rm` command to make sure that the analysis does not accidentally use unrelated variables. For consistency, we also set the seed of the random number generator.

We use simulated data to be able to compare the estimates to the true regression function β. We consider model (4.12) with $\alpha = 0$. We perform analysis separately for two regression functions:

$$\beta_1(t) = \sin(2\pi t), \quad \beta_2(t) = -f_1(t) + 3f_2(t) + f_3(t), \quad t \in [0,1],$$

where f_1, f_2, f_3 are normal densities with parameters specified in the following code:

```
n = 1000;   grid = seq(0, 1, length = 101)
beta = sin(grid * 2 * pi)
 beta = -dnorm(grid, mean=.2, sd=.03) +3*dnorm(grid, mean=.5,
     sd=.04)+dnorm(grid, mean=.75, sd=.05)
```

The functions β_1 and β_2 are shown as "Truth", respectively, in Figures 4.4 and 4.5.

The regressor curves are generated as

$$X(t_j) = Zt_j + U + \eta(t_j) + \varepsilon(t_j), \ Z \sim N(1, 0.2^2), \ U \sim \text{UNIF}[0, 5], \ \eta(t_j) \sim N(0, 1).$$

The iid variables $\eta(t_j)$ introduce a noisy component to the distribution of the regressor functions. The random curves, $\varepsilon(t)$, are generated as

$$\varepsilon(t) = \sum_{k=1}^{10} \frac{1}{k} \{Z_{1k} \sin(2\pi tk) + Z_{2k} \cos(2\pi tk)\},$$

where Z_{1k}, Z_{2k} are independent standard normal. The generation of $N = 1,000$ independent pairs (X_i, ε_i) of curves is accomplished with the following code:

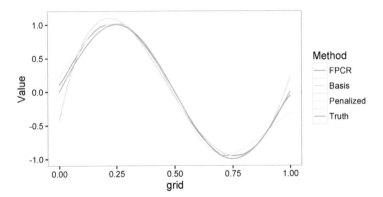

FIGURE 4.4: The regression function $\beta_1(t) = \sin(2\pi t)$ (Truth) and its estimates.

```
X <- matrix(0, nrow=n, ncol=length(grid))
for(i2 in 1:n){
  X[i2,]=X[i2,]+rnorm(length(grid), 0, 1)
  X[i2,]=X[i2,]+runif(1, 0, 5)
  X[i2,]=X[i2,]+rnorm(1, 1, 0.2)*grid
  for(j2 in 1:10){
    e=rnorm(2, 0, 1/j2^(2))
    X[i2,]=X[i2,]+e[1]*sin((2*pi)*grid*j2)
    X[i2,]=X[i2,]+e[2]*cos((2*pi)*grid*j2)
  }
}
```

We next generate the artificial data, estimate the model using the three methods and plot the estimates:

```
Y = X %*% beta * .01 + rnorm(n, 0, .4)

fit.fpcr = pfr(Y ~ fpc(X))
# without a penalty, use k= 3 for the first beta, and 15 for
    the second
# one can verify the efficacy of these choices by
# looking at the aic
fit.lin = pfr(Y ~ lf(X, bs = "ps", k = 15, fx = TRUE))
# "ps" stands for  "penalized splines", fx= TRUE means no
    penalty is used
fit.pfr = pfr(Y ~ lf(X, bs = "ps", k = 50))
# if sp is not specified, data driven smoothing is used
```

As noted toward the end of Section 4.5, the data driven approach to smoothing is based on REML in a mixed model formulation of the scalar–on–function regression. The `pfr` function uses mixed model software in the `mgcv` package, specifically the `gam` function, to address smoothness penalization.

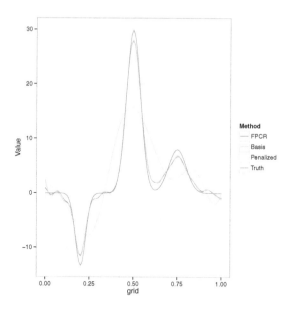

FIGURE 4.5: The regression function β_2 (Truth) constructed from normal densities and its estimates.

The remaining code plots the results.

```
coefs = data.frame(grid = grid,
                   FPCR = coef(fit.fpcr)$value,
                   Basis = coef(fit.lin)$value,
                   Penalized = coef(fit.pfr)$value,
                   Truth = beta)

coefs.m = melt(coefs, id = "grid")
colnames(coefs.m) = c("grid", "Method", "Value")

dev.new(width=6, height=3)
ggplot(coefs.m, aes(x = grid, y = Value, color = Method, group
    = Method), width=12, height=6) + geom_path() + theme_bw()
```

Goldsmith and Scheipl (2014) demonstrated on several real data that the performance of the various estimation methods, measured by prediction errors, depends on the data set. We can draw a similar conclusion from Figures 4.4 and 4.5. For the regression function β_1 in Figure 4.4, all three methods produce similar estimates. For the function β_2 in Figure 4.5, methods based on the simple basis expansions and on EFPC's produce almost identical and reasonably good estimates. The penalized approach oversmooths. Further details on the application of these methods can be found in the help files of the package refund, which also contains several other estimation methods for scalar–on–function regression.

4.8   Nonlinear scalar–on–function regression

Methods for scalar–on–function regression that do not rely on model (4.12) are referred to as nonlinear. We discuss two of them.

The first method is based on the *functional generalized additive model*, which is also called the *continuously additive model*. It takes the form

$$Y_i = \alpha + \int f(X_i(t), t)dt + \varepsilon_i, \tag{4.17}$$

where the bivariate function $(x, t) \mapsto f(x, t)$ is assumed to be smooth. It is estimated by an approximation $f(x, t) \approx \sum_{i=1}^{I} \sum_{\ell=1}^{L} f_{i\ell} B_i^\star(x) B_\ell(t)$, where the $B_i^\star$ and B_j are spline functions, which can be of different type, as the range and properties of the values $X_i(t_j)$ can be different from those of the domain points t_j. A penalty term is also introduced, following arguments analogous to those presented in Section 4.5. This method is implemented in the refund package with the default call af(X), see McLean *et al.* (2014) for further details.

A method that has gained a great deal of popularity and has been extensively studied is based on a fully nonparametric representation

$$Y_i = m(X_i) + \varepsilon_i, \tag{4.18}$$

where $m : L^2 \to \mathbb{R}$ is a functional that must be estimated. A large part of the monograph of Ferraty and Vieu (2006) is devoted to the study of this method, so we merely highlight one possible approach to the estimation of the functional m. For an arbitrary function $x \in L^2$, which however is thought of as being in some sense close to the observed regressor functions X_i, we define

$$\hat{m}(x) = \sum_{i=1}^{N} w_i(x)Y_i, \quad \sum_{i=1}^{N} w_i(x) = 1.$$

The idea is that a response to a new regressor function $X(t)$ should be a weighted average of known responses with coefficients $w_i(x)$ which give more weight to values Y_i which correspond to regressors close to x. A common approach is to use

$$w_i(x) = \frac{K(h^{-1}d(x, X_i))}{\sum_{i=1}^{N} K(h^{-1}d(x, X_i))},$$

where $d(\cdot, \cdot)$ is some measure of distance between two functions. The kernel K is defined on the nonnegative half–line and is some decreasing function, with $K(t)$ equal, or close, to zero for large t. The most critical way of selecting the Y_i to be effectively included in the estimation of $m(x)$ is through the selection of h. For any given kernel K, for the Y_i to be effectively included, $h^{-1}d(x, X_i)$ must be close to zero. Thus, if h is small, few Y_i will be included. As h increases, more and more Y_i are included. This will lead to estimates with a smaller variance but a larger bias (oversmoothing). Selecting the optimal h thus requires a balance between reducing variance and increasing the bias. Several methods of achieving it have been proposed. The simplest choice of $d(\cdot, \cdot)$ would be to consider the L^2 distance introduced in Chapter 3, i.e. $d(x, y) = \|x - y\|$, but this choice is not always optimal. If the regressors X_i are sufficiently smooth functions, the L^2 distance between the second derivatives is often useful, i.e. one measures distance between functions by $d(x, y) = \|x'' - y''\|$. Notice that this function is not a metric in the usual sense because it is possible that $d(x, y) = 0$ for some $x \neq y$. It is an example of a *semimetric*. The performance of predictors based on model (4.18) may depend quite a bit on a selection of the distance d.

A number of other approaches, both linear and nonlinear, and discussed in the review paper of Reiss *et al.* (2016).

4.9   Chapter 4 problems

4.1 Consider the design matrix $\mathbf{X}$ in (4.5). Show that if $\mathbf{X}$ has rank p, then $\mathbf{X}^T\mathbf{X}$ is nonsingular.

4.2 Consider the linear model (4.6) and the least squares estimator (4.7). Suppose $\mathbf{X}$ is a deterministic matrix of rank p and the errors ε_i are uncorrelated with variance σ_ε^2. Show that $E[\hat{\boldsymbol{\beta}}] = \boldsymbol{\beta}$ and $\mathrm{Var}[\hat{\boldsymbol{\beta}}] = \sigma_\varepsilon^2 (\mathbf{X}^T\mathbf{X})^{-1}$.

4.3 Verify relation (4.9).

4.4 Show that the matrix $\mathbf{C}_X$ in (4.9) is invertible if and only if relation (4.10) holds.

4.5 Suppose C is a covariance operator defined by (3.8). We will show in Chapter 11 that we must have $\sum_{j=1}^\infty \lambda_j < \infty$. Consider functions with expansions

$$f(t) = \sum_{j=1}^\infty f_j v_j(t), \quad g(t) = \sum_{j=1}^\infty g_j v_j(t).$$

Show that the equation $C(g) = f$ implies $\lambda_j g_j = f_j$. Find an example of eigenvalues λ_j and coefficients f_j which satisfy $\sum_{j=1}^\infty f_j^2 < \infty$, so that $f \in L^2$, but there is no $g \in L^2$ such that $C(g) = f$.

4.6 Verify relation (4.11).

4.7 Prove (4.16) by equating the partial derivatives to zero and solving a linear system.

4.8 Call the packages listed at the beginning of Section 4.7 and use the following code to reproduce Figures 4.1 and 4.2 related to the octane rating of spectrum regression.

```
plot(gasoline$octane, xlab="Gasoline sample",ylab="Octane
   rating", pch=15)

dev.new(height=6,width=7)
par(ps = 12, cex = 1, cex.lab=1.7, cex.axis=1.4, cex.main=1.7,
   cex.sub=1,mar=c(4.25,4.5,1,1))
plot.ts(gasoline$NIR[1,], lw=2, xlab="Wavelength", ylab="
   Spectrum")

plot.ts(gasoline$NIR[2,]- gasoline$NIR[1,], lw=2, lty=1, xlab=
   "Wavelength", ylab="Difference")
lines(gasoline$NIR[5,]- gasoline$NIR[1,], lw=2, lty=2, xlab="
   Wavelength", ylab="Difference")
```

4.9 Use the following code to reproduce the absorbance curves and their derivatives shown in Figure 4.3.

```
library("fda.usc"); data("tecator"); names(tecator)

absorp <- tecator$absorp.fdata
Fat20 <- ifelse(tecator$y$Fat < 20, 0, 1) * 2 + 2
plot(tecator$absorp.fdata, col = Fat20, ylab=" ",
xlab="Wavelength", main="Absorbances")
absorp.d1 <- fdata.deriv(absorp, nderiv = 1)
plot(absorp.d1, col = Fat20, ylab=" ",
xlab="Wavelength", main="Derivatives")
```

5

Functional response models

At the beginning of Chapter 4, we introduced a general classification of functional regression models. We divided them into three broad classes: scalar–on-function, function–on–scalar and function–on–function. In Chapter 4, we explained the main features and estimation principles of functional regression models using the scalar–on-function regression, and studied this type of regression in detail. In this chapter, we turn to the remaining two classes, both of them have functions as responses. We will focus on the practical aspects of estimation of these models, assuming that the reader is familiar with the general principles of penalized estimation introduced in Chapter 4.

We begin by introducing in Section 5.1 the function–on–scalar model and showing how it can be easily estimated using the least squares principle. Section 5.2 focuses on penalized estimation of this model. Penalized estimation of function–on–function regression is explained in Section 5.3. The implementation of these penalized estimation approaches in the `refund` package is illustrated in Section 5.4. The next two sections focus on inference for the function–on–function regression which is based on functional principal components. Section 5.5 is concerned with estimation, Section 5.6 with a hypothesis test. Section 5.7 explains how the validity of the assumption that the data follow a functional linear regression can be evaluated. We explain how to construct suitable diagnostic plots similar to the plots used in scalar linear regression models. We conclude with Section 5.8 which provides references to papers which elaborate on the topics studied in this chapter.

5.1 Least squares estimation and application to angular motion

The function–on–scalar regression model is

$$Y_i(t) = x_{i1}\beta_1(t) + x_{i2}\beta_2(t) + \ldots + x_{iq}\beta_q(t) + \varepsilon_i(t), \quad i = 1, 2, \ldots, N. \quad (5.1)$$

For each unit i, the response Y_i is a function. There are q scalar explanatory variables. The q functional regression parameters β_ℓ are often called the *effect functions.* Setting

$$
\mathbf{Y}(t) = \begin{bmatrix} Y_1(t) \\ Y_2(t) \\ \vdots \\ Y_N(t) \end{bmatrix}, \ \mathbf{X} = \begin{bmatrix} x_{11} & x_{12} & \cdots & x_{1q} \\ x_{21} & x_{22} & \cdots & x_{2q} \\ \vdots & \vdots & \vdots & \vdots \\ x_{N1} & x_{N2} & \cdots & x_{Nq} \end{bmatrix}, \ \boldsymbol{\beta}(t) = \begin{bmatrix} \beta_1(t) \\ \beta_2(t) \\ \vdots \\ \beta_q(t) \end{bmatrix}, \ \boldsymbol{\varepsilon}(t) = \begin{bmatrix} \varepsilon_1(t) \\ \varepsilon_2(t) \\ \vdots \\ \varepsilon_N(t) \end{bmatrix},
$$

equations (5.1) can be written as $\mathbf{Y}(t) = \mathbf{X}\boldsymbol{\beta}(t) + \boldsymbol{\varepsilon}(t)$. This section presents the simplest, yet often effective, approach to the estimation of the parameter functions β_ℓ in model (5.1), followed by an illustrative application of this model.

The least squares estimator of the functional parameter $\boldsymbol{\beta}$ is defined as the value minimizing

$$
\sum_{i=1}^{N} \left\| Y_i - \sum_{k=1}^{q} x_{ik}\beta_k \right\|^2 = \int \sum_{i=1}^{N} e_i^2(\boldsymbol{\beta}, t) \ dt, \tag{5.2}
$$

where

$$
e_i^2(\boldsymbol{\beta}, t) = \left(Y_i(t) - \sum_{k=1}^{q} x_{ik}\beta_k(t) \right)^2.
$$

For each fixed t, $\sum_{i=1}^{N} e_i^2(\boldsymbol{\beta}, t)$ is minimized if

$$
\hat{\boldsymbol{\beta}}(t) = (\mathbf{X}^T\mathbf{X})^{-1}\mathbf{X}^T\mathbf{Y}(t), \tag{5.3}
$$

see Section 4.2. Thus for any t, $\sum_{i=1}^{N} e_i^2(\hat{\boldsymbol{\beta}}, t) \le \sum_{i=1}^{N} e_i^2(\boldsymbol{\beta}, t)$, so the functions (5.3) minimize (5.2). The fitted (predicted) values and the residuals are computed using

$$
\widehat{\mathbf{Y}}(t) = \mathbf{X}\hat{\boldsymbol{\beta}}(t), \quad \hat{\boldsymbol{\varepsilon}}(t) = \mathbf{Y}(t) - \widehat{\mathbf{Y}}(t).
$$

The least squares estimation procedure above assumes that the functions Y_i are fully observed. The functional objects can be constructed from raw data $Y_i(t_{ij})$ in a way most suitable to a problem at hand. For example, basis expansions or interpolation can be used. For response functions defined on a grid which does not depend on i, and without any missing values, one can work directly with the raw observations $Y_i(t_j)$.

We illustrate the above estimation approach using data collected for a study aimed at designing an ergonomic car dashboard. Figure 5.1 shows 20 sets of 3 curves. Each set corresponds to a certain location that a driver would reach by taking his hand off the steering wheel. These data are part of a larger data set, and correspond only to one driver reaching each location 3 times, with appropriate randomization and rest intervals. There are thus 60 response curves, but one outlier curve (in the left.rear.shifter box) has been removed (the driver changed his mind half way through the motion). The sample size is thus $N = 59$. The values $Y_i(t), t \in [0, 1]$, represent the angle formed by the shoulder, elbow and wrist. The physical time required to reach a target is different for every motion; the time has been rescaled to the unit interval to represent the portion of the motion between start and end. Each of the 20 targets is described by its Cartesian coordinates $[x, y, z]$.

The initial regression that can be entertained is

$$Y_i(t) = \beta_0(t) + x_i \beta_x(t) + y_i \beta_y(t) + z_i \beta_z(t) + \varepsilon_i(t).$$

In this regression $q = 4$. Using the estimates $\hat{\boldsymbol{\beta}}(t) = [\hat{\beta}_0(t), \hat{\beta}_x(t), \hat{\beta}_y(t), \hat{\beta}_z(t)]^T$ computed according to (5.3), it has been found that the fitted curves $\widehat{\mathbf{Y}}(t) = \mathbf{X}\hat{\boldsymbol{\beta}}(t)$ provide a poor and inconsistent approximation to the responses. A better fit can be obtained by using the regression

$$\begin{aligned} Y_i(t) = {} & \beta_0(t) + x_i \beta_x(t) + y_i \beta_y(t) + z_i \beta_z(t) \\ & + x_i y_i \beta_{xy}(t) + y_i z_i \beta_{yz}(t) + z_i x_i \beta_{zx}(t) \\ & + x_i^2 \beta_{x^2}(t) + y_i^2 \beta_{y^2}(t) + z_i^2 \beta_{z^2}(t) + \varepsilon_i(t), \end{aligned}$$

with $q = 10$. The above regression can be used to predict the angular curve $Y_{x_0, y_0, z_0}(t), t \in [0, 1]$, for any location $[x_0, y_0, z_0]$ which may be different from any of the original 20 locations, but which is not too far away from the overall range of the experimental locations. Such information may aid the car cockpit design with the goal of reducing the need for excessive and difficult motions.

Figure 5.2 shows the first four EFPC's of the residual curves $\hat{\varepsilon}_i, 1 \leq i \leq 59$, computed using the formula $\hat{e}(t) = \mathbf{Y}(t) - \widehat{\mathbf{Y}}(t)$ together with the corresponding eigenvalues. Note that $\hat{\lambda}_4 = 5.3$ is over 30 times smaller than the largest eigenvalue $\hat{\lambda}_1 = 182.6$. The shape of the first EFPC $\hat{v}_1$ indicates that there is more variability in the middle part of the motion than at the start and end. This can be understood by realizing that the motion must start and end at a specific location, whereas the spatial path in the middle of the motion is less constrained. The shapes of the remaining EFPC's are to a large extent determined by their orthogonality. For example, the shape of $\hat{v}_2$ not only reflects the second most important mode of variability, but also the condition that $\hat{v}_2$ must also be orthogonal to $\hat{v}_1$, and this mathematical condition places a constraint on its shape. For this reason, the subsequent $\hat{v}_j$ have more and more zero crossings. The shapes of the EFPC's in Figure 5.2 are thus typical of those obtained from other "well–behaved" functional samples.

5.2   Penalized least squares estimation

In Section 5.1, the regression functions β_ℓ were estimated pointwise. This is justified if the response functions are smooth, as is the case for the angular

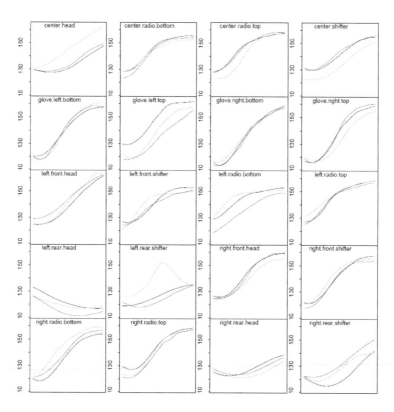

FIGURE 5.1: Angle curves for 20 different locations with 3 replications per location. Source: Faraway (1997).

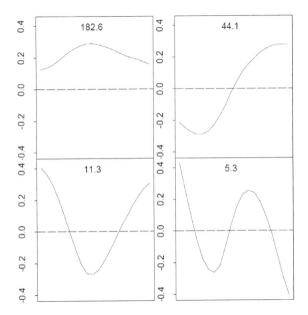

FIGURE 5.2: The first four EFPC's of the residual curves with corresponding eigenvalues. Source: Faraway (1997).

motion data. If the responses are noisy, the above approach is modified by introducing a penalty that favors smooth estimates $\hat{\beta}_\ell$. This reflects the assumptions that the functions β_ℓ in (5.1) are smooth, but the error functions ε_i are noisy, leading to noisy response functions Y_i.

The starting point of estimation with a penalty is to assume that the functions β_ℓ admit the expansion

$$\beta_\ell(t_j) = \sum_{k=1}^{K} b_{\ell k} \varphi_k(t_j), \qquad (5.4)$$

where $\varphi_1, \varphi_2, \ldots, \varphi_K$ are some basis functions, whose number should be large. These functions do not have to be orthogonal or normalized. Assuming that the responses are available at points $t_j, 1 \leq j \leq J$, model (5.1) can be written as

$$\mathbf{Y} = \mathbf{XB\Phi} + \varepsilon,$$

using the following matrices:

$$\mathbf{Y} = [Y_i(t_j), \ 1 \leq i \leq N, \ 1 \leq j \leq J], \quad (N \times J),$$
$$\mathbf{X} = [x_{i\ell}, \ 1 \leq i \leq N, \ 1 \leq \ell \leq q], \quad (N \times q),$$
$$\mathbf{B} = [b_{\ell k}, \ 1 \leq \ell \leq q, \ 1 \leq k \leq K], \quad (q \times K),$$
$$\mathbf{\Phi} = [\varphi_k(t_j), 1 \leq k \leq K, \ 1 \leq j \leq J], \quad (K \times J),$$
$$\varepsilon = [\varepsilon_i(t_j), \ 1 \leq i \leq N, \ 1 \leq j \leq J], \quad (N \times J).$$

The penalized sum of squared residuals is then defined as

$$\sum_{i=1}^{N} \sum_{j=1}^{J} (Y_i(t_j) - [\mathbf{XB\Phi}]_{ij})^2 + \sum_{\ell=1}^{q} \lambda_\ell \int \{(L\beta_\ell)(t)\}^2 \, dt. \qquad (5.5)$$

The integrand in the penalty term can be expressed in terms of the elements of the matrix $\mathbf{B}$ and derivatives of the basis functions φ_k. For example, if the second derivative is used, then

$$(L\beta_\ell)(t) = \beta_\ell''(t) = \sum_{k=1}^{K} b_{\ell k} \varphi_k''(t).$$

Similarly as in Section 4.5, the matrix $\mathbf{B}$ minimizing (5.5) can be expressed in terms of the matrices $\mathbf{X}, \mathbf{\Phi}, \mathbf{Y}$ and the matrices

$$\mathbf{\Lambda} = \operatorname{diag}(\lambda_1, \lambda_2, \ldots, \lambda_q), \quad \text{and} \quad \mathbf{R} = \left[\int (L\varphi_\ell)(t)(L\varphi_k)(t)dt, \ 1 \leq \ell, k \leq K \right].$$

To state the formula, we must introduce the concept of a vectorized matrix and of the Kronecker product of two matrices. Suppose $\mathbf{M}$ is a $p \times q$ matrix. Then $\operatorname{vec}(\mathbf{M})$ is a column vector of length pq constructed by stacking the columns of

$\mathbf{M}$, placing the first column on top. To define the Kronecker product, suppose $\mathbf{A}$ and $\mathbf{B}$ are two matrices with the following dimensions

$$\dim(\mathbf{A}) = p_A \times q_A, \quad \dim(\mathbf{B}) = p_B \times q_B.$$

Then, the Kronecker product $\mathbf{A} \otimes \mathbf{B}$ is the $p_A p_B \times q_A q_B$ matrix

$$\mathbf{A} \otimes \mathbf{B} = \begin{bmatrix} a_{11}\mathbf{B} & a_{12}\mathbf{B} & \cdots & a_{1q_A}\mathbf{B} \\ a_{21}\mathbf{B} & a_{22}\mathbf{B} & \cdots & a_{2q_A}\mathbf{B} \\ \vdots & \vdots & \vdots & \vdots \\ a_{p_A1}\mathbf{B} & a_{p_A2}\mathbf{B} & \cdots & a_{p_Aq_A}\mathbf{B} \end{bmatrix}.$$

Using this notation, the entries of the matrix $\mathbf{B}$ which minimizes (5.5) are given by

$$\text{vec}(\mathbf{B}^T) = \left[\mathbf{U}^T\mathbf{U} + \mathbf{\Lambda} \otimes \mathbf{R}\right]^{-1} \mathbf{U}^T \text{vec}(\mathbf{Y}^T), \quad \mathbf{U} = \mathbf{X} \otimes \mathbf{\Phi}^T. \tag{5.6}$$

The derivation of (5.6) is complex and is omitted. An alternative expression for $\text{vec}(\mathbf{B}^T)$ can be obtained by noticing that $\mathbf{U}^T\mathbf{U} = (\mathbf{X}^T\mathbf{X}) \otimes (\mathbf{\Phi}\mathbf{\Phi}^T)$. This follows from the following properties of the Kronecker product:

$$(\mathbf{A} \otimes \mathbf{B})^T = \mathbf{A}^T \otimes \mathbf{B}^T, \quad (\mathbf{A} \otimes \mathbf{B})(\mathbf{C} \otimes \mathbf{D}) = (\mathbf{AC}) \otimes (\mathbf{BD}).$$

The above properties can be established by direct verification that uses the definition.

In many applications, it is common to work with the response functions Y_i expressed in terms of a basis expansion:

$$Y_i(t) = \sum_{k=1}^{K} c_{ik}\varphi_k(t), \quad 1 \le i \le N. \tag{5.7}$$

Expansion (5.7), even with a large K, is typically an approximation of the data, but in the following we assume it holds exactly. It is advisable to use the same basis in expansions (5.4) and (5.7) because then the predicted functions have the appearance and the degree of smoothness similar to the responses, cf. (5.1). Introducing the matrix

$$\mathbf{C} = [c_{ik}, \ 1 \le i \le N, \ 1 \le k \le K], \quad (N \times K),$$

we can write relation (5.7) as $\mathbf{Y}(t) = \mathbf{C}\varphi(t)$, with the vector $\mathbf{Y}(t)$ defined following (5.1) and $\varphi(t)$ defined by

$$\varphi(t) = [\varphi_1(t), \varphi_2(t), \ldots, \varphi_K(t)]^T.$$

Using the relation $\boldsymbol{\beta}(t) = \mathbf{B}\varphi(t)$, we see that the penalized least squares loss function takes the form

$$\int \|\mathbf{C}\varphi(t) - \mathbf{X}\mathbf{B}\varphi(t)\|^2 \, dt + \sum_{\ell=1}^{q} \lambda_\ell \int \{(L\beta_\ell)(t)\}^2 \, dt, \tag{5.8}$$

where the norm in the first integral is the Euclidean norm in $\mathbb{R}^N$. The penalty term in (5.8) is the same as in (5.5).

To state the formula for the entries of the matrix $\mathbf{B}$ which minimizes (5.8), introduce the matrix

$$\mathbf{I}_\varphi = \left[\int \varphi_k(t)\varphi_\ell(t)dt, \ 1 \le k, \ell \le K \right], \quad (K \times K).$$

Note that if the functions φ_k are orthonormal, then $\mathbf{I}_\varphi$ is an identity matrix. For any φ_k, $\mathbf{I}_\varphi$ is nonnegative definite, Problem 5.1, so its square root $\mathbf{I}_\varphi^{1/2}$ exists. It can be shown that

$$\text{vec}(\mathbf{B}^T) = \left[\mathbf{U}_\varphi^T\mathbf{U}_\varphi + \mathbf{\Lambda} \otimes \mathbf{R}\right]^{-1}\mathbf{U}_\varphi^T\text{vec}(\mathbf{I}_\varphi^{1/2}\mathbf{C}^T), \quad \mathbf{U}_\varphi = \mathbf{X} \otimes \mathbf{I}_\varphi^{1/2}. \quad (5.9)$$

5.3   Functional regressors

If both responses and regressors are functions, the required model is called the function–on–function regression or a fully functional linear model. In its simplest formulation it takes the form

$$Y_i(t) = \alpha(t) + \int \psi(t, s)X_i(s)ds + \varepsilon_i(t), \quad i = 1, 2, \ldots, N. \quad (5.10)$$

The regressors X_i are now functions and the regression coefficient is a bivariate function, or kernel, ψ. We will consider a more general model which can include more functional regressors as well as scalar regressors. Since we want to focus on the functional regressors, we will now denote the scalar regressors using various fonts of the letter w. To lighten the notation, we will consider only two functional regressors. The methodology explained in this section can be extended to the case of more than two regressors without any substantive change. The model we consider thus has the form

$$Y_i(t) = \alpha(t) + \mathbf{w}_i^T\boldsymbol{\gamma} + \int_\mathcal{S} \psi_1(t, s)X_{i1}(s)ds + \int_\mathcal{R} \psi_2(t, r)X_{i2}(r)dr + \varepsilon_i(t). \quad (5.11)$$

For each subject i, $1 \le i \le N$, we observe

$$Y_i(t_{ij}), \ 1 \le j \le J_i, \quad X_{i1}(s_{ik}), \ 1 \le k \le K_i, \quad X_{i2}(r_{i\ell}), \ 1 \le \ell \le L_i,$$

as well as the values of the scalar regressors $\mathbf{w}_i = [w_{i1}, w_{i2}, \ldots, w_{iq}]^T$. The unknown parameters are $\boldsymbol{\gamma} = [\gamma_1, \gamma_2, \ldots, \gamma_q]^T$ and the bivariate functions ψ_1 and ψ_2. The response functions are defined on a compact interval $\mathcal{T}$, while the regressor functions on possibly different compact intervals $\mathcal{S}$ and $\mathcal{R}$. The error functions ε_i are assumed to be independent and identically distributed Gaussian random functions.

To ensure identifiability, i.e. to allow the estimators to distinguish between the kernels ψ_1, ψ_2 and the intercept function α, we assume that the regression functions are centered before the estimation. This means that they must be replaced by the functions

$$X_{i1}^c(s) = X_{i1}(s) - \frac{1}{N}\sum_{i=1}^N X_{i1}(s), \quad X_{i2}^c(r) = X_{i2}(r) - \frac{1}{N}\sum_{i=1}^N X_{i2}(r).$$

In the exposition that follows, we assume that this has been done, and we do not use the superscript ^{c}.

In this chapter, we assume that, for each i, the points $t_{ij}, s_{ik}, r_{i\ell}$ are dense in their respective domains. If these points are sparse, the methodology of Chapter 7 must be used. We now derive a discrete approximation to model (5.11) which is used to obtain estimates of the functional parameters α, ψ_1, ψ_2 and the scalar parameters $\gamma = [\gamma_1, \gamma_2, \ldots, \gamma_q]^T$. We consider the most common setting in which the functions are observed on the same dense grids for each subject, i.e. $t_{ij} = t_j, s_{ik} = s_k, r_{i\ell} = r_\ell$. Analogously to (5.7), in the following, the equalities stated in terms of basis expansions and Riemann sums are only approximations, but we use the equality sign, $=$, to be able to manipulate these expressions without adding additional error terms.

The intercept function is represented as $\alpha(t) = \sum_{m=1}^M \alpha_m B_m(t)$. The count M of the basis functions $B_1, B_2, \ldots, B_M$, as well as the count of the other basis systems used in the following, must be large. Analogously as in Section 5.2, smoothness of the estimated functions is achieved by applying suitable roughness penalties, as will be explained below. The regression kernels ψ_1 and ψ_2 are expanded using bivariate basis systems. Such systems are generally constructed as products of univariate basis functions, Problem 5.2. We thus use the approximations

$$\psi_1(t,s) = \sum_{g=1}^G \psi_{1,g} B_{1,g}(t,s), \quad \psi_2(t,r) = \sum_{h=1}^H \psi_{2,h} B_{2,h}(t,r).$$

Using these expansions and approximating the integral by a Riemann sum, we obtain

$$\int_S \psi_1(t,s)X_{i1}(s)ds = \sum_{k=1}^K (s_k - s_{k-1})\psi_1(t,s_k)X_{i1}(s_k)$$

$$= \sum_{k=1}^K (s_k - s_{k-1})\sum_{g=1}^G \psi_{1,g}B_{1,g}(t,s_k)X_{i1}(s_k)$$

$$= \sum_{g=1}^G B_{1,g,i}^\star(t)\psi_{1,g},$$

where

$$B_{1,g,i}^\star(t) = \sum_{k=1}^K (s_k - s_{k-1})B_{1,g}(t,s_k)X_{i1}(s_k).$$

Similarly,

$$\int_{\mathcal{R}} \psi_2(t,r)X_{i2}(r)dr = \sum_{h=1}^{H} B^{\star}_{2,h,i}(t)\psi_{2,h},$$

where

$$B^{\star}_{2,h,i}(t) = \sum_{\ell=1}^{L}(r_\ell - r_{\ell-1})B_{2,h}(t,r_\ell)X_{i2}(r_\ell).$$

Observe that once the bivariate basis functions and their count have been selected, the coefficients $B^{\star}_{1,g,i}$ and $B^{\star}_{2,h,i}$ are known. The counts G and H are large, they are typically products of large counts of basis functions for each of the two components. As a result of the above approximations, model (5.11) is approximated by the model

$$Y_i(t_j) = \mathbf{w}_i^T\boldsymbol{\gamma} + \sum_{m=1}^{M} B_m(t_j)\alpha_m + \sum_{g=1}^{G} B^{\star}_{1,g,i}(t_j)\psi_{1,g} + \sum_{h=1}^{H} B^{\star}_{2,h,i}(t_j)\psi_{2,h} + \varepsilon_i(t_j).$$

The unknown parameters are now scalars $\alpha_m, 1 \le m \le M$, $\psi_{1,g}, 1 \le g \le G$, $\psi_{2,h}, 1 \le h \le H$. We organize them into high–dimensional vectors

$$\boldsymbol{\alpha} = [\alpha_1, \ldots, \alpha_M]^T, \quad \boldsymbol{\psi}_1 = [\psi_{1,1}, \ldots, \psi_{1,G}]^T, \quad \boldsymbol{\psi}_2 = [\psi_{2,1}, \ldots, \psi_{2,H}]^T.$$

Using this notation, we now specify a penalized least squares criterion analogous to (5.5). The sum of squared residuals is

$$S(\boldsymbol{\gamma}, \boldsymbol{\alpha}, \boldsymbol{\psi}_1, \boldsymbol{\psi}_2) = \sum_{i=1}^{N}\sum_{j=1}^{J}(Y_i(t_j) - \mu_i(t_j; \boldsymbol{\gamma}, \boldsymbol{\alpha}, \boldsymbol{\psi}_1, \boldsymbol{\psi}_2))^2, \qquad (5.12)$$

where

$$\mu_i(t_j; \boldsymbol{\gamma}, \boldsymbol{\alpha}, \boldsymbol{\psi}_1, \boldsymbol{\psi}_2) = \mathbf{w}_i^T\boldsymbol{\gamma} + \sum_{m=1}^{M} B_m(t_j)\alpha_m$$

$$+ \sum_{g=1}^{G} B^{\star}_{1,g,i}(t_j)\psi_{1,g} + \sum_{h=1}^{H} B^{\star}_{2,h,i}(t_j)\psi_{2,h}.$$

We proceed with the derivation of the penalty terms. We motivate it by specific roughness penalties which will lead us to a general form. Suppose we want to penalize intercept function, α, whose second derivative is cumulatively too large over the interval $\mathcal{T}$. A penalty in this case is

$$P_0(\boldsymbol{\alpha}) = \int_{\mathcal{T}} \{\alpha''(t)\}^2 dt.$$

Using the basis expansion, we see that

$$P_0(\boldsymbol{\alpha}) = \sum_{m,m'=1}^{M} \alpha_m\alpha_{m'}\int_{\mathcal{T}} B''_m(t)B''_{m'}(t)dt = \boldsymbol{\alpha}^T\mathbf{D}_0\boldsymbol{\alpha},$$

where $\mathbf{D}_0$ is a nonnegative definite matrix. The roughness of the bivariate functions ψ_1 and ψ_2 can be quantified in many ways. Suppose we require that the Laplacian of ψ_1, i.e.

$$(L\psi_1)(t,s) = \frac{\partial^2}{\partial t^2}\psi_1(t,s) + \frac{\partial^2}{\partial s^2}\psi_1(t,s)$$

be cumulatively small over $\mathcal{T} \times \mathcal{S}$. Using the basis expansion, we obtain for any linear operator L, of which the Laplacian is an example,

$$(L\psi_1)(t,s) = \sum_{g=1}^{G} \psi_{1,g}(LB_{1,g})(t,s).$$

This leads to a penalty of the form

$$P_1(\boldsymbol{\psi}_1) = \int_{\mathcal{T}}\int_{\mathcal{S}} \{(L\psi_1)(t,s)\}^2 \, dt ds = \boldsymbol{\psi}_1^T \mathbf{D}_1 \boldsymbol{\psi}_1,$$

where $\mathbf{D}_1$ has entries

$$D_1(g,g') = \int_{\mathcal{T}}\int_{\mathcal{S}} (LB_{1,g})(t,s)(LB_{1,g'})(t,s) dt ds.$$

We conclude that a rich class of penalized least squares estimators is obtained by minimizing

$$S(\boldsymbol{\gamma}, \boldsymbol{\alpha}, \boldsymbol{\psi}_1, \boldsymbol{\psi}_2) + \lambda_0 \boldsymbol{\alpha}^T \mathbf{D}_0 \boldsymbol{\alpha} + \lambda_1 \boldsymbol{\psi}_1^T \mathbf{D}_1 \boldsymbol{\psi}_1 + \lambda_2 \boldsymbol{\psi}_2^T \mathbf{D}_2 \boldsymbol{\psi}_2, \tag{5.13}$$

where the sum of squared residuals, $S(\boldsymbol{\gamma}, \boldsymbol{\alpha}, \boldsymbol{\psi}_1, \boldsymbol{\psi}_2)$, is given by (5.12), and $\mathbf{D}_0, \mathbf{D}_1, \mathbf{D}_2$ are nonnegative definite matrices. The minimization can be conveniently carried out using the function `pffr` in the `refund` package, as explained in Section 5.4. As is the case with all penalized methods, the choice of the smoothing parameters $\lambda_0, \lambda_1, \lambda_2$ requires care. The function `pffr` provides default values, but other methods discussed in this book, including generalized cross–validation, can also be used.

5.4 Penalized estimation in the `refund` package

In this section, we provide examples of R code which implements the methods of Sections 5.2 and 5.3. We focus on the implementation in the `refund` package which offers the greatest flexibility at the time of writing of this textbook. Implementation in the `fda` package is discussed in Ramsay *et al.* (2009).

Function–on–scalar regression can be fitted by two different functions in `refund`: `fosr` and `pffr`. While `pffr` is more powerful and flexible overall, the following example illustrates two features currently available only in `fosr`: inputting the functional responses as `fd` objects from the `fda` package, and fully automated generalized least squares. (A third feature offered only by `fosr` is permutation testing, via the `fosr.perm` function, see Problem 5.14.)

To illustrate the interface between the function `fosr` and the `fda` package, we work with the Canadian temperature data available in the `fda` package. This data set will be revisited in a different context in Chapter 9. We consider 35 locations in Canada shown in Figure 9.2. Each of them belongs to one of four climatic regions: Arctic, Atlantic, Continental and Pacific. The temperature curves themselves are shown in Figure 9.1. We denote by $Y_i(t)$ the temperature at location i on the tth day of the year. (These are temperatures averaged over many years.) We postulate that

$$Y_i(t) = \mu(t) + \alpha_{r(i)}(t) + \varepsilon_i(t).$$

The function μ represents the overall annual pattern of temperature in Canada, we refer to it as the intercept function. The function $i \mapsto r(i)$ assigns one of the four regions to the location i. The function α_r describes the effect of the location being in region r, with $r = 2$ for Arctic, $r = 3$ for Atlantic, $r = 4$ for Continental and $r = 5$ for Pacific. The above model is a special case of model (5.1), with $\beta_1(t) = \mu(t)$, $\beta_r(t) = \alpha_r(t), r = 2, 3, 4, 5$. The 35×5 matrix $\mathbf{X}$ is defined by $x_{i1} = 1$ and

$$x_{ir} = \begin{cases} 1 & \text{if } i \text{ is in region } r, \\ 0 & \text{otherwise} \end{cases} \quad , \quad r = 2, 3, 4, 5.$$

Our objective is to estimate the functions $\mu(t)$ and $\alpha_r(t), r = 2, 3, 4, 5$. To ensure identifiability, we assume that $\sum_{r=2}^{5} \alpha_r(t) = 0$.

Due to the spacial dependence between the temperature curves at neighboring locations, the error curves ε_i are not independent, they are correlated. The ordinary least squares estimator (OLS) described in Section 5.2 is optimal if the error curves are uncorrelated. In case of dependence, an estimator that takes the dependence into account may be better. Such estimators are generally referred to as generalized least squares (GLS) estimators, and an estimator of this type is implement in the function `fosr`. We do not discuss the details, which are presented in Reiss *et al.* (2010).

The first commands, adapted from the help file for the `fRegress` function in the `fda` package, set up a functional data object representing mean temperatures (in degrees Celsius) for each day of the year at the 35 sites in Canada.

```
require(fda); require(refund)
daybasis25 <- create.fourier.basis(rangeval=c(0,365), nbasis
    =25, axes=list('axesIntervals'))
Temp.fd <- with(CanadianWeather, smooth.basisPar(day.5,
              dailyAv[,,'Temperature.C'], daybasis25)$fd)
```

We wish to estimate an intercept function representing an overall average of the mean temperature, along with coefficient functions that capture deviations from this average for each of four regions. We manually set up a model matrix **X** and sum-to-zero constraints for these four coefficient functions.

```
modmat=cbind(1,model.matrix(~factor(
    CanadianWeather$region)-1))
constraints = matrix(c(0,1,1,1,1),1)
```

The following command fits the model by penalized OLS, with smoothing parameter chosen by grid search:

```
olsmod = fosr(fdobj = Temp.fd, X = modmat, con = constraints,
    method="OLS", lambda=100*10:30)
```

Penalized GLS is implemented as follows:

```
glsmod = fosr(fdobj = Temp.fd, X = modmat, con = constraints,
    method="GLS")
```

We can then display the two sets of coefficient function estimates:

```
par(mfrow=c(2,5), mar=c(5,2,4,1))
plot(olsmod, split=1, set.mfrow=FALSE, titles=c("OLS: Intercept
    ", levels(factor(CanadianWeather$region))),ylab="", xlab="
    Day")
plot(glsmod, split=1, set.mfrow=FALSE, titles=c("GLS: Intercept
    ", levels(factor(CanadianWeather$region))),ylab="", xlab="
    Day")
```

The estimates are quite similar, but in this case GLS yields somewhat smoother coefficient functions than OLS.

We now turn to R code that illustrates how the estimation method of Section 5.3 is implemented. We begin by defining a convenience function which plots a matrix containing one curve in each row.

```
library(refund)

fplot <- function(x, y, lty=1, col=rgb(0, 0, 0, max(.1,sqrt(1/
    nrow(y)))),
    lwd=.5, ...) {
  if (missing(x)) {
    if (missing(y))
      stop("must specify at least one of 'x' and 'y'")
    else x <- 1L:NROW(y)
  } else {
    if (missing(y)) {
      y <- t(x)
      x <- 1L:NROW(y)
    } else {
      y <- t(y)
    }
```

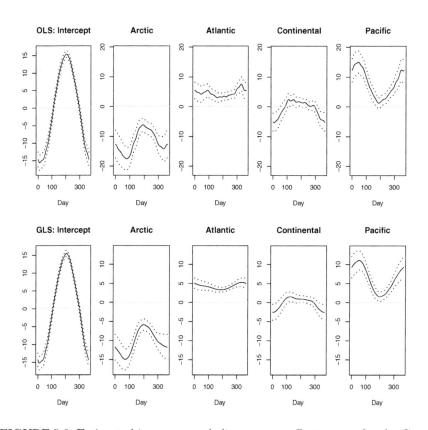

FIGURE 5.3: Estimated intercept and climate zone effect curves for the Canadian temperature data.

```
    }
    matplot(x, y, type="l", lty=lty, col=col, lwd=lwd, ...)
}
```

The simulated data follow model (5.10) with $N = 100$. Simulation is performed using the function pffrSim:

```
set.seed(9312)
data_ff <- pffrSim(scenario=c("int", "ff"), n=100)
```

The kernel ψ, denoted beta in the original source code of pffrSim, is hard coded as follows:

```
psi_st <- function(s, t) {s*cos(pi * abs(s - t)) - .19}
s <- seq(0, 1, length = 40);  t <- seq(0, 1, length = 60)
psi_true <- outer(s, t, psi_st)
```

This part of the code must be changed by the user to define a different kernel. The functions X_i are defined on a grid of 40 equidistant points in $[0, 1]$, the responses Y_i on a grid of 60 such points. The X_i are generated as linear combinations of cubic B-splines with random normal coefficients. The values $\varepsilon_i(t_j)$ are iid normal (noisy error curves). The function α is defined by 1 + dbeta(t, 2, 7) (a beta density). These hard coded defaults are easy to modify. The following code plots Figure 5.4 which shows the curves X_i and the effect of the action of the integral operator with kernel ψ.

```
par(mfrow=c(1,2))
fplot(s, data_ff$X1, xlab = "s", ylab = "", main="X(s)")
# highlight first three regressor functions X_i
matlines(s, t(data_ff$X1[1:4,]),  col=1, lwd=2)

fplot(t, attr(data_ff, "truth")$etaTerms$X1, xlab = "t",
    main=expression(integral(psi(t,s)*X(s)* ds)))
matlines(t, t(attr(data_ff, "truth")$etaTerms$X1[1:4,]),  col
    =1, lwd=2)
```

Next, we estimate the model on the simulated data using the function pffr, and plot the estimated surface $\hat{\psi}$ next to the true surface ψ. The plots are shown in Figure 5.5.

```
m_ff <- pffr(Y ~ ff(X1), data = data_ff)

psi_plot <- plot(m_ff, select = 2, pers=TRUE)[[2]]
layout(t(1:2))
# true psi surface
par(mar=c(0, 1, opar$mar[3], 1))
persp(s, t, psi_true, xlab = "s", ylab = "t", main=expression(
    psi(t,s)), phi=40, theta = 30, ticktype="detailed", zlab =
    "", border= NA, col="grey", shade = .7, zlim = range(
    beta_true))
# estimated psi surface
```

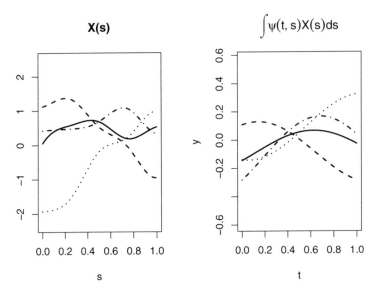

FIGURE 5.4: Regressor curves X_i and the transformed curves $\int \psi(t,s)X_i(s)ds$. The first four pairs are shown.

```
persp(psi_plot$x, psi_plot$y, matrix(psi_plot$fit, 40, 40),
    xlab = "s", ylab = "t", phi=40, theta = 30, ticktype="
    detailed", main=expression(hat(psi)(t,s)),   zlim = range(
    beta_true), zlab = "", border= NA, col="grey", shade = .7)
```

The following code produces Figure 5.6.

```
par(mar = opar$mar); layout(t(1:3))

fplot(t, data_ff$Y, xlab = "t",   ylab="", main="Observations",
    ylim=range(data_ff$Y))
matlines(t, t(data_ff$Y[1:4,]),   col =1, lty = 1:4, lwd=2)

fplot(t, attr(data_ff, "truth")$eta, xlab = "t", ylab="", main
    ="True predictor", ylim=range(data_ff$Y))
matlines(t, t(attr(data_ff, "truth")$eta[1:4,]),   col = 1, lty
    = 1:4, lwd=2)

fplot(t, fitted(m_ff), xlab = "t", ylab="", main="Estimates",
    ylim=range(data_ff$Y))
matlines(t, t(fitted(m_ff)[1:4,]),   col = 1, lty = 1:4, lwd=2)
```

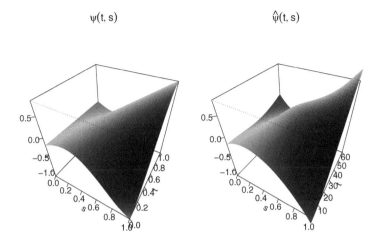

FIGURE 5.5: The true and estimated kernels.

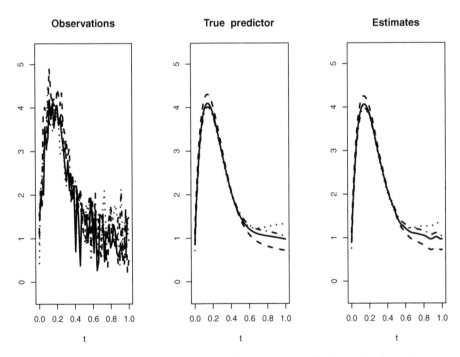

FIGURE 5.6: Illustration of the performance of fitting the function-on-function regression (5.10). The left panel shows the functional responses Y_i, the middle panel shows true predictor curves $E[Y_i(t)|X_i] = \alpha(t) + \int \psi(t,s)X_i(s)ds$, the right panel shows the fitted response curves $\hat{Y}_i(t) = \hat{\alpha}(t) + \int \hat{\psi}(t,s)X_i(s)ds$. The first four curves out of $N = 100$ are shown.

5.5   Estimation based on functional principal components

In this section, and the following two sections, we consider inference for the fully functional model based on FPC's. This section is concerned with estimation in model (5.10). The error functions ε_i are assumed to be iid, square integrable, mean zero and independent of the regressors X_i, which are iid with the same distribution as a square integrable random function X in L^2. The concept of square integrable random functions is introduced in Chapter 3, and elaborated on in Chapter 11.

We first show how to reduce the estimation of the kernel ψ to the case of the zero intercept function α. Taking the expectations of both sides of (5.10), we obtain $\mu_Y(t) = \alpha(t) + \int \psi(t,s)\mu_X(s)ds$, which leads to

$$Y_i(t) - \mu_Y(t) = \int \psi(t,s)\left(X_i(s) - \mu_X(s)\right)ds + \varepsilon_i(t).$$

We assume that the population mean functions μ_Y and μ_X can be estimated well enough. In the case of fully observed functions, or functions transformed to functional objects, e.g by spline smoothing of individual trajectories, we can use the estimators

$$\hat{\mu}_Y = \frac{1}{N}\sum_{i=1}^{N} Y_i(t), \quad \hat{\mu}_X = \frac{1}{N}\sum_{i=1}^{N} X_i(t).$$

The construction of suitable estimators in the case of sparsely observed functions is described in Chapter 7. In either case, we replace the curves X_i and Y_i, respectively, by $X_i^c(t) = X_i(t) - \hat{\mu}_X$ and $Y_i^c(t) = Y_i(t) - \hat{\mu}_Y$ and obtain an estimator $\hat{\psi}$ in the model

$$Y_i^c(t) = \int \psi(t,s)X_i^c(s)ds + \varepsilon_i(t), \quad i = 1, 2, \ldots, N. \tag{5.14}$$

The intercept function is then estimated by $\hat{\alpha}(t) = \hat{\mu}_Y(t) - \int \hat{\psi}(t,s)\hat{\mu}_X(s)ds$.

In the remainder of this section, we therefore assume that the data follow the population model

$$Y(t) = \int \psi(t,s)X(s)ds + \varepsilon(t) \tag{5.15}$$

and that $EX = 0$, which implies $EY = 0$. Our objective is to derive an estimator of the kernel ψ in (5.15). The starting point is to find a suitable expression for the kernel ψ. Since we assume zero mean functions, we have the expansions

$$X(s) = \sum_{i=1}^{\infty} \xi_i v_i(s), \quad Y(t) = \sum_{j=1}^{\infty} \zeta_j u_j(t), \tag{5.16}$$

where the v_i are the FPC's of X and the u_j the FPC's of Y, cf. Chapter 3. Assuming that $\iint \psi^2(t,s)dtds < \infty$, we will show that

$$\psi(t,s) = \sum_{k=1}^{\infty}\sum_{\ell=1}^{\infty} \frac{E[\xi_\ell\zeta_k]}{E[\xi_\ell^2]}u_k(t)v_\ell(s). \qquad (5.17)$$

Since the functional principal components of a square integrable random function form a basis in $L^2([0,1])$, the bivariate functions $\{v_i(s)u_j(t), 0 \le s,t \le 1, i,j \ge 1\}$ form a basis in $L^2([0,1] \times [0,1])$, cf. Problem 5.2. Therefore, the kernel ψ admits the expansion

$$\psi(t,s) = \sum_{k,\ell=1}^{\infty} \psi_{k\ell}u_k(t)v_\ell(s).$$

We must show that $\psi_{k\ell} = E[\xi_\ell\zeta_k]/E[\xi_\ell^2]$. Using (5.15) and (5.16), we get

$$\sum_{j=1}^{\infty} \zeta_j u_j(t) = \int \left(\sum_{k,\ell=1}^{\infty} \psi_{k\ell}u_k(t)v_\ell(s)\right)\left(\sum_{i=1}^{\infty} \xi_i v_i(s)\right) ds + \varepsilon(t)$$

$$= \sum_{k,\ell=1}^{\infty} \psi_{k\ell}u_k(t) \sum_{i=1}^{\infty} \xi_i \int v_\ell(s)v_i(s)ds + \varepsilon(t)$$

The orthonormality of the v_i thus implies that

$$\sum_{j=1}^{\infty} \zeta_j u_j(t) = \sum_{k=1}^{\infty}\sum_{i=1}^{\infty} \psi_{ki}\xi_i u_k(t) + \varepsilon(t).$$

Multiplying both sides by $u_\ell(t)$ and integrating with respect to dt, we obtain

$$\zeta_\ell = \sum_{i=1}^{\infty} \psi_{\ell i}\xi_i + \int u_\ell(t)\varepsilon(t)dt. \qquad (5.18)$$

Since $\xi_k = \langle X, v_k\rangle$, it is independent of $\int u_\ell(t)\varepsilon(t)dt$, and since both have mean zero, we conclude that $E\xi_k \int u_\ell(t)\varepsilon(t)dt = 0$. Therefore, since the scores are uncorrelated,

$$E[\xi_k\zeta_\ell] = \sum_{i=1}^{\infty} \psi_{\ell i}E[\xi_k\xi_i] = \psi_{\ell k}E[\xi_k^2].$$

This completes the verification of (5.17).

Since $E[\xi_\ell^2] = \lambda_\ell$, setting $\sigma_{\ell k} = E[\xi_\ell\zeta_k]$, relations (5.17) leads to the expression

$$\psi(t,s) = \sum_{k=1}^{\infty}\sum_{\ell=1}^{\infty} \frac{\sigma_{\ell k}}{\lambda_\ell}u_k(t)v_\ell(s)$$

and an estimator

$$\hat{\psi}(t,s) = \sum_{k=1}^{q} \sum_{\ell=1}^{p} \frac{\hat{\sigma}_{\ell k}}{\hat{\lambda}_{\ell}} \hat{u}_k(t) \hat{v}_{\ell}(s). \tag{5.19}$$

The specific estimators $\hat{\sigma}_{\ell k}$ and $\hat{\lambda}_{\ell}$ depend on the structure of the data. The estimates $\hat{\lambda}_{\ell}$ are generally available as output of any FPC software, either in the dense or sparse cases. The parameters $\sigma_{\ell k} = E[\xi_{\ell}\zeta_k]$ can be estimated by

$$\hat{\sigma}_{\ell k} = \frac{1}{N} \sum_{i=1}^{N} \langle X_i, \hat{v}_{\ell} \rangle \langle Y_i, \hat{u}_k \rangle .$$

where the X_i can be functions observed densely, or their basis expansion approximations, or reconstructions in case of sparse data. The truncation levels p and q can be selected using the cumulative percentage variance approach.

We conclude this section by showing how the above estimator can be computed in R using the `fda` package. We work with the Canadian weather data. We first define functional objects which contain curves of average daily precipitation and average daily temperature at the 35 sites in Canada, and their attributes.

```
library(fda); data(daily)
precav    <- daily$precav; tempav <- daily$tempav
daytime   <- (1:365)-0.5; dayrange   <- c(0,365); dayperiod <-
    365
```

The daily log–precipitation curves are noisy, even after averaging over the years 1960 to 1994, see Figure 2.3, so we smooth them before proceeding with the computation of the estimator. We also smooth the temperature curves, so that the responses Y_i (precipitation) and the regressors X_i (temperature) have the same degree of smoothness. The same degree of smoothness is however not necessarily required. We use the harmonic acceleration penalty and generalized cross validation (GEV) to find the smoothing parameter, see Section 2.2.

```
#  ----------- set up the fourier basis  -----------
daybasis = create.fourier.basis(dayrange, 365)
#  ----------- set up the harmonic acceleration operator
    -----------
Lcoef <- c(0,(2*pi/dayperiod)^2,0)
harmaccelLfd <- vec2Lfd(Lcoef, dayrange)
#  Do final smooth
lambda    = 1e6;  #  minimum GCV estimate, corresponding to 12
    .3 df
fdParobj = fdPar(daybasis, harmaccelLfd, lambda)
precfd = smooth.basis(daytime, precav, fdParobj)$fd
tempfd = smooth.basis(argvals=daytime, tempav, fdParobj)$fd
```

Next, we center the smooth functions and compute the EFPC's, four for temperature and six for precipitation.

```
precfd.c = center.fd(precfd); tempfd.c = center.fd(tempfd)
prec.pca = pca.fd(precfd.c,       =6); temp.pca = pca.fd(
    tempfd.c,       =4)
```

Using the output of `pca.fd` we can compute the coefficients $\hat{\sigma}_{\ell k}/\hat{\lambda}_\ell$ in (5.19).

```
sigmas = t(temp.pca$scores)%*%prec.pca$scores
cs = diag(1/temp.pca$values[1:4])%*%sigmas
```

Now we translate this into a bivariate functional data object, in terms of the basis expansions used to express the eigenfunctions (called harmonics), and create a bivariate functional data object

```
beta.coefs = temp.pca$harmonics$coefs%*%cs%*%t(prec.pca$
    harmonics$coefs)
beta.bifd = bifd(beta.coefs,sbasisobj = temp.pca$harmonics$
    basis,
  tbasisobj = prec.pca$harmonics$basis)
```

Problem 5.13 explains how to plot the function `beta.bifd`.

5.6   Test of no effect

In this section, we further illustrate inference based on FPC's expansions by deriving a test of $\psi = 0$, where ψ is the regression kernel in model (5.15). The testing problem thus is

$$H_0 : \ \psi(\cdot, \cdot) = 0, \quad \text{vs.} \quad H_A : \ \psi(\cdot, \cdot) \neq 0.$$

The function ψ is treated as an element of $L^2([0, 1] \times [0, 1])$, so H_0 is equivalent to $\iint \psi^2(t, s)dtds = 0$ or, equivalently $\psi(t, s) = 0$ for almost all $t, s \in [0, 1]$. We assume that the $(Y_i, X_i, \varepsilon_i)$ are independent copies of (Y, X, ε), the $\{\varepsilon_i, i \geq 1\}$ are independent of the $\{X_i, i \geq 1\}$, and all random functions are square integrable.

Assume that model (5.15) holds and consider the following operators:

$$C(x) = E[\langle X, x \rangle X], \quad \Gamma(x) = E[\langle Y, x \rangle Y], \quad \Delta(x) = E[\langle X, x \rangle Y].$$

The covariance operators C and Γ have nonnegative, summable eigenvalues defined by

$$C(v_i) = \lambda_i v_i, \quad \Gamma(u_j) = \gamma_j u_j.$$

Denote by $\widehat{C}, \widehat{\Gamma}$ and $\widehat{\Delta}$ the sample counterparts of C, Γ and Δ defined by

$$\widehat{C}(x) = \frac{1}{N} \sum_{n=1}^{N} \langle X_n, x \rangle X_n,$$

$$\widehat{\Gamma}(x) = \frac{1}{N} \sum_{n=1}^{N} \langle Y_n, x \rangle Y_n,$$

$$\widehat{\Delta}(x) = \frac{1}{N} \sum_{n=1}^{N} \langle X_n, x \rangle Y_n.$$

The eigenvalue/eigenfunction pairs of the $\widehat{C}$ are denoted by $\hat{\lambda}_i, \hat{v}_i$, and those of $\widehat{\Delta}$ by $\hat{\gamma}_j, \hat{u}_j$.

The test exploits the interpretation of the EFPC's as an orthonormal system that best approximates a set of functions, cf. Section 12.2 and Section 11.4. The test is also based on the relation

$$\lambda_i \Psi(v_i) = \Delta(v_i), \tag{5.20}$$

where Ψ is the integral operator with kernel $\psi(\cdot, \cdot)$, see Problem 5.4.

By (5.16), $\Psi(X) = \sum_{i=1}^{\infty} \xi_i \Psi(v_i)$. The sum involves only the indexes i such that $\lambda_i = E\xi_i^2 > 0$, so by (5.20), $\Psi(X) = 0$ if and only if $\Delta(v_i) = 0$ for every i such that $\lambda_i > 0$. Since the $u_j, j \geq 1$, form a basis, H_0 is thus equivalent to the condition $\langle \Delta(v_i), u_j \rangle = 0$ for any $i, j \geq 1$. We cannot test an infinite number of conditions, so we test

$$\langle \Delta(v_i), u_j \rangle = 0, \quad \text{for } 1 \leq i \leq p, \ 1 \leq j \leq q, \tag{5.21}$$

where p and q are such that the span of $v_1, v_2, \ldots, v_p$ approximates the support of X sufficiently well, and the span of $u_1, u_2, \ldots, u_q$ approximates the support of Y sufficiently well. We thus reduce $\Psi(X) = 0$ to $\Psi(v_i) = 0$ i.e., by (5.20), to $\Delta(v_i) = 0$, for $i = 1, 2, \ldots, p$. Note that

$$\Delta(v_i) \approx \widehat{\Delta}(v_i) = \frac{1}{N} \sum_{n=1}^{N} \langle X_n, v_i \rangle Y_n,$$

so the functions $\Delta(v_i)$ are well approximated by the span of the functions $Y_1, Y_2, \ldots Y_N$, which, in turn, is well approximated by the span of the FPC's $u_1, u_2, \ldots, u_q$.

A suitable test statistic is

$$\hat{T}_N(p, q) = N \sum_{k=1}^{p} \sum_{j=1}^{q} \hat{\lambda}_k^{-1} \hat{\gamma}_j^{-1} \left\langle \widehat{\Delta}(\hat{v}_k), \hat{u}_j \right\rangle^2. \tag{5.22}$$

It can be shown that if H_0 is true, then $\hat{T}_N(p, q)$ converges to the chi–square distribution with pq degrees of freedom, see Problem 5.8. The following algorithm summarizes the test procedure.

ALGORITHM 5.6.1 [Chi–square test based on the statistic $\hat{T}_N(p,q)$]

1. Select the number of important FPC's, p and q, using the scree plot and/or the cumulative variance method.

2. Center the X_n and the Y_n, i.e. replace them by the X_n^c and the Y_n^c defined in Section 5.5.

3. Compute the test statistics $\hat{T}_N(p,q)$ (5.22). Note that

$$\langle \hat{\Delta}(\hat{v}_k), \hat{u}_j \rangle = \left\langle \frac{1}{N} \sum_{n=1}^{N} \langle X_n, \hat{v}_k \rangle Y_n, \hat{u}_j \right\rangle = \frac{1}{N} \sum_{n=1}^{N} \langle X_n, \hat{v}_k \rangle \langle Y_n, \hat{u}_j \rangle,$$

where $\langle X_n, \hat{v}_k \rangle$ is the kth score of the X_n, and $\langle Y_n, \hat{u}_j \rangle$ is jth score of the Y_n. These scores and the eigenvalues $\hat{\lambda}_k$ and $\hat{\gamma}_j$ are available as output of suitable R functions.

4. If $\hat{T}_N(p,q) > \chi^2_{pq}(\alpha)$, reject the null hypothesis of no linear effect. The critical value $\chi^2_{pq}(\alpha)$ is the $(1-\alpha)$th quantile of the chi-squared distribution with pq degrees of freedom.

5.7 Verification of the validity of a functional linear model

Before using a functional linear regression model, its approximate validity should be checked. This can be done in a similar way as the verification that a straight line regression, $y_i = a + bx_i + \varepsilon_i$, is suitable to describe the dependence between scalar responses y_i and scalar regressors x_i. In this simple case, the most common approach is to examine the scatter plot of the pairs $(x_i, y_i), 1 \leq i \leq N$. This plot should not exhibit any obvious nonlinear pattern of the dependence of the y_i on the x_i; the points should scatter along a line. If the pairs (x_i, y_i) follow a jointly normal distribution, the plot should be football shaped, with one of the axes being a line with slope b. This approach carries over to the functional linear models considered in this chapter and in Chapter 4. Basically, two extensions are needed: 1) in case of more than one regressors, scatter plots of responses vs. each of the regressors must be examined, 2) if responses or regressors are functions, they must be replaced by their scores with respect to the estimated FPC's. Extension 1) is not related to functional data, it is a standard approach valid for the scalar model (4.1). The functional modification is that functions must be replaced by vectors of scores. Before justifying this approach, we show how it works. Consider a non–linear functional regression: $Y_i(t) = H_2(X_i(t)) + \varepsilon_i(t)$, where $H_2(x) = x^2 - 1$. For this model, the plot in the top left corner of Figure 5.7

exhibits a quadratic trend. In this model, the responses Y_i and the regressors X_i are functions. We can produce a large number of the scatter plots of the scores $\langle Y_i, \hat{u}_k \rangle$ vs. the scores $\langle X_i, \hat{v}_j \rangle$ for various choices of $k = 1, 2, \ldots$ and $j = 1, 2, \ldots$. In this example, already the plot for $k = 1$ and $j = 1$ exhibits a non–linear pattern.

We now justify the approach explained above in case of the fully functional linear model (5.10). The derivation will lead us additional useful insights into this technique. Consider the FPC expansions already used in Sections 5.5 and 5.6, but include the mean functions for completeness. We thus expand the response and regressor functions in (5.10) as

$$Y_i(t) = \mu_Y(t) + \sum_{k=1}^{\infty} \zeta_{ik} u_k(t), \quad X_i(s) = \mu_X(s) + \sum_{j=1}^{\infty} \xi_{ij} v_j(s).$$

This leads to the identity

$$\mu_Y(t) + \sum_{k=1}^{\infty} \zeta_{ik} u_k(t) = \alpha(t) + \int \psi(t, s) \left(\mu_X(s) + \sum_{j=1}^{\infty} \xi_{ij} v_j(s) \right) ds + \varepsilon_i(t)$$

$$= g(t) + \sum_{j=1}^{\infty} \xi_{ij} h_j(t) + \varepsilon_i(t),$$

where

$$g(t) = \alpha(t) + \int \psi(t, s) \mu_X(s) ds, \quad h_j(t) = \int \psi(t, s) v_j(s) ds.$$

Taking the inner product with u_ℓ, and using the fact that the u_k are orthonormal, we obtain

$$\langle \mu_Y, u_\ell \rangle + \zeta_{i\ell} = \langle g, u_\ell \rangle + \sum_{j=1}^{\infty} \xi_{ij} \langle h_j, u_\ell \rangle + \langle \varepsilon_i, u_\ell \rangle.$$

Setting $a_\ell = \langle g - \mu_Y, u_\ell \rangle$, $b_{\ell j} = \langle h_j, u_\ell \rangle$, we rewrite the above relation as

$$\zeta_{i\ell} = a_\ell + b_{\ell j} \xi_{ij} + \eta_{i,\ell j}, \tag{5.23}$$

where, setting $\varepsilon_{i\ell} = \langle \varepsilon_i, u_\ell \rangle$,

$$\eta_{i,\ell j} = \sum_{j' \neq j} b_{\ell j'} \xi_{ij'} + \varepsilon_{i\ell}.$$

To obtain a visually clearer interpretation of (5.23), we suppress the indexes ℓ and j; in each scatter plot, like each of the four scatter plots in Figure 5.7, we consider a fixed ℓ and a fixed j. Doing so, leads to the relation $\zeta_i = a + b \xi_i + \eta_i$, which is a scalar straight line regression equation. The errors η_i have mean

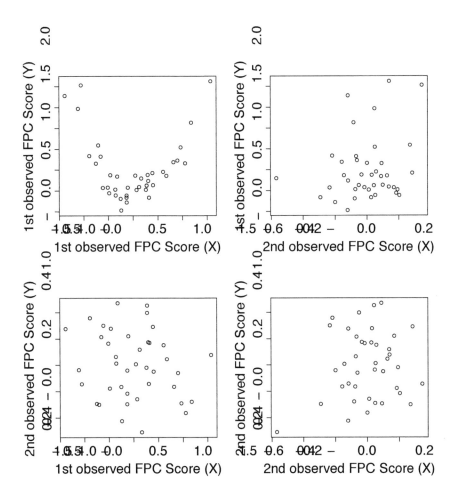

FIGURE 5.7: Predictor-response scatter plots of the first two scores of response functions versus the first two scores of the regressor functions for the non–linear regression $Y_i(t) = H_2(X_i(t)) + \varepsilon_i(t)$, where $H_2(x) = x^2 - 1$, $i = 1, \ldots, 40$.

zero and are uncorrelated with the regressors ξ_i, and uncorrelated with each other, Problem 5.10. If the functions X and Y are jointly Gaussian, then their projections, respectively, onto u_ℓ and v_j are also jointly Gaussian (Chapter 11 discusses Gaussian random functions in detail). Therefore, if model (5.10) is true with the X_i and Y_i being replications of jointly Gaussian random functions, then the scatter plots discussed in this section should be roughly football shaped. Problems 5.11 and 5.12 are concerned with analogous straight line regressions, respectively, for function–on–scalar and scalar–on–function regressions.

5.8   Extensions and further reading

Morris (2015) provides an extensive review of literature on functional regression. Section 5.1 is based on Faraway (1997). Formulas (5.6) and (5.9) are derived in Reiss *et al.* (2010). Section 5.3 is based on Ivanescu *et al.* (2015) who discuss connections to mixed models, provide a comparison to other estimation and prediction methods, and explain how to apply the function `pffr` if the curves are irregularly and sparsely observed.

The methods described in Sections 5.3 and 5.4 are substantially extended in Scheipl *et al.* (2015) who consider models of the form $Y_i(t) = \mu_i(t) + \varepsilon_i(t)$. As in (5.11), the errors ε_i are assumed to be iid Gaussian random functions, but the mean functions μ_i can depend in a complex way on scalar and functional regressors and categorical variables. To illustrate the kind modeling this more general framework can accommodate, we use the example of a medical study considered by the authors. The index i has two components, ι and κ. The index ι refers to a patient, and κ to a medical examination visit. The index κ introduces temporal dependence of the regressors; in the previous sections of this chapter the regressors were treated either as deterministic or as realizations of iid random vectors or functions. For every patient ι, two categorical variables are considered: $d(\iota) = 1$ if the patient has a disease in question, $d(\iota) = 0$ otherwise; $g(\iota) = 1$ if the patient is a man, $g(\iota) = 0$ otherwise. By $u(\iota, \kappa)$ we denote the age of patient ι at his/her κth visit. The mean function $\mu_i(t) = \mu_{\iota\kappa}(t)$ is assumed to have the following form

$$\mu_{\iota\kappa}(t) = \alpha_1(d(\iota), t) + \alpha_2(g(\iota), t) + \alpha_3(u(\iota, \kappa), t)$$
$$+ \int_S \psi_1(d(\iota), t, s) X_{1,\iota\kappa}(s) ds + \int_{\mathcal{R}} \psi_2(d(\iota), t, r) X_{2,\iota\kappa}(r) dr.$$

What could be called the intercept structure now has a more complex form than the single function α in (5.11). We must estimate the four functions $\alpha_1(0, \cdot), \alpha_1(1, \cdot), \alpha_2(0, \cdot), \alpha_2(1, \cdot)$ and the bivariate function α_3. There are also four regression kernels to be estimated. In the application considered by Scheipl *et al.* (2015), the regressors $X_{1,\iota\kappa}$ and $X_{2,\iota\kappa}$ are functions obtained from measurements along two tracts in the human brain. In medical research, it is stipulated that these functions have the potential to explain the response functions $Y_{\iota\kappa}$ constructed from measurements along a different tract, and that the pattern of dependence is different for patients with the disease.

The estimation approach based on FPC's, Section 5.5, is developed in the context of sparsely observed functions in Yao *et al.* (2005) and Yao *et al.* (2005b). Section 5.6 is based on Kokoszka *et al.* (2008) who provide an application of this test to space physics data, see also Chapter 9 of Horváth and Kokoszka (2012). The approach to verifying the linearity assumption presented in Section 5.7 was proposed by Chiou and Müller (2007) who also consider different types of scatter plots derived from FPC scores. Gabrys *et al.* (2010) derived hypothesis tests whose null hypothesis is that the error function ε_i in (5.10) are iid, and the alternative that they are correlated.

5.9 Chapter 5 Problems

5.1 Show that for any functions $\varphi_1, \varphi_2, \ldots, \varphi_K$, the $K \times K$ matrix $\mathbf{I}_\varphi$ with the entries $\varphi_{k\ell} = \int \varphi_k(t)\varphi_\ell(t)dt$, $1 \leq k, \ell \leq K$, is nonnegative definite, i.e. for any real numbers $x_1, x_2, \ldots, x_K$,

$$\sum_{k,\ell=1}^{K} \varphi_{k\ell} x_k x_\ell \geq 0.$$

The matrix $\mathbf{I}_\varphi$ is obviously symmetric. It is known, e.g. Seber and Lee (2003), Appendix A, that a symmetric nonnegative definite matrix, say $\mathbf{A}$, can be decomposed as $\mathbf{A} = \mathbf{U}\boldsymbol{\Lambda}\mathbf{U}^T$, where $\boldsymbol{\Lambda}$ is a diagonal matrix with nonnegative entries, and $\mathbf{U}$ satisfies $\mathbf{U}^T\mathbf{U} = \mathbf{I}$. Then $\mathbf{A}^{1/2} = \mathbf{U}^T\boldsymbol{\Lambda}^{1/2}\mathbf{U}$ satisfies $\mathbf{A}^{1/2}\mathbf{A}^{1/2} = \mathbf{A}$. Many software packages compute the matrix $\mathbf{A}^{1/2}$.

5.2 Show that if $\{u_j, j \geq 1\}$ and $\{v_i, i \geq 1\}$ are bases in $L^2([0,1])$ (not necessarily orthonormal), then

$$\{v_i(s)u_j(t), 0 \leq s, t \leq 1, \ i, j \geq 1\} \tag{5.24}$$

is a basis in $L^2([0,1] \times [0,1])$. Show that if $\{u_j, j \geq 1\}$ and $\{v_i, i \geq 1\}$ are both orthonormal systems, then (5.24) is an orthonormal system as well.

5.3 Consider estimator (5.19). We know that the EFPC's $\hat{u}_k$ and $\hat{v}_\ell$ are determined only up to a sign. Show that if $\hat{u}_k$ is replaced by $d_k\hat{u}_k$ and $\hat{v}_\ell$ by $c_\ell\hat{v}_\ell$, with $d_k^2 = 1, c_\ell^2 = 1$, then the estimator $\hat{\psi}(t,s)$ does not change. Show that the statistic (5.22) is also invariant to such a substitution.

5.4 Assuming model (5.15) holds and X is independent of ε, verify identity (5.20).

In Problems 5.5, 5.6 and 5.7, which require background of Chapter 11, assume that the null hypothesis of Section 5.6 holds and the following two assumptions hold as well.

ASSUMPTION 5.9.1 *The triples $(Y_n, X_n, \varepsilon_n)$ form a sequence of independent identically distributed random elements such that ε_n is independent of X_n and*

$$EX_n = 0 \quad \text{and} \quad E\varepsilon_n = 0; \tag{5.25}$$

$$E\|X_n\|^4 < \infty \quad \text{and} \quad E\|\varepsilon_n\|^4 < \infty. \tag{5.26}$$

ASSUMPTION 5.9.2 *The eigenvalues of the operators C and Γ satisfy, for some $p > 0$ and $q > 0$, $\lambda_1 > \lambda_2 > \ldots \lambda_p > \lambda_{p+1}$, $\gamma_1 > \gamma_2 > \ldots \gamma_q > \gamma_{q+1}$.*

5.5 Establish the following joint convergence:

$$\sqrt{N} \left\{ \left\langle \widehat{\Delta}(v_k), u_j \right\rangle, \ 1 \le j \le q, 1 \le k \le p \right\}$$
$$\xrightarrow{d} \left\{ \eta_{kj} \sqrt{\lambda_k \gamma_j}, \ 1 \le j \le q, 1 \le k \le p \right\}, \tag{5.27}$$

with $\eta_{kj} \sim N(0, 1)$ such that η_{kj} and $\eta_{k'j'}$ are independent if $(k, j) \ne (k', j')$.

5.6 Verify that $E\|\widehat{\Delta}\|_{\mathcal{S}}^2 = N^{-1} E\|X\|^2 E\|\varepsilon_1\|^2$.

5.7 Show that

$$\sqrt{N} \left\{ \left\langle \widehat{\Delta}(\hat{v}_k), \hat{u}_j \right\rangle, \ 1 \le j \le q, 1 \le k \le p \right\}$$
$$\xrightarrow{d} \left\{ \eta_{kj} \sqrt{\lambda_k \gamma_j}, \ 1 \le j \le q, 1 \le k \le p \right\} \tag{5.28}$$

with η_{kj} equal to those in Problem 5.5.

5.8 Using the results in Problems 5.5, 5.6 and 5.7 show that if the null hypothesis of Section 5.6 holds, then, under Assumptions 5.9.1 and 5.9.2, $\hat{T}_N(p, q) \xrightarrow{d} \chi_{pq}^2$, as $N \to \infty$.

5.9 This problem establishes the consistency of the asymptotic test based on statistic (5.22). Show that if Assumptions 5.9.1 and 5.9.2 hold, and $\langle \Psi(v_k), u_j \rangle \ne 0$ for some $k \le p$ and $j \le q$, then $\hat{T}_N(p, q) \xrightarrow{P} \infty$, as $N \to \infty$.

5.10 Consider the errors in (5.23) for fixed ℓ and j, i.e. $\eta_i = \sum_{j' \ne j} b_{\ell j'} \xi_{ij'} + \varepsilon_{i\ell}$. Assume that the pairs (X_i, ε_i) are iid and $\{\varepsilon_i, i \ge 1\}$ independent of $\{X_i, i \ge 1\}$. Show that

$$E\eta_i = 0, \quad E[\eta_i \xi_{ij}] = 0, \quad E[\eta_i \eta_m] = 0, \ i \ne m.$$

Show that $\mathrm{Var}[\eta_i] = \sum_{j' \ne j} b_{\ell j'}^2 \lambda_{j'} + E\varepsilon_{i\ell}^2$, where $\lambda_{j'}$ is the j'th eigenvalue of the covariance operator of X.

5.11 Consider the function–on–scalar regression (5.1) with a deterministic matrix $\mathbf{X}$. The scores of the ith response function are $\zeta_{i\ell} = \langle Y_i - \mu_Y, u_\ell \rangle$, $\ell = 1, 2, \ldots$ Identify the coefficients a, b and the errors η_i in the regression $\zeta_{i\ell} = a + b x_{ij} + \eta_i$.

5.12 Consider the scalar–on–function regression (4.12) and the scores $\xi_{ij} = \langle X_i - \mu_X, v_j \rangle$. Identify the coefficients a, b and the errors η_i in the regression $Y_i = a + b \xi_{ij} + \eta_i$.

5.13 At the time of writing of this book, the fda package does not provide a custom function for plotting bivariate functional data objects .bifd. However plotting them is easy by first extracting the values with the function eval.bifd followed by the application of the plotting commands used in Section 1.2, like contour and persp. Generate the contour and perspective plots of the regression function beta.bifd computed in Section 5.5.

5.14 This problem explains the application of the function fosr.perm in the refund package. Consider model (5.1). For $m < q$, we wish to test

$$H_0 : \ \beta_{m+1}(t) = \ldots = \beta_q(t) = 0, \quad \forall \ t \in [0,1]. \tag{5.29}$$

For each t, we compute the F–statistic given by (4.8) (with $p = q$), we denote its value by $\widehat{F}(t)$. The estimators $\hat{\beta}_j^{(q)}(t)$ and $\hat{\beta}_j^{(m)}(t)$ generally involve a penalty and so depend all values $Y_i(s), s \in [0,1]$. Therefore, even for a fixed t, $\hat{\beta}_j^{(q)}$ will not have an F–distribution. Moreover, the null hypothesis (5.29) postulates that the *functions* are zero.

The following approach to testing (5.29) has been used. Consider the test statistic

$$\widehat{F}_{\mathrm{sup}} = \sup_{t \in [0,1]} \widehat{F}(t).$$

The distribution of $\widehat{F}_{\mathrm{sup}}$ under H_0 is unknown, so it is approximated as follows. For each of the $N!$ permutations, π, of $\{1, 2, \ldots, N\}$ consider the permuted data set

$$\mathcal{D}_\pi = (Y_{\pi(i)}(t), \ t \in [0,1], \ x_{i1}, \ldots x_{iq}).$$

Using $\mathcal{D}_\pi$, we compute the $\hat{\beta}_{\pi,j}^{(m)}(t), j = 1, 2, \ldots, m$ and the $\hat{\beta}_{\pi,j}^{(q)}(t), j = 1, 2, \ldots, q$, and the statistic $\widehat{F}_{\mathrm{sup}}^\pi$. The $N!$ values $\widehat{F}_{\mathrm{sup}}^\pi$ are arranged from the smallest to the largest, and their 95th percentile is denoted by $c_{0.95}(N)$. The null hypothesis (5.29) is rejected if $\widehat{F}_{\mathrm{sup}} > c_{0.95}(N)$. The justification of this approach (for vector regression) is provided, for example, in Manly (1991), pp. 156-162.

The function fosr.perm produces the output similar to the one shown in Figure 5.8. The continuous blue curve is the function $t \mapsto \widehat{F}(t)$. The dashed red line is the critical level $c_{0.95}(N)$. The gray lines are the $N!$ functions $t \mapsto \widehat{F}^\pi(t)$. If the maximum of the continuous blue curve is above the dashed critical level, we reject H_0.

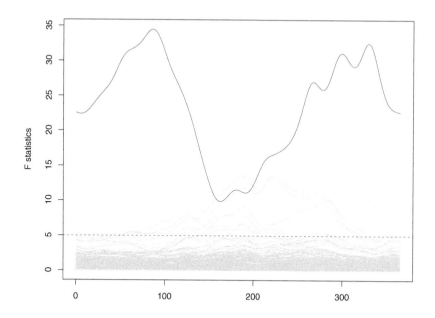

FIGURE 5.8: Graphical output of the function `fosr.perm` applied to the regression of annual temperature in Canada on the four climatic regions.

Use the following code, adapted from the help file, to assess the significance of the effects of the climatic regions on the annual temperature curves in Canada, cf. Figure 5.3. Clearly formulate the null and alternative hypotheses. Explain what all arguments of `fosr.perm` are.

```
require(fda); require(refund)
smallbasis  <- create.fourier.basis(c(0, 365), 25)
tempfd <- smooth.basis(day.5, CanadianWeather$dailyAv[,,"
    Temperature.C"], smallbasis)$fd
        Xreg = cbind(1, model.matrix(~factor(CanadianWeather
    $region)-1))
conreg = matrix(c(0,1,1,1,1), 1)    # constrain region effects
    to sum to 0

regionperm = fosr.perm(fdobj=tempfd, X=Xreg, con=conreg,
method="OLS", nperm=200, prelim=30)
```

5.15 This problem further explains the idea of a permutation test using a two sample testing problem. It is based on Section 10.5.1 of Ramsay *et al.* (2009).

We observe two samples of curves defined on the same interval: $X_1, X_2, \ldots X_N$ and $X_{N+1}, X_{N+2}, \ldots X_{N+M}$. The first N curves are iid draws from a population with mean function μ_1, the last M curves are iid draws from a population with mean function μ_2. We want to test

$$H_0 : \mu_1(t) = \mu_2(t), \quad \forall \ t \in [0, 1].$$

We want to use the test statistics

$$\widehat{T}_{\mathrm{sup}} = \sup_{t \in [0,t]} \widehat{T}(t), \quad \widehat{T}(t) = \frac{|\bar{X}_N(t) - \bar{X}_M(t)|}{\left\{ N^{-1} \widehat{V}_N(t) + M^{-1} \widehat{V}_M(t) \right\}^{1/2}},$$

where

$$\bar{X}_N(t) = \frac{1}{N} \sum_{i=1}^{N} X_i(t), \quad \bar{X}_M(t) = \frac{1}{M} \sum_{i=N+1}^{N+M} X_i(t),$$

and where $\widehat{V}_N(t)$ and $\widehat{V}_M(t)$ are the corresponding sample variances.

The null hypothesis is rejected if $\widehat{T}_{\mathrm{sup}}$ is large, i.e. if the normalized difference between the sample mean functions is large for some t. To compute exact critical values, we would need to know the distribution of $\widehat{T}_{\mathrm{sup}}$ under H_0. This distribution is approximated as follows. Denote by π one of the $(N+M)!$ permutations of the index set $\{1, \ldots, N, N+1, \ldots, N+M\}$. If H_0 is true, the two samples

$$X_{\pi(1)}, \ldots, X_{\pi(N)} \quad \text{and} \quad X_{\pi(N+1)}, \ldots, X_{\pi(N+M)}$$

have the same mean. The empirical distribution (histogram) of the $(N+M)!$ values of $\widehat{T}_{\mathrm{sup}}^{\pi}$ is used as an approximation to the null distribution of $\widehat{T}_{\mathrm{sup}}$. (In numerical implementations, not all $(N+M)!$ values are used, as this number can be very large; a random sample of them is used.)

Using the help file of the function tperm.fd in the fda package, and the Berkeley growth data, test if the mean growth curves for boys and girls are the same. Obtain a graph analogous to Figure 5.8 and interpret it.

5.16 Perform the test described in Problem 5.14 using the function Fperm.fd in the fda package.

5.17 [Concurrent functional model] Dynamic biometric measurements motivated the early development of functional data analysis. A classical example, explained in some detail in Section 10.2.3. of Ramsay *et al.* (2009), is the gait data. Figure 5.9 shows suitably defined angles made at matching times by knees and hips of 39 boys. A question of interest is the extent to which the hip angle can explain the knee angle.

A model proposed to study this question takes the form

$$Y_i(t) = \beta_0(t) + \beta_1(t) X_i(t) + \varepsilon_i(t).$$

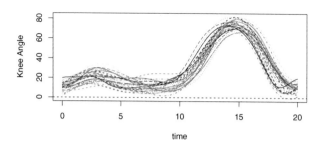

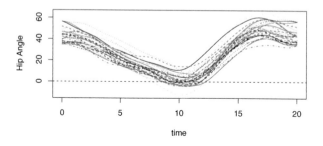

FIGURE 5.9: The curves show suitably defined knee and hip angles for a sample of children.

It is know as the concurrent model because the value $Y_i(t)$ of the response curve Y_i depends only on the value of the regressor X_i at the same (concurrent) time t. This should be contrasted with the fully functional model (5.10). The coefficient functions β_0 and β_1 can be estimated in several ways, saturated basis expansion with a penalty term being the most popular one.

(a) Use the following code to create functional objects containing hip and knee angles:

```
require(fda)
gaittime <- as.numeric(dimnames(gait)[[1]])*20
gaitrange <- c(0,20)
gaitbasis <- create.fourier.basis(gaitrange, nbasis=21)
harmaccelLfd <- vec2Lfd(c(0, (2*pi/20)^2, 0), rangeval=
    gaitrange)
gaitfd <- smooth.basisPar(gaittime, gait,
        gaitbasis, Lfdobj=harmaccelLfd, lambda=1e-2)$fd
hipfd  <- gaitfd[,1];  kneefd <- gaitfd[,2]
```

Using the objects defined above, reproduce Figure 5.9.

(b) Use the call

```
knee.hip.f <- fRegress(kneefd ~ hipfd)
```

to estimate the concurrent model to predict the knee angle from the hip angle. Plot the functions β_0 and β_1 and comment on the information contained in their shape.

(c) We want to compare the predicted curve for boy 1 in the sample with his observed knee angle curve. Use the the following calls to plot these two curves in separate panels:

```
plot(knee.hip.f$yhatfd$y[,1],type='l',lwd=2,lty=2, ylab="Knee
    Angle")
plot(kneefd['boy1'], lwd=2)
```

Notice and explain the different ranges of the x–arguments in the two plots.

Plot the two curves in one panel, as shown in Figure 5.10. Extract the x–axis values of the predicted curve with argvals. Use legend to identify the curves.

5.18 Estimation of the function–on–function regression (5.10) is also implemented in the fda function linmod. Reproduce Figure 5.5 using linmod rather than the refund function pffr.

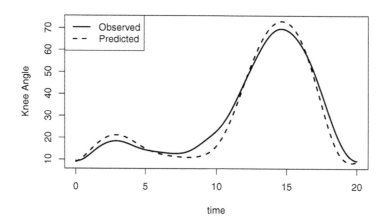

FIGURE 5.10: Observed and predicted knee angle curves for boy 1.

6

Functional generalized linear models

In applications it is quite common to encounter data which are non-normal. For example, counts, data which are strictly positive, or skewed data. To model such data, distributions such as Bernoulli, Binomial, Poisson, Exponential, etc are often appropriate. However, with such distributions, the assumptions made in linear models are generally not satisfied. Generalized linear models provide effective tools to handle such data.

At the heart of GLM's is the view that, for non-Gaussian data, the mean of the response is usually not linearly related to the predictors. This is maybe most clearly illustrated for binary outcomes. In that case, the mean of the response is the probability of "success", and therefore must be bounded between 0 and 1. Assuming that this parameter is a linear combination of predictors can easily lead to mean estimates/predictions which fall outside this region. GLM's address this problem by assuming that the mean of the response is linearly related to the predictors only after a suitable transformation. We will first review the core concepts of scalar GLM's in Section 6.1. In Section 6.2 we will discuss how to handle functional predictors, while in Section 6.3 we will examine functional response models. In Section 6.4 we will discuss how the models can be fit using the `refund` package in R, and an application to DTI data is presented in Section 6.5. We conclude with a brief discussion of more advanced theoretical topics in Section 6.6.

6.1   Background

The brief background we provide here follows the outline and notation given in McCullagh and Nelder (1989), and we refer to them for a more detailed discussion. The theory for generalized linear models is built upon exponential families, though this can be generalized. Defining an exponential family is usually done in one of two equivalent ways. The first defines a set of densities that factor in a particular way, while the second is more constructive and built on exponential tilting. Here we take latter approach as it helps shed light on why densities are so difficult to work with in infinite dimensions, but the former is also quite common in introductory statistics texts, e.g. Casella and Berger (2002). Consider a density $f_0(y)$ with respect to a base measure,

μ, on $\mathbb{R}$. Recall that if the base measure is Lebesgue, we call the density continuous, if it is the counting measure then we say the density is discrete. Such generality is useful so that both cases can be considered simultaneously. Let $M(\theta)$ denote the moment generating function of f_0,

$$M(\theta) = \int e^{\theta x} f_0(x) \, d\mu(x),$$

and assume that it is finite for $\theta \in \Theta$, an interval of $\mathbb{R}$ containing the origin. A second density $f(y|\theta)$ can then be constructed via *exponential tilting*:

$$f(y|\theta) = f_0(y) \exp\{\theta y - b(\theta)\}, \tag{6.1}$$

where

$$b(\theta) = \log M(\theta) = \log \int f_0(y) \exp\{\theta y\} \, d\mu(y),$$

is the cumulant generating function, i.e. the log of the moment generating function. The family of densities $\{f(\cdot|\theta) : \theta \in \Theta\}$ is called an *exponential family*.

It is common to allow for a *dispersion parameter*, ϕ, and the typical form for expressing the densities is given by

$$f(y|\theta, \phi) = \exp\left\{ \frac{\theta y - b(\theta)}{a(\phi)} + c(y, \phi) \right\}, \tag{6.2}$$

where ϕ can either be known or estimated from data. In this context $f_0(y) = \exp\{c(y, \phi)\}$ would also include a dispersion parameter.

EXAMPLE 6.1.1 Consider the standard normal density and cumulant generating function:

$$f_0(x) = \frac{1}{\sqrt{2\pi}} \exp\left\{ -\frac{x^2}{2} \right\} \quad \text{and} \quad b(\theta) = \log(M(\theta)) = \frac{\theta^2}{2}.$$

Generating a new density via exponential tilting yields

$$f(x|\theta) = f_0(x) \exp\{\theta x - b(\theta)\} = \frac{1}{\sqrt{2\pi}} \exp\left\{ -\frac{x^2}{2} + \theta x - \frac{\theta^2}{2} \right\}$$

$$= \frac{1}{\sqrt{2\pi}} \exp\left\{ -\frac{(x - \theta)^2}{2} \right\},$$

which is just the normal density with mean θ and variance 1. So, the exponential family consists of all normal densities with unit variance. By including a dispersion parameter for the variance, we can get any member of the normal family, i.e. any mean and variance. Using μ for the mean and σ^2 for the variance, any normal density can be expressed as

$$f(x|\mu, \sigma^2) = \frac{1}{\sqrt{2\pi\sigma^2}} \exp\left\{-\frac{(x-\mu)^2}{2\sigma^2}\right\}$$

$$= \frac{1}{\sqrt{2\pi\sigma^2}} \exp\left\{-\frac{x^2}{2\sigma^2} + \frac{\mu x}{\sigma^2} - \frac{\mu^2}{2\sigma^2}\right\}$$

$$= \exp\left\{\frac{\mu x - \mu^2/2}{\sigma^2} - \frac{x^2}{2\sigma^2} - \frac{1}{2}\log(2\pi\sigma^2)\right\}.$$

We can now express the density using the form given in (6.2). Let $\theta = \mu$ and $\phi = \sigma^2$, then

$$f(x|\theta, \phi) = \exp\left\{\frac{\theta x - \theta^2/2}{\phi} - \frac{x^2}{2\phi} - \frac{1}{2}\log(2\pi\phi)\right\}.$$

We therefore have that

$$b(\theta) = \theta^2/2, \quad a(\phi) = \phi, \quad \text{and} \quad c(x, \phi) = -\frac{x^2}{2\phi} - \frac{1}{2}\log(2\pi\phi).$$

$\square$

Maybe the most commonly used GLM is logistic regression, which involves binomial outcomes. Logistic regression involves transforming the success parameter p onto the *logit* scale, defined as $\text{logit}(p) = \log(p/(1-p))$, $0 < p < 1$. This is illustrated in the next example.

EXAMPLE 6.1.2 Let $X \sim Bin(n, p)$ be a binomial random variable. Then the density (with respect to counting measure on $\{1, \ldots, n\}$) is given by

$$f(x|p) = \binom{n}{x} p^x (1-p)^{n-x} = \exp\left\{x\log p + (n-x)\log(1-p) + \log\binom{n}{x}\right\}$$

$$= \exp\left\{x\,\text{logit}\,p + n\log(1-p) + \log\binom{n}{x}\right\}.$$

To express the distribution in a canonical form, we would take $\text{logit}(p) = \log(p/(1-p)) = \theta$, which yields

$$f(x|\theta) = \exp\left\{x\theta + n\log\left(1 - \frac{\exp(\theta)}{1+\exp(\theta)}\right) + \log\binom{n}{x}\right\}$$

$$= \exp\left\{x\theta - n\log\left(1 + \exp(\theta)\right) + \log\binom{n}{x}\right\}.$$

This implies that

$$b(\theta) = n\log\left(1 + \exp(\theta)\right), \qquad c(x) = \log\binom{n}{x}.$$

$\square$

Suppose now that a random variable Y has density $f(y|\theta, \phi)$, then one can verify the following properties

$$E[Y] = \mu = b'(\theta), \qquad \text{Var}[Y] = a(\phi)b''(\theta). \qquad (6.3)$$

In a generalized linear model, GLM, one no longer assumes that μ is a linear combination of the predictors, call them $x_1, \ldots, x_p$. Instead, one introduces a *link function*, $g(\cdot)$, and assumes that the linear relationship holds only after transforming μ by g, i.e.

$$\eta := g(\mu) = \sum_{i=1}^{p} x_i \beta_i = \mathbf{X}^\top \boldsymbol{\beta}. \qquad (6.4)$$

Throughout we will assume that g is a known user chosen function. Common examples include logistic or probit (inverse normal cdf) transformations for binary data, the logarithm for Poisson data (i.e. a log-linear model), and the inverse for exponential data. It may be a bit confusing at this point as we now have three different parameters which are all related through transformations, μ, θ, and η:

$$\eta = g(\mu) \quad \text{and} \quad \mu = b'(\theta) \quad \text{imply} \quad \eta = g(b'(\theta)).$$

It is common to use a link function g which results in $\eta = \theta$, i.e. $g = (b')^{-1}$, in which case g is called the *canonical link function*.

For a given sample $\{(\mathbf{X}_n, Y_n)\}$, one can estimate $\boldsymbol{\beta}$ using maximum likelihood, which leads to a particular set of *estimating equations*. The likelihood can be expressed as

$$\prod_{n=1}^{N} f(Y_n|\theta_n) = \exp\left\{ \sum_{n=1}^{N} \frac{\theta_n Y_n - b(\theta_n)}{a(\phi)} + \sum_{n=1}^{N} c(Y_n, \phi) \right\},$$

and therefore the derivative of the log-likelihood (wrt $\boldsymbol{\beta}$) is given by

$$\sum_{n=1}^{N} \frac{\partial \theta_n}{\partial \boldsymbol{\beta}} \frac{(Y_n - b'(\theta_n))}{a(\phi)}.$$

We can express $\partial \theta_n / \partial \boldsymbol{\beta}$ as

$$\theta_n = b'^{-1}(g^{-1}(\mathbf{X}_n^\top \boldsymbol{\beta})) \implies \frac{\partial \theta_n}{\partial \boldsymbol{\beta}} = [b'^{-1}]'(g^{-1}(\mathbf{X}_n^\top \boldsymbol{\beta}))[g^{-1}]'(\mathbf{X}_n^\top \boldsymbol{\beta})\mathbf{X}_n.$$

To ease the notation, one usually introduces the *mean* and *variance* functions

$$\mu(\eta) = g^{-1}(\eta) \qquad V(\mu) = \text{Var}(Y) = a(\phi)b''(b'^{-1}(\mu)),$$

and one can show that

$$\frac{\partial \theta_n}{\partial \boldsymbol{\beta}} = \frac{\mu'(\mathbf{X}_n^\top \boldsymbol{\beta})}{V(\mu(\mathbf{X}_n^\top \boldsymbol{\beta}))}\mathbf{X}_n. \qquad (6.5)$$

Finding the MLE's is now equivalent to solving the following equations,

$$S(\boldsymbol{\beta}) := \sum_{n=1}^{N} \frac{\mu'(\mathbf{X}_n^\top \boldsymbol{\beta})}{V(\mu(\mathbf{X}_n^\top \boldsymbol{\beta}))} \mathbf{X}_n (Y_n - \mu(\mathbf{X}_n^\top \boldsymbol{\beta})) = 0, \qquad (6.6)$$

which must be solved in terms of $\boldsymbol{\beta}$ (typically via Newton-Raphson) to find the maximum likelihood estimates. Given the context, the above are often referred to as *estimating equations*. The estimating equations take into account two aspects which are not present in least squares estimates. The first is the nonlinearity inherent in the problem when g is not the identity function. The second is that the variance of Y changes with its mean, which holds for all families commonly used in GLM's except the normal.

6.2   Scalar-on-function GLM's

When X_n is replaced by a function, $X_n(t)$, (6.4) must be modified. In particular, we use

$$\eta_n := g(\mu_n) = \alpha + \int X_n(t)\beta(t) \, dt = \alpha + \langle X_n, \beta \rangle.$$

However, as with the case of scalar-on-function regression, one cannot estimate $\beta(t)$ without imposing additional structure. One can again use a small number of FPC's of X, and then fit a scalar GLM, or one can use basis functions and add a smoothness penalty on the β. We outline the penalized basis approach here. We approximate β using K basis functions:

$$\beta(t) \approx \sum_{k=1}^{K} \langle \beta, B_k \rangle B_k(t) = \sum_{k=1}^{K} \beta_k B_k(t).$$

This gives the familiar relationship

$$\langle X_n, \beta \rangle = \sum_{k=1}^{K} \beta_k X_{nk},$$

where $X_{nk} = \langle X_n, B_k \rangle$. We then utilize existing methods for scalar GLM's by noticing that

$$\eta_n \approx \alpha + \sum_{k=1}^{K} x_{nk} \beta_k.$$

At present, there are then two primary approaches to fitting the model. The first is to use a small number K, selected by BIC, AIC, or cross-validation, in conjunction with the glm function in R. The second approach is to take K large and include a penalty on the smoothness of β when fitting the model. This can be done using the refund package in R.

6.3   Functional response GLM

Fitting a GLM with a functional response is, at least conceptually, more challenging than with a scalar response. This is primarily due to the currently incomplete development of non-Gaussian functional models. For example, what is the functional version of an exponential random variable or a binomial? The answers to these questions are not immediately obvious, though important for fully generalizing the scalar GLM to functional settings.

To get beyond this issue, it has become common to avoid functional densities and instead focus on moment estimates and estimating equations of a form analogous to (6.6). We do not discuss the theoretical problems in this section (see Section 6.6), but rather move directly to the formulation of currently used GLM's with functional responses. To fit a function-on-scalar GLM, we assume that at each point t, the random variable $Y_n(t)$ satisfies a GLM with scalar predictor x_n and the same density at each time point, except for a parameter $\eta_n(t) = \alpha(t) + x_n\beta(t)$, which is allowed to vary with time. In other words, we assume that the outcome has a density of the form $Y_n(t) \sim f(y|\eta_n(t))$, where $\eta_n(t)$ is now allowed to vary with time. As before

$$\eta_n(t) := g(\mathrm{E}[Y_n(t)]) = \alpha(t) + x_n\beta(t),$$

where g is a known link function. We do not make assumptions about the joint distribution of the $Y_n(t)$ across t. One can then estimate $\alpha(t)$ and $\beta(t)$ at each time point, usually assuming that it is smooth across t.

The function-on-function GLM is nearly the same, but we have

$$\eta_n(t) = \alpha(t) + \int X_n(s)\beta(t,s) \ ds. \tag{6.7}$$

To estimate $\beta(t,s)$ in (6.7), we fit the model at a finite number of time points, possibly including a penalty on the overall smoothness of β. This can be done in the `refund` package. To fit (6.7) we expand $\beta(t,s)$ using a basis:

$$\beta(t,s) \approx \sum_{k=1}^{K} \beta_k(t)B_k(s),$$

i.e. we expand $\beta(t,s)$ at each time point t. We then have that

$$\eta_n(t) \approx \alpha(t) + \sum_{k=1}^{K} \beta_k(t)X_{nk},$$

which can now be fit using the function-on-scalar methodology.

6.4 Implementation in the `refund` package

We begin with a simulation study to help illustrate the discussed methods and highlight the R tools. We demonstrate how to fit a function-on-scalar and function-on-function GLM. We then proceed to several data examples.

Function-on-Scalar Regular Design: We simulate data from a function-on-scalar GLM using a latent variable approach. We assume here that the design is regular, meaning that all functions are observed at the same time points. First we generate data from a function-on-scalar linear model, then we threshold the resulting outcome:

$$Z_n(t) = \alpha(t) + x_n \beta(t) + \varepsilon_n(t) \quad \text{and} \quad Y_n(t) = 1_{Z_n(t)>0}.$$

We will take the errors, $\varepsilon_n(t)$ to be Mátern Gaussian (see Problem 1.5) processes whose point-wise variance is equal to 1 so that the resulting GLM is a function-on-scalar probit model. Recall that the link function for a probit model is the standard normal cdf:

$$E[Y_n(t)] = g((\alpha(t) + x_n \beta(t)) = \Phi(\alpha(t) + x_n \beta(t)).$$

For this example we will take

$$\alpha(t) = \cos(\pi t + \pi) \qquad \beta(t) = 2t. \tag{6.8}$$

First let us define some basic data generating parameters.

```
library(refund); library(MASS)
N<-200; M<-50; time = seq(0,1, length=M)
# Mean Function
mu_f<-function(t){cos(pi*t + pi)}
# Coefficient Function
beta_f<-function(t){2*t}
# Matern Covariance Function
C_f<-function(t,s){
        sig2<-1; rho<-0.5
        d<-abs(outer(t,s,"-"))
        tmp2<-sig2*(1+sqrt(3)*d/rho)*exp(-sqrt(3)*d/rho)}
```

Next we generate the data from a functional probit model. Here Z represents an unobserved latent process. Generating a functional GLM can be surprisingly challenging and using latent processes is an effective strategy since Gaussian processes are much easier to define. The predictor is X while the outcome is Y.

```
set.seed(2000)
Sigma<-C_f(time,time)
mu<-mu_f(time)
```

```
beta<-beta_f(time)
X<-rnorm(N,mean=0)
Xdata<- data.frame(X = X)
Z<-mvrnorm(N,mu,Sigma) + X%*%t(beta)
Y<-matrix(Z>0,nrow=N)
```

We now fit a functional GLM with $Y(t)$ as the functional outcome and X as
the scalar predictor using the pffr function from the refund package in R.
The parameters and their estimates are plotted in figure 6.1.

```
pffr_fit<-pffr(Y~X ,family=binomial(link="probit"),
               yind=time, data=Xdata)
par(mfrow=c(1,2))
plot(pffr_fit,select=1,xlab="",ylab="Intercept",ylim=c(-1.25,1
    .5),cex.lab=1.25)
points(time,mu_f(time),typ="l",lty=4,lwd=4)
plot(pffr_fit,select=2,xlab="t",ylab="Slope",ylim=c(-0.25,2.75
    ),cex.lab=1.25)
points(time,beta_f(time),typ="l",lty=4,lwd=4)
```

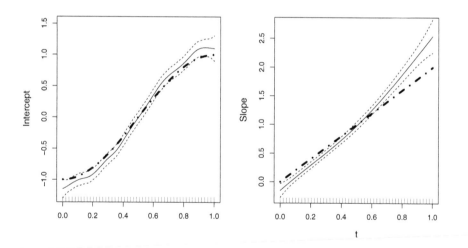

FIGURE 6.1: Plot of the slope and intercept parameters, (6.8), and their
estimates for a function-on-scalar probit model with a regular design. The
parameters functions are plotted in dot-dashed lines while the estimates are
plotted using solid lines. Point-wise confidence intervals, are plotted using
dashed lines.

Function-on-Scalar Irregular Design: Here we show how a slight modification can be used to handle irregular designs, where the number of time
points per subject may vary and/or the curves can be observed at different

times. We simulate data in much the same way, but the domain points are randomly generated. Again we define the same parameters, but reduce the number of points per curve to 5 instead of 50.

```
library(refund); library(MASS)
N<-200; M<-10; time = seq(0,1,length=M)
# Mean Function
mu_f<-function(t){cos(2*pi*t/2 + pi)}
# Coefficient Function
beta_f<-function(t){2*t}
# Matern Covariance Function
C_f<-function(t,s){
        sig2<-1; rho<-0.5
        d<-abs(outer(t,s,"-"))
        tmp2<-sig2*(1+sqrt(3)*d/rho)*exp(-sqrt(3)*d/rho)}
```

We now generate the data, but we also generate the observed domain points according to a uniform.

```
set.seed(2000)
X<-rnorm(N,mean=0)
Xdata= data.frame(X = X)
Z<-numeric(0)
time_all<-numeric(0)
for(i in 1:N){
  time = sort(runif(M))
  Sigma<-C_f(time,time)
  mu<-mu_f(time)
  beta<-beta_f(time)
  Z<-c(Z,mvrnorm(1,mu,Sigma) + beta*X[i])
  time_all<-c(time_all,time)}
Y = as.numeric(Z>0)
Y_all<-data.frame(.obs=rep(1:N,each=M),.index=time_all,.value=
    Y)
```

When the data are irregular, then Y needs to be in the form of a data frame with each observation put on a different row. The first column indicates which subject the observation is from, the second column indicates the domain point of the observation, and the third column indicates the actual value. The model is then fit much the same way as before, however, because the domain locations are included in Y, we no longer need to include them as part of the yind option.

```
pffr_fit<-pffr(Ydummy~X, family=binomial(link="probit"),
              data=Xdata, ydata=Y_all)
par(mfrow=c(1,2))
t_pts = seq(0,1,length=50)
plot(pffr_fit,select=1,xlab="",ylab="Intercept",ylim=c(-1.25,1
    .5),cex.lab=1.25)
points(t_pts,mu_f(t_pts),typ="l",lty=4,lwd=4)
```

```
plot(pffr_fit,select=2,xlab="",ylab="Slope",ylim=c(-0.5,2),
    cex.lab=1.25)
points(t_pts,beta_f(t_pts),typ="l",lty=4,lwd=4)
```

Since the Y values are actually input via the ydata option, the left side of the formula used to define the model plays no role; one could put any variable/label there and it would not be used. To stress this, in the code we used the label Ydummy to demonstrate that it is not used.

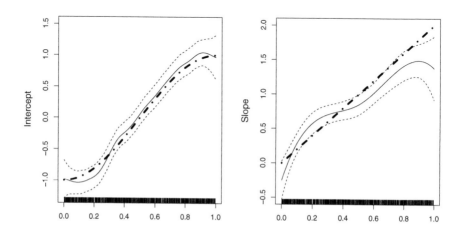

FIGURE 6.2: Plot of the parameters and their estimates for a function-on-scalar probit model with an irregular design. The parameters functions are plotted in dot-dashed lines while the estimates are plotted using solid lines. Point-wise confidence intervals, are plotted using dashed lines.

Function-on-function: In this section we show how to fit a function-on-function GLM by modifying our previous simulation scheme. We generate the X using the same Matern process and we make a slight adjustment to β, from (6.8), to make it bivariate

$$\beta(t, s) = 4|ts|.$$

We also increase the sample size, N, as the parameter estimation for function-on-function models tends to be very noisy, and in practice a visual inspection of the parameter estimates is important.

```
library(refund); library(MASS)
N<-1000; M<-50; time = seq(0,1,length=M)
# Mean Function
mu_f<-function(t){cos(2*pi*t/2 + pi)}
# Coefficient Function
```

```
beta_f<-function(t,s){
  d<-abs(outer(t,s,"*"))
  return(4*d)}
# Matern Covariance Function
C_f<-function(t,s){
        sig2<-1; rho<-0.5
        d<-abs(outer(t,s,"-"))
        tmp2<-sig2*(1+sqrt(3)*d/rho)*exp(-sqrt(3)*d/rho)}
```

Everything else looks very similar. However, notice that when multiplying the matrices X and beta we divide M, the number of points. This is to emulate taking an L^2 inner product.

```
set.seed(2000)
Sigma<-C_f(time,time)
mu<-mu_f(time)
beta<-beta_f(time,time)
X=mvrnorm(N,mu,Sigma)
Xdata<- data.frame(X = X)
Z<-mvrnorm(N,mu,Sigma) + X%*%t(beta)/M
Y<-matrix(Z>0,nrow=N)
```

Fitting the model is now nearly the same as for the Function-on-scalar setting, but we have to indicate to pffr that X should be treated as functional, in which case a function-on-function GLM will be fit. The defaults for the ff function can be used, but it is usually advisable to check alternative bases to make sure the fit seems reasonable. Here we fit using the default (basistype = "te") which uses a "tensor basis", that is, constructs a bivariate basis by taking tensor products of univariate bsplines.

```
pffr_fit<-pffr(Y~ff(X,basistype="te",xind=time),
                family=binomial(link="probit"),
                yind=time, data=Xdata)
par(mfrow=c(1,3),mar=c(4,4,0,0))
plot(pffr_fit,select=1,xlab="",ylab="intercept",cex.lab=1.5)
points(time,mu_f(time),typ="l",lty=4,lwd=4)
plot(pffr_fit,select=2,pers=TRUE,xlab="",ylab="",main="
    Estimated Slope")
persp(time,time,beta_f(time,time),xlab="",ylab="",zlab="True
    Slope",theta=30,phi=30)
```

6.5 Application to DTI

We return the DTI dataset in the Refund package (see Section 1.5). We consider the the corpus collosum tracts, CCA, as predictor functions, $X_n(t)$, and compare them between MS patients and controls, where the outcome Y_n

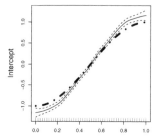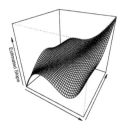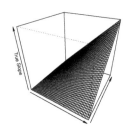

FIGURE 6.3: Plot of the parameters and their estimates for a function-on-function probit model with a regular design. The intercept parameter (left) is plotted in dot-dashed lines while the estimates are plotted using solid lines. Point-wise confidence intervals, are plotted using dashed lines. The bivariate slope function is estimated using a tensor product of univariate splines (middle panel), while the true slope is given in the right panel.

is 1 if case (MS) and 0 if control (healthy):

$$E[Y_n(t)] = \Phi^{-1}(\alpha + \langle X_n, \beta \rangle).$$

In this case we are fitting a scalar-on-function probit GLM, and we swap from using `pffr` to `pfr`. Since the outcome is a binary scalar, the intercept is a scalar and the only parameter function to estimate and plot is the intercept $\beta(t)$. Calling on `pfr` and `pffr` are very similar.

```
Y<-DTI$case; X<-DTI$cca
N<-dim(X)[1]; M<-dim(X)[2]; time = seq(0,1,length=M)
Xdata<- data.frame(X = X)
dti_fit<-pfr(Y~lf(X,argvals=time),
            family=binomial(link="probit"), data=Xdata)
plot(dti_fit,xlab="t",ylab=expression(paste(beta(t))))
```

Here we see a key problem in handling models with functional predictors; it is often very difficult to interpret the parameters because (1) the noise is typically high and (2) the effects must be considered jointly across t. Here we see that the parameter estimate is mostly negative, meaning that the thinner CCA tracts are associated with MS. There is a pronounced dip towards the end suggesting that thickness right before the end of the tract might play an important role in MS. As a last point, the ends of the parameter estimate seem to be positive (or are nearly so). One might be tempted to interpret this as saying that thicker ends of the CCA are associated with MS. However, this is incorrect as one must consider the estimates jointly across the tract. Namely, since the ends are positive and the middle is negative, this indicates that it is the relative difference between the thickness at the ends and the thickness in the middle which is associated with MS. This is an important point to keep in mind, when dealing with a functional predictor that has both positive and

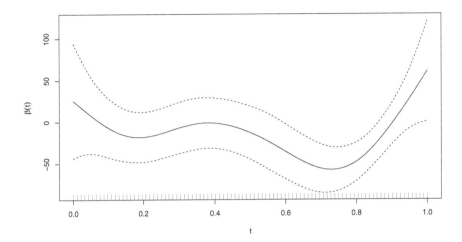

FIGURE 6.4: Estimated slope paramter for DTI data.

negative estimates, it means that it is the contrast at different points which is important for the outcome.

6.6 Further reading

Densities play a central role in the theory of scalar and multivariate GLM's. However, densities have proven a very difficult concept to fully integrate into functional data methods. In this section we briefly discuss why and provide some additional references for the interested reader. The exponential tilting method in Section 6.1 can be generalized to a functional setting, that is, a base density, f_0, over a Hilbert space $\mathcal{H}$ can be defined and then tilted to create a family of densities. However, this family will always be highly incomplete. For example, one cannot generate the family of all Gaussian processes using this approach. This has to do with the *orthogonality* of measures in infinite dimensional spaces (Ibragimov and Rozanov, 1978; Stein, 1999). Two probability measures, P_1 and P_2 are orthogonal if there exists a set A such that $P_1(A) = 1$ and $P_2(A) = 0$. We say they are equivalent if for any (measurable) set A, $P_1(A) = 0$ implies $P_2(A) = 0$ and vice versa.

To help visualize this issue, compare standard Brownian motion, $\{W(t) : t \in [0,1]\}$, to a standard Brownian bridge, $\{B(t) : t \in [0,1]\}$. Recall $B(t)$ can be obtained from $W(t)$ through the relation $B(t) \stackrel{D}{=} W(t) - tW(1)$. In other words, $B(t)$ is a Brownian motion which has been "tied down" at the ends. Since $B(1) = 0 \neq W(1)$ with probability 1, we have that the two probability measures or orthogonal and we can't define a density of one with respect to the other. This problem goes far beyond this simple example as, in some sense, most Gaussian processes are orthogonal to each other, especially when comparing processes with different covariance functions. Thus, densities and likelihoods over function spaces are currently very difficult to successfully implement.

Another view of this problem is discussed in Li and Linde (1999); Delaigle and Hall (2010); Dai *et al.* (2016). In particular, they focus on "small ball probabilities", that is, the probability of a random function falling into a small ball in the respective function space. They focus on how quickly the probability tends to zero as these balls shrink. In the finite dimensional case, these probabilities shrink like the volume of the ball multiplied by the density of the random vector evaluated at the center of the ball. However, this does not occur in the infinite dimensional setting, and further supports the difficulty in defining densities.

6.7   Chapter 6 problems

6.1 Show that the density defined in (6.1) is proper probability density.

6.2 Verify (6.3).

6.3 Verify (6.5).

6.4 Assume that Y_n are independent normals with mean $\mathrm{E}[Y_n] = \mathbf{X}_n^\top \boldsymbol{\beta}$ and variance $\mathrm{Var}(Y_n) = \sigma^2$, as in Example 6.1.1. Find the estimating equations (6.6), i.e. replace μ' etc with their corresponding values.

6.5 Assume that Y_n are independent Bernoullis with mean $\mathrm{E}[Y_n] = p_n = \mathrm{logit}^{-1}(\mathbf{X}_n^\top \boldsymbol{\beta})$ and variance $\mathrm{Var}(Y_n) = p_n(1 - p_n)$, as in Example 6.1.2. Find the estimating equations (6.6), i.e. replace μ' etc with their corresponding values.

6.6 Consider a Gaussian process $Z(t)$ in $L^2[0,1]$ with mean 0 and covariance C. Suppose we also have a second process $X(t) := \mu(t) + Z(t)$. Let $v_j(t)$ be the eigenfunctions of C and λ_j the eigenvalues.

a. Write down the joint density of $\{\langle Z, v_1 \rangle, \ldots, \langle Z, v_m \rangle\}$ for some fixed $m \in \mathbb{N}$, call it $g(x_1, \ldots, x_m)$. Write down the joint density of $\{\langle X, v_1 \rangle, \ldots, \langle X, v_m \rangle\}$, $f(x_1, \ldots, x_m)$.

b. You can obtain the density of $\{\langle X, v_i \rangle\}$ with respect to $\{\langle Z, v_i \rangle\}$, by taking their ratio, i.e. $f(x_1, \ldots, x_m)/g(x_1, \ldots, x_m)$. Write down this ratio and simplify it.

c. Suppose you tried to take the limit $m \to \infty$ of the ratio you obtained in (b). What requirement on μ do you need to ensure the limit exists and is finite? (hint: think about the inner product from Reproducing Kernel Hilbert Spaces)

d. Based on the above, form a hypothesis about when the distribution of X is orthogonal/equivalent to the distribution of Z.

7

Sparse FDA

Longitudinal studies often produce data which can be viewed as functional data observed at a small number of time points. In such a setting, custom methods are needed to incorporate the sparse structure. In this chapter, we present the core methods for analyzing such data, which we call *sparse functional data analysis*.

In Section 7.1 we introduce the core concepts and ideas behind sparse FDA. In Section 7.2 we present several options for estimating the mean function, including local linear regression, basis expansions, and reproducing kernel Hilbert spaces. In Section 7.3 we discuss how the methods from Section 7.2 can be used for covariance function estimation. In Section 7.4 we illustrate how to carry out FPCA, and how it can be used for individual curve estimation/prediction. We complete this Chapter with Section 7.5, which discusses how to fit functional regression models with sparsely observed curves.

7.1 Introduction

In this chapter, we explore methods for functions which are sparsely observed. Such data arise quite often in longitudinal studies in which researchers are only able to observe subjects at a relatively small number of time points, which can be different for different patients. For example, patients may arrive for diagnostic examinations only at a handful of irregularly and sparsely distributed time points. For this reason, we will refer to the methods in this section as *Sparse Functional Data Analysis* or S-FDA. An entire textbook could be devoted to this topic, thus in one chapter we will only be able to outline key methodologies and differences with methods for densely observed functions. In S-FDA, smoothing is not applied to individual sparse trajectories. Imputed smooth trajectories can be obtained only after information from the whole sample has been suitably combined. A distinguishing feature from approaches discussed in previous chapters is thus that one does not usually directly embed each unit into a function space. To do so could produce very unreliable curve estimates and potentially introduce a substantial amount of bias into the observations. Instead, most sparse FDA methods rely heavily on pooling across subjects and utilizing *nonparametric smoothing*, also called *scatterplot smoothing* or *nonparametric regression*.

Throughout this chapter, we assume that the observed data are of the form

$$Y_{nm} = Y_n(t_{nm}) = \mu(t_{nm}) + \varepsilon_n(t_{nm}) + \delta_{nm}, \qquad (7.1)$$

for $1 \leq n \leq N$, $1 \leq m \leq M_n \leq M < \infty$, and $t_{nm} \in [0,1]$. The index n refers to a unit or subject, for example a patient; there are M_n observations at times t_{nm} for subject n. The mean function μ is required to be smooth, and indeed, the discussed methods work better for smoother functions. The error functions $\varepsilon_n(t)$ are subject specific and induce correlation between observations on the same subject. The errors δ_{nm} are assumed to be measurement noise and are iid across both n and m. While the number of observations per subject, M_n, is allowed to vary with n, it is assumed to be bounded. The currently available methods have been developed under the assumption that the observations on the units, Y_n, are iid, we have a simple random sample of the units. It is also assumed that the times $t_{nm}, 1 \leq n \leq N, 1 \leq m \leq M$, cover the interval $[0,1]$ relatively densely.

We first discuss a longitudinal data example. The data concerns the Comparison of Age-Related Macular Degeneration Treatments Trials, CATT. The data are freely available at www.med.upenn.edu/cpob/studies/CATT.shtml.

EXAMPLE 7.1.1 The Comparison of Age-Related Macular Degeneration Treatments Trials, CATT, is an ongoing multicenter clinical trial with the aim of understanding the impact of different treatments on Age Related Macular Degeneration, AMD, which impacts eyesight. Each subject made clinical visits every four weeks with up to 27 total visits (including a baseline visit). After their initial visit, subjects were randomly assigned to one of four treatment groups involving different administrations of the drugs Lucentis and Avastin. The primary outcome we consider here is Visual Acuity Score, VAS. A VAS of 0.5 is 20/40 vision, i.e., the subject reads at 20 feet, what a person with sound eyesight could read at 40 feet. Thus, the lower the VAS, the worse a subject's vision. A subject is considered legally blind when their VAS is less than 0.1 or 20/200, while a VAS of 0 is actually blind.

While the study design is balanced (all subjects are observed at the same times), there are many missing values. Only 19% of subjects are completely observed, while 54% are missing 1-3 observations. We will use study time (as opposed to subject age) for the domain, but either could be used.

A plot of the data is given in the left panel of Figure 7.1 and point-wise mean estimates and confidence intervals in the right panel.

□

Before a systematic exposition of the chief aspects of sparse FDA, we present an introductory example which illustrates the differences between nonparametric curve smoothing and the type of smoothing used in sparse FDA.

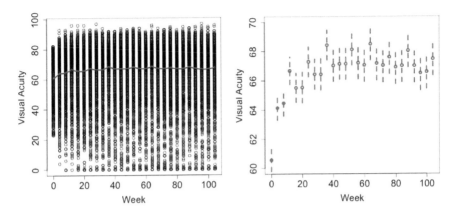

FIGURE 7.1: Plot of raw visual acuity data (multiplied by 100) against the week of the trial in the left panel (red line indicates the point-wise mean). In the right panel we plot the point-wise estimates by week along with 95% point-wise confidence intervals. No smoothing is preformed. When pooled across subjects, the domain is densely sampled, but the majority of subjects do not contribute to every clinical visit.

In nonparametric smoothing, theoretical considerations focus on the interplay between the number of points, say M, at which the curve is observed, and the bandwidth h over which the observations are, in some sense, averaged. In asymptotic theory for sparse FDA, we consider the interplay of the number of subjects, N, the number of observations per subject, $M_n \sim M$, and the bandwidth, h. As noted above, the main conceptual difference is that in sparse FDA, smoothing is based on pooling information from N subjects, rather than using just one curve. Thus, when N is large, potentially very precise estimators can be constructed. However, if the number of observations per subject is small, the precision, as measured by the MSE, will deteriorate. Example 7.1.2 quantifies these arguments in a simple setting.

EXAMPLE 7.1.2 For simplicity, we assume that the number of observations per subject, M_n, is the same, $M_n \equiv M$. Consider the model

$$Y_{nm} = \mu(t_{nm}) + \varepsilon_n + \delta_{nm},$$

where t_{nm} are iid $U(0,1)$, $\{\varepsilon_n\}$ are iid with mean zero and finite variance τ^2, and $\{\delta_{nm}\}$ are iid measurement errors with mean zero and variance σ^2. We also assume that the three sequences, $\{t_{nm}\}$, $\{\varepsilon_n\}$, and $\{\delta_{nm}\}$ are independent of each other. Such a model is a called a random effects model with a block design, and is also one of the simplest models for longitudinal data. In the context of (7.1), the subject specific error curves are constant over time: $\varepsilon_n(t_{mn}) \equiv \varepsilon_n$.

Our goal is to estimate the function $\{\mu(t) : t \in [0,1]\}$. The simplest form of local polynomial (kernel) smoothing would be to estimate $\mu(t)$ by averaging those points which are close to t. Taking an interval around t (assume that t is an interior point for simplicity) of radius h and averaging all points within that interval yields

$$\hat{\mu}_h(t) = \frac{\sum_n \sum_m Y_{nm} 1_{t_{nm} \in [t-h, t+h]}}{\sum_n \sum_m 1_{t_{nm} \in [t-h, t+h]}}.$$

Defining a kernel function $K(\cdot)$ by

$$K(x) = \begin{cases} 1 & \text{if } |x| \leq 1 \\ 0 & \text{otherwise} \end{cases},$$

we can then express the estimator as

$$\hat{\mu}_h(t) = \frac{\sum_n \sum_m K\left(\frac{t - t_{nm}}{h}\right) Y_{nm}}{\sum_n \sum_m K\left(\frac{t - t_{nm}}{h}\right)}.$$

As is always the case in kernel smoothing, the estimator is sensitive to the choice of h. For larger values of h, one is averaging more values and getting a smoother estimate. For smaller values of h, one is averaging more locally, and getting a rougher estimate. This kind of analysis leads to what is commonly known as the *bias-variance tradeoff*. A smaller value of h will use only points close to t and so $\hat{\mu}_h(t)$ will be centered (i.e. in expected value) closer to $\mu(t)$. However, since for a smaller h, fewer points t_{nm} will be used, the variance of $\hat{\mu}_h(t)$ will be larger. To illustrate this point quantitatively, let us plug in the expression for Y_{nm} into the estimator:

$$\hat{\mu}_h(t) = \frac{\sum_n \sum_m K\left(\frac{t - t_{nm}}{h}\right) \mu(t_{nm})}{\sum_n \sum_m K\left(\frac{t - t_{nm}}{h}\right)} + \frac{\sum_n \sum_m K\left(\frac{t - t_{nm}}{h}\right)(\varepsilon_n + \delta_{nm})}{\sum_n \sum_m K\left(\frac{t - t_{nm}}{h}\right)}.$$

The expected value of the estimator can then be approximated as follows:

$$E[\hat{\mu}_h(t) | \{t_{nm}\}] \approx (2h)^{-1} \int_{t-h}^{t+h} \mu(x) \, dx, \tag{7.2}$$

see Problem 7.1. To get a handle on the bias, we have to control how far $(2h)^{-1} \int_{t-h}^{t+h} \mu(x) \, dx$ can be from $\mu(t)$. So let

$$F(h) = \int_{t-h}^{t+h} \mu(x) \, dx = \int_{t}^{t+h} \mu(x) \, dx + \int_{t-h}^{t} \mu(x) \, dx.$$

Carrying out a one term Taylor expansion of F about zero we get

$$F(h) = 2\mu(t)h + O(h^3), \tag{7.3}$$

if the second derivative of μ is bounded on $[0, 1]$ (see Problem 7.2). We then have that

$$\left| \mu(t) - \frac{1}{2h} \int_{t-h}^{t+h} \mu(x) \, dx \right| \leq Ch^2,$$

uniformly in t. We see that the bias term tends to zero, as $h \to 0$, at a rate of h^2. Turning to the variance, we have that

$$\text{Var}(\hat{\mu}_h(t)|\{t_{nm}\}) = \frac{\sum_n \text{Var}\left(\sum_m K\left(\frac{t-t_{nm}}{h}\right)(\varepsilon_n + \delta_{nm})|\{t_{nm}\}\right)}{\left(\sum_n \sum_m K\left(\frac{t-t_{nm}}{h}\right)\right)^2}. \tag{7.4}$$

Examining the numerator we have that

$$\text{Var}\left(\sum_m K\left(\frac{t-t_{nm}}{h}\right)(\varepsilon_n + \delta_{nm})|\{t_{nm}\}\right) \approx Mh(\tau^2 + \sigma^2) + M^2 h^2 \tau^2, \tag{7.5}$$

see Problem 7.3. The double sum in the denominator of (7.4) is proportional to $N^2 M^2 h^2$. The asymptotic variance is therefore given by

$$\frac{NMh(\tau^2 + \sigma^2) + NM^2 h^2 \tau^2}{N^2 M^2 h^2} = \frac{(\tau^2 + \sigma^2)}{NMh} + \frac{\tau^2}{N},$$

and we see that it is larger for smaller values of h.

Now that we have an expression for both the variance and the bias, we can find the h which leads to an optimal estimate. This means making the MSE as small as possible, which in turn, means choosing h such that the square of the bias and the variance are roughly equal (see Problem 7.4). However, the second term in the variance does not involve h, thus to balance the variance with the bias, we equate only those terms which do, namely,

$$\frac{1}{NMh} = h^4 \Rightarrow h = (NM)^{-1/5},$$

which gives

$$Bias^2 \sim (NM)^{-4/5} \quad \text{and} \quad Var \sim (NM)^{-4/5} + N^{-1}.$$

We thus have three scenarios: (1) where $(NM)^{4/5}/N \to 0$, (2) where $(NM)^{4/5}/N \to \infty$, and (3) where $(NM)^{4/5}/N \to c > 0$. This implies that the tipping point occurs when

$$M \sim N^{1/4}.$$

If M grows faster than this rate, then one has what is known as a *parametric rate of convergence*, namely the MSE of $\hat{\mu}(t)$ converges to zero at a rate of N^{-1}, which is what one commonly finds in parametric models. If M is fixed (i.e. doesn't go to infinity with N) then the MSE of $\hat{\mu}(t)$ converges to zero at the slower rate of $N^{-4/5}$, this rate is commonly called the *nonparametric rate of convergence*, as it comes up often in the theory of nonparametric statistics. If M tends to infinity, but slower than $N^{1/4}$, then one ends up with a rate which is somewhere in the middle. For this reason, the conditions $M \to \infty$ and $M/N^{1/4} \to 0$ can be viewed as a sort of informal *rule of thumb* for determining if a sparse methodology is appropriate.   □

The above example illustrates, in the simplest possible setting, the basic behavior of kernel smoothers applied to sparse functional data. One can estimate nonlinear mean functions, but the convergence rates can be slower than for parametric estimates. However, it is rare to be in a setting where one has a nonparametric rate of convergence for functional data, furthermore, it takes a surprisingly small number of observations to reach a parametric rate.

7.2   Mean function estimation

In this section we will discuss several options for estimating mean functions in S-FDA, including local polynomial regression, basis functions, and reproducing kernel Hilbert spaces. We will then illustrate these methods on data from CATT.

Local polynomial regression

Local polynomial regression is a special case of *kernel smoothing* or *scatter plot smoothing*. A number of texts now exist on this topic and we refer to Wand and Jones (1994); Fan and Gijbels (1996) for more details. The intuition behind local polynomial regression is that one does not attempt to fit a linear or polynomial trend globally, but instead fits a separate model at each individual time point, t. Observations which are far from the current t are down weighted using a kernel function so that they have less impact on the fit of the model. To illustrate this point, consider the following simple example, which leads to a kernel estimator, called a Nadarya–Watson estimator, that locally fits constant functions. Suppose the observations Y_{nm} follow model (7.1). We want to construct an estimate of $\mu(t)$. Define it as the solution to the weighted least squares estimation problem:

$$\hat{\mu}(t) = \operatorname{argmin}_\alpha \sum_{n=1}^{N} \sum_{m=1}^{M_n} K\left(\frac{t - t_{nm}}{h}\right) (Y_{nm} - \alpha)^2, \qquad (7.6)$$

where $K(\cdot)$ is a kernel function. In principle, any function which satisfies $0 < \int_{-\infty}^{\infty} K(t) < \infty$ can be used as a kernel function. However, it is common to also require certain properties for the "moments" of the kernel. In particular, one nearly always uses a kernel which is symmetric, hence satisfying $\int_{-\infty}^{\infty} tK(t) =$

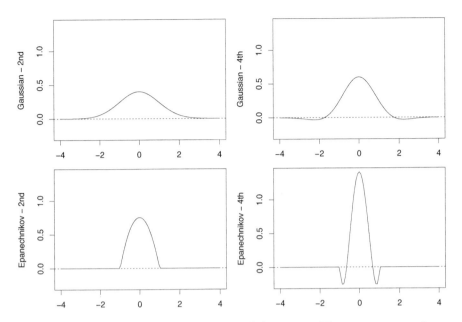

FIGURE 7.2: Plot of four different kernel functions. The top two panels are Gaussian kernels and the bottom two are Epanechnikov. The left two panels are 2nd order kernels while the right two are 4th order.

0. Such a kernel is called a *second order kernel*. A $\nu \in \mathbb{N}$ order kernel is one which satisfies $\int_{-\infty}^{\infty} t^j K(t) = 0$ for $j = 1, 2, \ldots, \nu - 1$, i.e. ν is the first nonzero moment. Notice that this implies that the kernels must take both positive and negative values when $\nu \geq 3$, which, at first, might seem counter intuitive. Higher order kernels better control the asymptotic estimation bias, but also require more smoothness assumptions. Most 2nd order kernel functions can be adjusted to achieve higher orders. As an illustration, Figure 7.2 plots two popular 2nd order and 4th order versions of the Gaussian and Epanechnikov kernels.

Simple calculations (Problem 7.9) show that the solution, $\hat{\mu}(t)$, is the same as in Example 7.1.2, namely,

$$\hat{\mu}(t) = \frac{\sum_n \sum_m K\left(\frac{t-t_{nm}}{h}\right) Y_{nm}}{\sum_n \sum_m K\left(\frac{t-t_{nm}}{h}\right)}. \tag{7.7}$$

Such an estimator is called a *Nadarya–Watson* estimator and can be thought of as fitting a constant function locally at each point t. Once this point of view is adopted, it is easy to imagine fitting higher order polynomials. Consider the loss function

$$L(\beta) = \sum_{n=1}^{N} \sum_{m=1}^{M_n} K\left(\frac{t - t_{nm}}{h}\right) \left(Y_{nm} - \sum_{i=0}^{P} \beta_i (t - t_{nm})^i\right)^2, \qquad (7.8)$$

where $\beta = (\beta_0, \ldots, \beta_p)^\top$ is the unknown parameter vector. After finding the minimum, $\hat{\beta}$, one would set $\hat{\mu}(t) = \hat{\beta}_0$. Note that the other parameters are also useful as they can be used to estimate derivatives (see Problem 7.16). Finding the minimizer, for a fixed t, turns out to be a relatively straightforward regression problem. Define the following

$$\mathbf{Y} = \begin{pmatrix} Y_{11} \\ Y_{12} \\ \vdots \\ Y_{NM_N} \end{pmatrix} \qquad \mathbf{Z} = \begin{pmatrix} (t_{11} - t)^0 & \cdots & (t_{11} - t)^P \\ (t_{12} - t)^0 & \cdots & (t_{12} - t)^P \\ \vdots & & \\ (t_{NM_N} - t)^0 & \cdots & (t_{NM_N} - t)^P \end{pmatrix}$$

and

$$\mathbf{K}_h = \operatorname{diag}\left(K\left(\frac{t_{11} - t}{h}\right), K\left(\frac{t_{12} - t}{h}\right), \ldots, K\left(\frac{t_{NM_N} - t}{h}\right)\right).$$

We can then express the loss function as

$$L(\beta) = (\mathbf{Y} - \mathbf{Z}\beta)^T \mathbf{K}_h (Y - \mathbf{Z}\beta).$$

Using multivariate calculus (Problem 7.5), one can show that

$$\hat{\beta} = (\mathbf{Z}^\top \mathbf{K}_h \mathbf{Z})^{-1} \mathbf{Z}^\top \mathbf{K}_h \mathbf{Y}. \qquad (7.9)$$

As with the basis expansions seen in the first chapters, we have that the kernel smoothers are also linear estimators since they depend linearly on $\mathbf{Y}$.

Local polynomial regression is a very effective tool for estimating nonlinear mean functions. Sampling designs which are highly unbalanced will result in point-wise variances which are larger in some regions, and smaller in others. However, one potential drawback is that the estimators are often not as computationally convenient to work with as basis function expansions, which are often represented using some relatively small number of basis coefficients. For example, to compute the $L^2[0, 1]$ norm of an estimate from local polynomial regression, one would evaluate the estimator on some large grid and calculate the norm numerically, while with a basis representation, one would simply use the sum of the squared coefficients (for an orthonormal basis).

One caution about implementation is the choice of the bandwidth h. A popular method for choosing h is to use k−fold cross-validation, which involves splitting the data into k groups, and then applying cross-validation to the groups treating them as units (so leave one group out, fit the model, predict the left out group, etc.). However, in FDA applications, one must make sure that entire subjects, not individual points, are divided amongst the groups. In other words, while traditional cross-validation is "leave one point out at time", for this setting it must be "leave one subject out at a time". The reason for this is due to the potentially very strong within subject correlations which will result in a poor choice of h. Additionally, cross-validation has a tendency to under smooth, thus a visual inspection is always recommended to see if the smoothing parameter should be further adjusted.

Basis function regression

We have used basis functions for functional parameter estimation extensively in previous chapters. We specialize these methods to the setting of mean function estimation in the context of sparsely observed functions. One assumes that the mean function can be expressed as a linear combination of functions $e_1(t), \ldots, e_J(t)$:

$$\mu(t) = \sum_{j=1}^{J} \mu_j e_j(t) = \boldsymbol{\mu}^\top \mathbf{e}(t) \qquad \text{where} \qquad \boldsymbol{\mu} = (\mu_1, \ldots, \mu_J)^\top.$$

One then estimates $\boldsymbol{\mu}$ via least squares, i.e. by minimizing

$$L(\boldsymbol{\mu}) = \sum_{n=1}^{N} \sum_{m=1}^{M_n} (Y_{nm} - \boldsymbol{\mu}^\top \mathbf{e}(t_{nm}))^2.$$

If we once again stack Y_{nm} into a large vector $\mathbf{Y}$ and the $e(t_{nm})$ into a large matrix

$$\mathbf{E} = \begin{pmatrix} \mathbf{e}(t_{11})^\top \\ \vdots \\ \mathbf{e}(t_{NM_N})^\top \end{pmatrix},$$

then the LS loss function is expressed as

$$L(\boldsymbol{\mu}) = (\mathbf{Y} - \mathbf{E}\boldsymbol{\mu})^\top (\mathbf{Y} - \mathbf{E}\boldsymbol{\mu}),$$

which leads to the estimator

$$\hat{\boldsymbol{\mu}} = (\mathbf{E}^\top \mathbf{E})^{-1} \mathbf{E}^\top \mathbf{Y}.$$

A penalty on the resulting function can be added to produce smoother functions. As we have seen in previous chapters, putting a penalty on some linear combination of derivatives of $\mu(t)$ always results in a form like the following

$$L_\lambda(\boldsymbol{\mu}) = (\mathbf{Y} - \mathbf{E}\boldsymbol{\mu})^\top (\mathbf{Y} - \mathbf{E}\boldsymbol{\mu}) + \lambda \boldsymbol{\mu}^\top \mathbf{R} \boldsymbol{\mu},$$

where $\mathbf{R}$ is some symmetric matrix. For example, if we placed a penalty on the second derivative then

$$R_{ij} = \int e_i^{(2)}(t) e_j^{(2)}(t) \, dt.$$

This results in an estimator of the form

$$\hat{\mu} = (\mathbf{E}^\top \mathbf{E} + \lambda \mathbf{R})^{-1} \mathbf{E}^\top \mathbf{Y}.$$

Reproducing kernel Hilbert spaces

In some ways reproducing kernel Hilbert spaces, RKHS, are a form of penalized basis regression. However, the penalty and the basis both come from a specified kernel function (which should not be confused with the kernels and kernel smoothing in Section 7.2). RKHS methods provide a very flexible framework for including assumptions about the underlying parameter functions. One can tailor the kernel to reflect smoothness and boundary assumptions. The theory for RKHS's is quite rich with many fascinating mathematical results. There are many texts available for the interested reader, we have found Berlinet and Thomas-Agnan (2011) to be especially useful. We refer to this monograph for further details. In this section we merely state some relevant concepts.

Let $K : [0,1] \times [0,1] \to \mathbb{R}$ be a positive definite, symmetric, square integrable kernel, i.e.

1. for any distinct points $t_1, \ldots, t_n \in [0,1]$, the matrix $\{K(t_i, t_j) : i = 1, \ldots, n \ j = 1, \ldots, n\}$ is positive definite,

2. $K(t,s) = K(s,t)$;

3. and $\iint K(t,s)^2 \, dt \, ds < \infty$.

Define the set

$$A_K = \left\{ f : f(t) = \sum_{j=1}^J \alpha_j K(t, s_j) \text{ for some } 1 \leq J < \infty \text{ and } s_j \in [0,1], \alpha_j \in \mathbb{R} \right\},$$

that is, we assume that A_K is the set of finite linear combinations of $K(t, s_j)$ for some points s_j in the domain. As we explain in Chapters 3 and 10, the function $K(t,s)$ can be viewed as the kernel of a Hilbert–Schmidt operator, and thus has a *spectral decomposition* as in (3.8)

$$K(t,s) = \sum_{i=1}^\infty \lambda_i v_i(t) v_i(s),$$

where $\{v_i\}$ is an orthonormal basis in $L^2[0,1]$. Using the v_i and λ_i, we can define an inner product on A_K as

$$\langle f, g \rangle_{H_K} = \sum_{i=1}^\infty \frac{\int f(t) v_i(t) \, dt \int g(s) v_i(s) \, ds}{\lambda_i},$$

and inner product norm $\|f\|^2_{H_K} = \langle f, f \rangle_{H_K}$. It can be shown (Problem 7.6) that

$$\|f\|^2_{H_K} = \sum_{j=1}^{J} \sum_{k=1}^{J} \alpha_j \alpha_k K(s_j, s_k), \qquad (7.10)$$

which, in the RKHS literature, is the more common way to phrase the norm. The above is always positive for nonzero $\{\alpha_i\}$ since K is positive definite. Since we have an inner-product norm on A_K, we can complete the space with respect to the norm to obtain a Hilbert space (see Section 10.1). The completed space is denote as H_K and is called an RKHS.

The reasoning behind the term *reproducing kernel Hilbert space* is the following property, commonly known as the *Representer Theorem*.

THEOREM 7.2.1 *Let H_K be an RKHS of real valued functions over the domain $[0, 1]$, with kernel K. Then for any $x \in H_K$ and $t \in [0, 1]$ one has*

$$x(t) = \langle x, K_t \rangle_{H_K},$$

where $K_t(s) := K(t, s)$.

The Representer Theorem gives a direct connection between point-wise evaluation and the inner product. It also implies that point-wise evaluation is a continuous linear functional over H_K, a property which does not hold in L^2.

We now return to the problem at hand, i.e. to the estimation of the mean function. Suppose the observations Y_{nm} follow model (7.1). The RKHS approach is to estimate μ using least squares, but including a penalty on μ using the $\| \cdot \|_{H_K}$ norm:

$$L_\lambda(\mu) = \sum_{n=1}^{N} \sum_{m=1}^{M_n} \|Y_{nm} - \mu(t_{nm})\| + \lambda \|\mu\|^2_{H_K}.$$

Using the Representer Theorem it can be shown that the minimizer of the above takes on a very particular form. Notice that if $\mu \in H_K$ then we can express $\mu(t_{nm}) = \langle \mu, K_{t_{nm}} \rangle$. This means that $\hat{\mu} \in \text{span}\{K_{t_{nm}}\}$, since adding anything that is in the orthogonal compliment would increase the penalty without reducing the sum of squares (Problem 7.12). We therefore know that the solution is of the form

$$\hat{\mu}(t) = \sum_{n=1}^{N} \sum_{m=1}^{M_n} \hat{\alpha}_{nm} K(t, t_{n,m}), \qquad (7.11)$$

and solving for μ is equivalent to solving for the α_{nm}. Grouping the $\boldsymbol{\alpha} = \{\alpha_{n'm'}\}$ into a vector and the $\mathbf{K} = \{K(t_{n,m}, t_{n',m'})\}$ into a matrix, we can then express

$$L_\lambda(\mu) = L_\lambda(\boldsymbol{\alpha}) = (\mathbf{Y} - \mathbf{K}\boldsymbol{\alpha})^\top (\mathbf{Y} - \mathbf{K}\boldsymbol{\alpha}) + \lambda \boldsymbol{\alpha}^\top \mathbf{K}\boldsymbol{\alpha},$$

which gives that

$$\hat{\alpha} = (\mathbf{K}^{\top}\mathbf{K} + \lambda\mathbf{K})^{-1}\mathbf{K}^{\top}Y.$$

We see that RKHS and penalized basis regression take a very similar form. There is good reason for this, as they are very similar procedures. Primary differences between the two include (1) the eigenfunctions of K are not usually used as the coordinate system for RKHS estimates, though they can be, and (2) the penalty and basis are decoupled in penalized regression, while for an RKHS they both come from the choice of kernel. For this reason, it can, at first, be a bit unclear what assumptions are being made about the parameters when using different kernels, however, one can usually carefully tailor the kernel to reflect smoothness and/or boundary assumptions. We conclude this section with three examples.

EXAMPLE 7.2.1 [Sobolev Space] Consider $L^2(\mathcal{D})$, where $\mathcal{D}$ is a compact subset of $\mathbb{R}^d$. Define H_K to be the subset of functions in $L^2(\mathcal{D})$ that have up to and including m^{th} order derivatives that are also in $L^2(\mathcal{D})$. A family of norms can be defined on H_K as

$$\|x\|_{H_K}^2 = \sum_{|\alpha| \leq m} \frac{1}{\tau_\alpha^2} \int_{\mathcal{D}} |x^{(\alpha)}(\mathbf{s})|^2 \, d\mathbf{s}.$$

Here α is a vector of integers whose sum is less than or equal to m, while the τ_α are nonzero weights. Equipped with this norm, H_K is an RKHS if and only if $m > d/2$. The kernel cannot always be written down explicitly, but in the case where $\mathcal{D} = [0,1]$ and $m = 1$, we have that

$$K(t,s) = \begin{cases} \frac{\tau}{\sinh(\tau)} \cosh(\tau(1-s)) \cosh(\tau t)) & t \leq s \\ \frac{\tau}{\sinh(\tau)} \cosh(\tau(1-t)) \cosh(\tau s)) & t > s. \end{cases}$$

These details are nontrivial and can be found on Page 281 of Berlinet and Thomas-Agnan (2011).

□

EXAMPLE 7.2.2 [Gaussian Kernel] Again, consider $L^2(\mathcal{D})$, then the Gaussian kernel is given by

$$K(\mathbf{s},\mathbf{s}') = \exp\left\{-\sigma|\mathbf{s}-\mathbf{s}'|^2\right\}.$$

While the Sobolev spaces contain functions which are differentiable up to a given order, the space H_K here contains functions which are infinitely differentiable (see Problem 7.17 for an illustration).

□

EXAMPLE 7.2.3 [Exponential Kernel] The exponential kernel is on the other end of the "smoothness" spectrum compared to the Gaussian kernel. In this case we have

$$K(\mathbf{s},\mathbf{s}') = \exp\left\{-\sigma|\mathbf{s}-\mathbf{s}'|\right\}.$$

This seemingly minor adjustment to the power in the exponent produces a space consisting of continuous functions which need not be differentiable (see Problem 7.17).

□

Examples and coding

Kernel smoothers are readily fit using a few functions in R. Local constant smoothers, i.e. Nadarya-Watson, can be fit using `ksmooth`, which is in the `stats` package, and is included in the base installation of R. Also part of the `stats` package is `lowess` which can be used for local linear, $P = 1$, smoothing, which is maybe the most common form of kernel smoothing since it does not have the boundary problems of $P = 0$. Finally, for a general P, one can use the function `loess`, and it will be the option we take here in our examples.

While the use of basis functions can always be programmed manually in R, the `mgcv` package, and in particular the `gam` function, is a very useful and extensive R package for fitting such models using splines. *GAM* stands for generalized additive models, but the package is programmed in such generality that it can be readily used to fit functional models as well; much of the backbone of the `Refund` package is built on `mgcv`. An interested reader may find Wood (2006) a useful resource.

Lastly, at the time of this writing, there is currently no implementation in R for a general RKHS smoothing. Thus we will not implement it here, however, an RKHS smoother using a Gaussian kernel function is part of Problem 7.7

We illustrate these methods on the CATT data, which can be dowloaded from http://www.med.upenn.edu/cpob/studies/CATT.shtml. We begin by loading in the data, combining them, and converting into a list format.

```r
# VAS Longitudinal Measurements
Data<-read.csv("data tables/va.csv")
# VAS Baseline Measurements
Base<-read.csv("data tables/bv.csv")
# Other Subject Information
general_info<-read.csv("data tables/gi.csv")

# Each row corresponds to a particular subject,
# so we start by pulling off the subject ids for each dataset.
IdData<-Data[,1]
IdBase<-Base[,1]
IdGeneral<-general_info[,1]
# Unique ids
Code_levels<-levels(IdBase)
# Number of subects
n<-length(Code_levels)
# The "time" variable which is given by the week of the study.
Week<-Data[,"week"]
# Unique weeks
Week_levels<-levels(factor(Week))

# Select Visual Acuity values
VA<-Data[,"studyeye_va"]
VAB<-Base[,"studyeye_va"]
```

```
Age<-general_info[,"age"]

Y_l<-by(c(VAB,VA),c(IdBase,IdData),c)
T_l<-by(c(rep(0,times=n),Week),c(IdBase,IdData),c)
counts<-sapply(Y_l,length,simplify=TRUE)
# Drop those with only a baseline measurement
Y_l<-Y_l[counts!=1]
T_l<-T_l[counts!=1]
counts<-counts[counts!=1]
n<-length(counts)
```

We now fit a local linear smoother using loess. To choose the bandwidth, we
split the data randomly into two sets, one used for fitting the model and the
second used for testing the fit of the model. We then choose the one that fits
the data best.

```
# 2 fold CV
set.seed(2016)
n_fit<-floor(n/2)
r = 10 # number of lambdas
lambda_all<-.2*10^seq(0,2,length=r)

#Randomly divide sample into a testing a training set
Fit_sample<-sample(1:n,n_fit)
Fit<-cbind(unlist(T_l[Fit_sample]),
       unlist(Y_l[Fit_sample]))
Fit<-na.omit(Fit)
Test<-cbind(unlist(T_l[-Fit_sample]),
       unlist(Y_l[-Fit_sample]))
Test<-na.omit(Test)

# Check fit for each lambda
MPE<-numeric(0)  # Mean Prediction Error
for(lambda in lambda_all){
       loess_tmp<-loess(Fit[,2]~Fit[,1], span = lambda,
   degree=1)
          pred<-predict(loess_tmp, newdata=Test[,1])
          MPE<-c(MPE,mean((pred - Test[,2])^2))
}

# Select best lambda and refit with all data
lambda<-lambda_all[which.min(MPE)]
All<-rbind(Fit,Test)
LLF<-loess(All[,2]~All[,1],span = lambda, degree=1)
```

Next we illustrate how to use *gam* to fit a penalized splines model, which
is a specific example of the basis function methods from Section 7.2. The
preparation of the data is exactly the same as before, so we won't repeat
those commands.

```
library(mgcv)
```

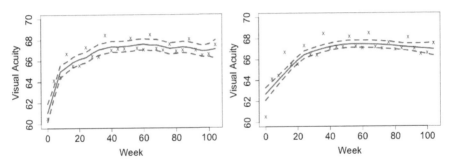

FIGURE 7.3: Estimated mean function of CATT data using local linear smoother. Dashed lines indicate 95% point-wise confidence intervals. In the left panel, the smoothing parameter was chosen by CV, while in the right panel this parameter is increased slightly after a visual inspection.

```
r = 20
lambda_all<-10^seq(-8,2,length=r)
MPE<-numeric(0)     # Mean Prediction Error
for(lambda in lambda_all){
        x = Fit[,1]; y = Fit[,2]
        GAMF<-gam(y~s(x,sp=lambda))
        pred<-predict(GAMF, newdata=data.frame(x=Test[,1]))
        MPE<-c(MPE,mean((pred - Test[,2])^2))
}
lambda<-lambda_all[which.min(MPE)]
All<-rbind(Fit,Test)
x = All[,1]; y = All[,2]
GAMF<-gam(y~s(x,sp=lambda))
```

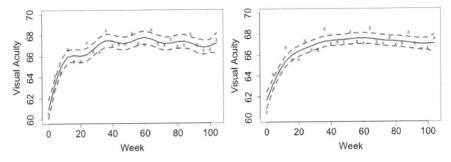

FIGURE 7.4: Estimated mean function of CATT data using a spline basis. Dashed lines indicate 95% point-wise confidence intervals. In the left panel, the smoothing parameter was chosen by CV, while in the right panel this parameter is increased slightly after a visual inspection.

All plots show roughly the same pattern: a significant increase in VAS as the study progresses, but the increase levels off. This is likely due to the fact that all subjects were assigned to some treatment, i.e. there was no control group. Furthermore, the subjects had not previously received any treatment for their AMD. Thus it is reasonable to expect an increase in visual acuity as the study progresses, with a leveling-off period after the treatments have reached their potential.

One potential point of concern is the baseline measurement. This measurement is made before applying any treatment, and as we move to the first observation which includes a treatment, we see a sharp jump. Subsequent increases seem to be substantially smaller, and there is a clear leveling off. Thus, there is a very real concern that there is a discontinuity at the beginning, or at the very least, a very rapid increase. Such artifacts can make choosing tuning parameters difficult, as different parameter values might produce estimates which fit well in certain regions/domains, but not others. As part of Problem 7.8, the reader is asked to explore a potential fix to this problem by omitting the baseline measurement, and instead subtracting it from each subsequent observation.

7.3   Covariance function estimation

Covariance estimation can be viewed as a bivariate version of mean estimation, at least in its implementation. Indeed, all of the discussed smoothing tools can be used for covariance function estimation as well. There are, however, a few differences between estimating a covariance function vs estimating a bivariate mean. First, the covariance function is positive definite, a property which is not a priori guaranteed to hold for smoothing estimates. Second, the data are assumed to be contaminated with independent noise, this means that one has to be very careful when estimating the diagonal of the covariance function.

As mentioned, there are two components contributing to the variance of the Y_{nm}, the first is the subject specific error function ε_n, which is assumed to be relatively smooth. We will denote the covariance function of $\varepsilon_n(t)$ as $c(t, s)$. The second is structural or measurement noise δ_{nm}. We will denote the variance of δ_{nm} as $\sigma^2(t_{nm})$, where $\sigma^2(t)$ is a continuous function. While we allow the variance of the structural noise to vary with time, we still assume that it is independent across both n and m. For many dense settings, this measurement noise is either negligible or can be smoothed out. For many longitudinal settings, this is not the case as the noise can be substantial. The covariance function of Y therefore has a discontinuity on the diagonal:

$$\text{Cov}(Y(t), Y(s)) = c(t, s) + \sigma^2(t)1_{t=s}.$$

To estimate $c(t, s)$ we can adapt the methodology presented in Section 7.2, and use a bivariate smoother on the pairs

$$\tilde{Y}_{nm_1}\tilde{Y}_{nm_2} = (Y_{nm_1} - \hat{\mu}(t_{nm_1}))(Y_{nm_2} - \hat{\mu}(t_{nm_2})),$$

with bivariate arguments (t_{nm_1}, t_{nm_2}). However, the discontinuity on the diagonal means that this must be done with care. The common approach is to remove the diagonal terms, and use only the off diagonal terms to estimate the covariance surface. In this case, one would create a large vector by multiplying observations

$$\tilde{\mathbf{Y}} = \{\tilde{Y}_{nm_1}\tilde{Y}_{nm_2} : 1 \le n \le N, 1 \le m_1 \le M_n, 1 \le m_2 \le M_n, m_1 \ne m_2\}$$

and then a large matrix with two columns for the time points

$$\mathbf{T} = \{(t_{nm_1}, t_{nm_2}) : 1 \le n \le N, 1 \le m_1 \le M_n, 1 \le m_2 \le M_n, m_1 \ne m_2\}.$$

We can then use any of the discussed smoothers to estimate the covariance surface. However, two properties must be kept in mind: symmetry and positive definiteness. The first property is usually not a problem as most smoothers will produce a symmetric estimate when used for covariance estimation. The second is more of an issue and it is common to end up with an estimate which is not positive definite. This is because there is nothing inherent in any of the smoothers which will guarantee a positive definite estimate. To remedy this, one usually expands the estimate using an eigenvalue/function decomposition and then sets any of the negative eigenvalues to zero (Problem 7.14).

Once $c(t, s)$ has been estimated, one can now estimate $\sigma^2(t)$ by running one of the univariate smoothers from Section 7.2 on the diagonal terms only, call it $\tilde{c}(t, t)$, and then forming the difference with $\hat{c}(t, s)$:

$$\hat{\sigma}^2(t) = \tilde{c}(t, t) - \hat{c}(t, t). \tag{7.12}$$

Or, equivalently, one could form the differences $\tilde{Y}_{nm}\tilde{Y}_{nm} - \hat{c}(t_{nm}, t_{nm})$ and run a univariate smoother through them. If the resulting $\hat{\sigma}^2(t)$ is negative for any t, then these values can be set to zero (since one cannot have a negative variance) or smoothers that incorporate positivity can be employed (Ramsay and Silverman, 2002, Section 6.2), though we do not explore this further. If $\sigma(t) \equiv \sigma^2$ is assumed to be constant, then it can be estimated more simply as

$$\hat{\sigma}^2 = \frac{1}{N}\sum_{n=1}^{N}\frac{1}{M_n}\sum_{m=1}^{M_n}(\tilde{Y}_{nm}\tilde{Y}_{nm} - \hat{c}(t_{nm}, t_{nm}))^2.$$

To illustrate these points, we return to the CATT data. In Figure 7.5 we plot the estimated covariance surface using gam. The left panel chose the tuning parameter using CV, while the right panel increases the tuning parameter slightly after a visual inspection. The plot indicates a similar pattern to the mean, the variability rapidly increases from baseline, peaking at $s = 1$ and

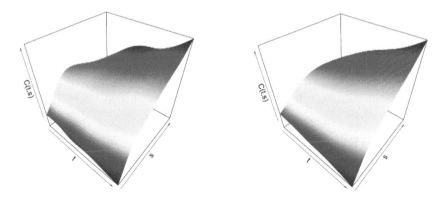

FIGURE 7.5: Estimated covariance function of CATT data using a spline basis. Left panel uses a tuning a parameter chosen with cross-validation. The right panel increases the left's tuning parameter to produce a slightly smoother estimate.

$t = 1$, but then levels off. Interestingly, the covariance seems to drop off pretty substantially as one moves away from the diagonal.

In Figure 7.6 we illustrate the effect of "contamination", i.e. the δ_{nm} from (7.1), on the estimation of the diagonal $c(t, t)$. First we take the gam estimates of $\widehat{c}(t, s)$ from Figure 7.5 and plot their diagonals using a red dashed line (using CV) and purple (increased slightly from CV). Next we plot, using blue x, the raw point-wise variances (computed at each time point) which have not been corrected. We then smooth (again using gam) these raw values to get the black line, which is then an estimate of $c(t, t) + \sigma^2(t)$. The difference between the black line and the red/purple indicates $\hat{\sigma}^2(t)$, and in this case the noise is quite noticeable. It is because of features like this that the diagonal terms are omitted when estimating $c(t, s)$.

7.4   Sparse functional PCA

Another key difference between nonparametric smoothing and sparse FDA occurs after covariance surface has been estimated as now FPCA can be used. Recall the Karhunen-Loéve expansion:

$$Y_n(t) = \mu(t) + \sum_{j=1}^{\infty} \xi_{nj} v_j(t),$$

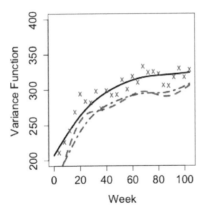

FIGURE 7.6: Estimated variance functions of CATT data using a spline basis. Purple and red lines are computed using (7.12) with the same smoothing parameters as in Figure 7.5, and represent estimates of $\widehat{c}(t,t)$. The blue x denote raw point-wise variances, and the black line is a smooth estimate $\widehat{c}(t,t)+\widehat{\sigma}^2(t)$ based on those values. The difference between the black and red/purple lines is $\widehat{\sigma}^2(t)$ and denotes the level of "contamination" in the data.

where $\xi_{n1}, \xi_{n2}, \dots$ are mean zero, uncorrelated, and have variances $\lambda_1, \lambda_2, \dots$. The goal of FPCA then is to replace the observed data $\{Y_n(t_{nm})\}$ with estimated/predicted principal component scores $\{\widehat{\xi}_{nj}\}$ or smoothed trajectories. FPCA accomplishes a few things:

1. irregular data are now on a "common scale" when working with scores,

2. the scores are typically of a lower dimension (though not necessarily for very sparse data),

3. the scores can be used in further multivariate statistical analyses,

4. the scores and FPC's can be used to reconstruct individual trajectories.

Estimating the eigenvalues and eigenfunctions is done as before, a spectral decomposition is applied to $\widehat{c}(t,s)$ (see Section 12.2). However, obtaining the scores must now be done with more care. Whereas for densely observed curves we could form inner products, such an approach cannot be applied to sparse data. In the case of densely observed curves, step 4 is interpreted as smoothing; the curves $\widehat{X}_n^{(p)}(t) := \widehat{\mu}(t) + \sum_{i=1}^{p} \widehat{\xi}_{nj}\widehat{v}_j(t)$ are smoother than the observed curves, and so may be closer to the underlying unobserved curves which are postulated to be smooth. In case of sparse data, the curve $\widehat{X}_n^{(p)}$ is, in addition, a remarkable reconstruction of a a curve X_n which is postulated to exist at any t, and be smooth, but which is observed only at a handful of time points.

To explain the approach used for sparse functional data, let us work at the population level. Assume that $\mu(t) = 0$ and both $\sigma^2(t)$ and $c(t,s)$ are known.

We can then ask, what is the conditional distribution of the first score, ξ_1, given the observations $Y(t_1), \ldots, Y(t_M)$? If the data are Gaussian, then the joint distribution of $(\xi_1, Y(t_1), \ldots, Y(t_M))^\top$ is multivariate normal. The mean vector is zero, and the covariance between the score, ξ_1, and the observation $Y(t_m)$ is given by

$$
\mathrm{Cov}(\xi_1, Y(t_m)) = \mathrm{E}\left[\left(\int Y(t) v_1(t) \, dt\right) Y(t_m)\right]
$$
$$
= \int c(t, t_m) v_1(t) \, dt = \lambda_1 v_1(t_m).
$$

The covariance matrix of $(\xi_1, Y(t_1), \ldots, Y(t_M))^\top$ is then given by

$$
\Xi =
\begin{bmatrix}
\lambda_1 & \lambda_1 v_1(t_1) & \cdots & \lambda_1 v_1(t_M) \\
\lambda_1 v_1(t_1) & c(t_1, t_1) + \sigma^2(t_1) & \cdots & c(t_1, t_M) \\
\vdots & & \ddots & \vdots \\
\lambda_1 v_1(t_M) & \cdots & \cdots & c(t_M, t_M) + \sigma^2(t_M)
\end{bmatrix}.
$$

Since the observations are jointly normal, this completely defines their joint distribution. We can now obtain the best linear unbiased predictor of ξ_1 by using its conditional expectation. Recall that if two vectors $\mathbf{Z}_1, \mathbf{Z}_2$ are jointly normal, then $\mathbf{Z}_1 | \mathbf{Z}_2$ is normal with mean

$$
\mathrm{E}[\mathbf{Z}_1 | \mathbf{Z}_2] = \boldsymbol{\mu}_1 + \boldsymbol{\Sigma}_{12} \boldsymbol{\Sigma}_{22}^{-1} (\mathbf{Z}_2 - \boldsymbol{\mu}_2)
$$

and covariance matrix

$$
\mathrm{Cov}(\mathbf{Z}_1 | \mathbf{Z}_2) = \boldsymbol{\Sigma}_{11} - \boldsymbol{\Sigma}_{12} \boldsymbol{\Sigma}_{22}^{-1} \boldsymbol{\Sigma}_{21}.
$$

This implies that $\xi_1 | Y(t_1), \ldots, Y(t_M)$ is normal with mean

$$
\begin{bmatrix}
\lambda_1 v_1(t_1) \\
\vdots \\
\lambda_1 v_1(t_M)
\end{bmatrix}^\top
\begin{bmatrix}
c(t_1, t_1) + \sigma^2(t_1) & \cdots & c(t_1, t_M) \\
\vdots & \ddots & \vdots \\
c(t_M, t_1) & \cdots & c(t_M, t_M) + \sigma^2(t_M)
\end{bmatrix}^{-1}
\begin{bmatrix}
Y(t_1) \\
\vdots \\
Y(t_M)
\end{bmatrix},
$$

and variance

$$
\lambda_1 -
\begin{bmatrix}
\lambda_1 v_1(t_1) \\
\vdots \\
\lambda_1 v_1(t_M)
\end{bmatrix}^\top
\begin{bmatrix}
c(t_1, t_1) + \sigma^2(t_1) & \cdots & c(t_1, t_M) \\
\vdots & \ddots & \vdots \\
c(t_M, t_1) & \cdots & c(t_M, t_M) + \sigma^2(t_M)
\end{bmatrix}^{-1}
\begin{bmatrix}
\lambda_1 v_1(t_1) \\
\vdots \\
\lambda_1 v_1(t_M)
\end{bmatrix}.
$$

From normal theory, the conditional expectation, call it $\hat{\xi}_1$, is the best linear unbiased predictor (BLUP), and thus we use it to predict ξ_1. We can predict $\xi_1, \ldots, \xi_p$ by adding more columns into the first term in the conditional expectation:

$$
\begin{bmatrix} \hat{\xi}_1 \\ \vdots \\ \hat{\xi}_p \end{bmatrix} = \begin{bmatrix} \lambda_1 v_1(t_1) & \cdots & \lambda_p v_p(t_1) \\ \vdots & \cdots & \vdots \\ \lambda_1 v_1(t_M) & \cdots & \lambda_p v_p(t_M) \end{bmatrix}^\top
$$

$$
\times \begin{bmatrix} c(t_1,t_1) + \sigma^2(t_1) & \cdots & c(t_1,t_M) \\ \vdots & \ddots & \vdots \\ c(t_M,t_1) & \cdots & c(t_M,t_M) + \sigma^2(t_M) \end{bmatrix}^{-1} \begin{bmatrix} Y(t_1) \\ \vdots \\ Y(t_M) \end{bmatrix}.
$$

Turning to the sample versions, one replaces the various parameters with their estimates, and we end up with $\{\hat{\xi}_{nj} : 1 \leq n \leq N, 1 \leq j \leq p\}$. The trajectories can now be reconstructed using

$$
\widehat{Y}_n(t) = \sum_{j=1}^p \hat{\xi}_{nj} \hat{v}_j(t),
$$

though one can also reconstruct Y_n without using FPCA, Problem 7.15.

7.5 Sparse functional regression

In this section we will focus on using the basis function methods of Section 7.2 to carry out functional regression. The methodology we use here is based on FPCA, though other methods are possible. In particular, functions `pfr` and `pffr` in the `refund` package can be used as well, see Section 6.4 for more details. In any of the settings, function-on-scalar, scalar-on-function, or function-on-function, we will carry out FPCA on any of the functional objects. Once converted to multivariate objects, we then carry out multivariate regression. Recall that Chapters 4 and 5 deal with functional regression for densely observed functions.

For function-on-scalar regression, as in Chapter 5, we assume that

$$
Y_n(t) = X_n \beta(t) + \varepsilon_n(t).
$$

We assume that both X_n and Y_n have been centered so that no intercept is required in the model. As before, we observe each function Y_n at time points $t_{n1}, \ldots, t_{nM_n}$. We carry out FPCA on the $Y_n(t_{nm})$ to obtain an $n \times q$ matrix of scores $\boldsymbol{\xi}$, and estimated FPCs $\hat{u}_j(t)$. We expand $\beta(t) \approx \sum_{j=1}^p \beta_j \hat{u}_j(t)$, and then use the least squares estimate

$$
\hat{\boldsymbol{\beta}} = \begin{pmatrix} \hat{\beta}_1 \\ \vdots \\ \hat{\beta}_p \end{pmatrix} = \frac{1}{\sum X_n^2} \begin{pmatrix} \sum X_n \xi_{n1} \\ \vdots \\ \sum X_n \xi_{nq} \end{pmatrix}.
$$

The estimate, $\hat{\beta}(t)$, is then given by

$$\hat{\beta}(t) = \sum_{j=1}^{p} \hat{\beta}_j \hat{u}_j(t) = \sum_{j=1}^{q} \frac{\sum X_n \xi_{nj}}{\sum X_n^2} \hat{u}_j(t).$$

In the scalar-on-function case, as in Chapter 4, the model is given by

$$Y_n = \int \beta(t) X_n(t) \, dt + \varepsilon_n.$$

We instead use FPCA on the X_n, resulting in an $n \times p$ matrix of scores ζ. Once again we approximate $\beta(t)$ using the FPC's of X, $\beta(t) \approx \sum_{j=1}^{p} \beta_j \hat{v}_j(t)$. Using least squares, we again have

$$\hat{\beta} = \begin{pmatrix} \hat{\beta}_1 \\ \vdots \\ \hat{\beta}_p \end{pmatrix} = \begin{pmatrix} \frac{\sum \xi_{n1} Y_n}{\sum \xi_{n1}^2} \\ \vdots \\ \frac{\sum \xi_{n1} Y_n}{\sum \xi_{np}^2} \end{pmatrix},$$

and the resulting estimate is given by

$$\hat{\beta}(t) = \sum_{j=1}^{q} \hat{\beta}_j \hat{v}_j(t) = \sum_{j=1}^{p} \frac{\sum \zeta_{jn} Y_n}{\sum \zeta_{jn}^2} \hat{v}_j(t).$$

Lastly, we turn to a full function-on-function linear model, as in Chapter 5, given by

$$Y_n(t) = \int \beta(t, s) X_n(s) \, ds + \varepsilon_n(t).$$

Here we carry out FPCA on both X and Y. In this case we approximate $\beta(t, s) \approx \sum \sum \beta_{ij} \hat{u}_i(t) \hat{v}_j(s)$. Using least squares we have that

$$\hat{\beta}_{ij} = \frac{\sum \zeta_{jn} \xi_{in}}{\sum \zeta_{jn}^2},$$

and the resulting estimate is given by

$$\sum_{i=1}^{q} \sum_{j=1}^{p} \frac{\sum \zeta_{jn} \xi_{in}}{\sum \zeta_{jn}^2} \hat{u}_i(t) \hat{v}_j(s).$$

7.6   Chapter 7 problems

7.1 Let $N = \sum_n M_n$. Justify (7.2). What conditions on h do you require?

7.2 Justify (7.3) assuming that $\mu(t)$ has a bounded second derivative.

7.3 Justify (7.5). What conditions on h do you require?

7.4 Consider the asymptotic MSE derived in Example 7.1.2:

$$MSE(h) = Ch^2 + \frac{\tau^2 + \sigma^2}{NMh} + \frac{\tau^2}{N}.$$

a) Using calculus, find the value of $h \geq 0$ that minimizes the above expression. What does this say about the size of the bias relative to the variance for the optimal h?

b) Plug the optimal value of h into the MSE. Suppose that $M = N^\delta$. What can you conclude about the asymptotic behavior of the MSE with respect to δ?

7.5 Verify (7.9).

7.6 Verify (7.10).

7.7 Estimate the mean function from Example 7.1.1 using an RKHS method. Use the Gaussian kernel $K(t, s) = \exp\{(t - s)^2\}$. Choose the penalty parameter, λ, using cross-validation.

7.8 Estimate the mean function from Example 7.1.1 using gam, however, do not include the baseline as a time point. Instead, subtract it from all of the subsequent observations. Does this help cross-validation in choosing a reasonable tuning parameter? If so, why do you think this is occurring?

7.9 Show that (7.7) minimizes (7.6).

7.10 Using the gam function, estimate $c(t, s)$ and $\sigma^2(t)$ from the CATT example, as in Figures 7.6 and 7.5.

7.11 Prove Theorem 7.2.1 in the case that $x \in A_K$.

7.12 Prove that the RKHS minimizer is of the form given if (7.11).

7.13 An advantage of RKHS methods is that different data structures can be built into the kernel. Here you will estimate the mean function of the Canadian Weather data, Section 2.2, using an RKHS method with the two different kernels described below. Make sure the domain of the functions is rescaled to be $[0, 1]$. Compare and discuss the results.

a) The Gaussian kernel:

$$K(t, s) = \exp\{-(t - s)^2\}$$

b) The Periodic kernel

$$K(t, s) = \exp\{-\sin^2(\pi|t - s|)\}.$$

7.14 Returning to Problem 7.13, estimate the covariance function using the method of your choosing. Evaluate the estimate on a grid. Is the estimate symmetric? What about positive definite? Explain.

7.15 Suppose that $\{Y(t) : t \in \mathcal{T}\}$ is a Gaussian process with a known mean function $\mu(t)$ and known covariance function $c(t, s) + \sigma^2(t)1_{t=s}$. Suppose that $Y(t)$ is observed at time points $t_1, \ldots, t_m$. What is the BLUP of $Y(t)$ given $Y(t_1), \ldots, Y(t_m)$?

7.16 Consider the classic Taylor expansion of a function $\mu(t)$ about a point t_0. Recall that if μ is P times differentiable at t then one has

$$\mu(t) = \sum_{i=1}^{P} \frac{\mu^{(i)}(t)[t - t_0]^i}{i!} + o(|t - t_0|^P).$$

a) Comparing this to the local polynomial regression equation (7.8), what do you think is the relationship between $\mu^{(i)}(t)$ and β_i?

b) Verify your conjecture via simulations using the random effects model from Example 7.1.2. Use a domain of $[0, 1]$, $\tau^2 = 1$, $\sigma^2 = 1/2$, and $\mu(t) = \sin(2\pi t)$. Check several different sample sizes to see how the estimates converge. Use local linear ($P = 1$) and take the bandwidth $h = N^{-1/5}$. Evaluate the estimates on an evenly spaced grid with 50 points.

7.17 Consider the following classic result from real analysis. Let $\{f_n(t) : t \in [0, 1], \ n = 1, 2, \ldots\}$ be a sequence of functions that (1) converges point-wise to a function $f(t)$ and (2) has derivatives $\{f_n'(t)\}$ that are uniformly convergent to a function g. Then it must be the case that $f' = g$.

a) Use the above result to show that the RKHS generated by the Gaussian kernel (Example 7.2.2) consists of functions which are at least one time continuously differentiable. Use the fact that every element of the RKHS can be expressed as

$$x(t) = \sum_{j=1}^{\infty} \alpha_j K(t, s_j),$$

where $\{\alpha_j\}$ and $\{s_j\}$ satisfy

$$\sum_j \sum_k \alpha_j \alpha_k K(s_j, s_k) < \infty.$$

b) Briefly explain why the RKHS for the exponential kernel contains functions which need not be differentiable.

8

Functional time series

This Chapter is concerned with functional data observed sequentially over time. For example, the curves $X_1, X_2, \ldots$ can represent pollution levels at some location; $X_n(t)$ is the level of pollution at time t of day n. Such curves cannot be treated as being independent with the same distribution; they do not form a simple random sample of curves. Pollution on a given day will be affected by pollution on previous days, so the curves will be dependent. A daily pollution pattern will be different on Sunday and on Monday, so the curves will not have the same distribution every day. We will discuss several more examples in this chapter.

To understand this chapter fully, some familiarity with basic tools and objectives of time series analysis is needed. We provide a concise introduction and list several textbooks in Section 8.1. In Section 8.2, we introduce the most extensively used functional time series model, the autoregressive process of order 1, the FAR(1) model. Sections 8.3 and 8.4 consider forecasting methods for functional time series which are implemented in easy to use R packages. In Section 8.5, we discuss the long–run covariance function. While the covariance function is appropriate for iid functions, many inferential procedures for functional time series are based on the long–run covariance function. Section 8.6 illustrates the application of this concept in the context of stationarity tests. The chapter concludes with two elaborations on the FAR(1) model. In Section 8.7, we show how to estimate it in R, in Section 8.8, we prove its existence. Some topics related to functional time series which are not discussed in this chapter are briefly discussed in Section 8.9.

8.1 Fundamental concepts of time series analysis

The purpose of this section is to review fundamental concepts of time series analysis in order to facilitate the understanding of the following sections of this chapter. This section is concerned with scalar time series. Time series analysis is a very large field which has been developed for over one hundred years. Modern time series analysis can be viewed as a subfield of statistics, due to its reliance on probability models. However, many of its fundamental ideas have been developed by researchers in other fields, most notably signal processing and econometrics, and it finds applications in practically all fields of science and engineering. Many excellent textbooks exist, for example Brockwell and Davis (2002), and its more advanced precursor Brockwell and Davis (1991), Shumway and Stoffer (2011) and Hamilton (1994).

A time series is a sequence of observations, $x_n, n = 1, 2, \ldots, N$, collected over time. In time series analysis, it is customary to use the index t, instead of n or i, to denote time. In this textbook, t is generally used to denote the argument of a function, so we use n to denote the time index. Two examples of time series are shown in Figures 8.1 and 8.2. Time series typically show trends, periodicities or abrupt changes in their general patterns. The objective of time series analysis is to quantitatively describe the random mechanism which generates the data. This is done by removing trends, periodicities, structural change points, so that the resulting time series looks stationary. Intuitively, every subsection of a stationary time series looks as if it was generated by the same random mechanism. The top panel of Figure 8.3 shows a simulated realization which is stationary. The time series in Figure 8.1 is not stationary because the mean and variability at the beginning and the end of the period are very different. The time series in Figure 8.2 is not stationary either because the variability in the first half of the period is smaller than in the second half.

After the original time series observations have been transformed in such a way that they can be assumed to be stationary, a suitable time series model is found. A *time series model* specifies the distribution of *random variables* $X_n, n \in \mathbb{Z}$. It is typically stated by formulas which describe how X_n depends on $X_{n-1}, X_{n-2}, \ldots$. The most commonly used model is the autoregressive process of order 1, AR(1), defined by the equations

$$X_n - \mu = \varphi(X_{n-1} - \mu) + \varepsilon_n.$$

The sequence $\{\varepsilon_n\}$ is the *white noise*, e.g. a sequence of random variables for which

$$E\varepsilon_n = 0, \quad \mathrm{Var}[\varepsilon_n] = \sigma^2, \quad \mathrm{Cov}(\varepsilon_n, \varepsilon_{n+h}) = 0, \text{ if } h \neq 0.$$

The concept of stationarity is actually defined for time series models, not for the time series observations. The broadest definition of stationarity is as follows:

DEFINITION 8.1.1 A time series model $\{X_n, n \in \mathbb{Z}\}$ is stationary if the expected value EX_n and the covariances $\mathrm{Cov}(X_n, X_{n+h})$ do not depend on n.

One can show that the AR(1) equations have a stationary solution if and only if $|\varphi| < 1$. A realization of this process with $\varphi = 0.6$ is shown in Figure 8.3.

Once a stationary model for the transformed observations is found, it is used to predict future values or test hypotheses about the data generating mechanism, e.g. whether there is a trend in the data. The most important tool that is used for such purposes is the *autocovariance function* defined by

$$\gamma_h = \mathrm{Cov}(X_n, X_{n+h}), \quad h = 0, 1, 2, \ldots$$

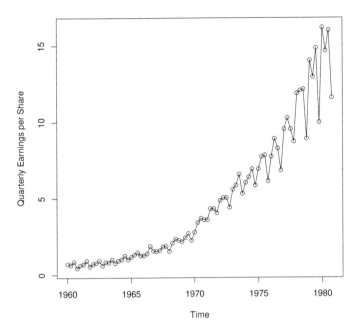

FIGURE 8.1: Time series of quarterly earnings of the Johnson & Johnson company. The circles show the actual observations. Reproduced from Shumway and Stoffer (2011).

We emphasize that the autocovariance function is defined only for a stationary model. Only nonnegative *lags* h are used because

$$\gamma_{-h} = \text{Cov}(X_n, X_{n-h}) = \text{Cov}(X_{n-h}, X_n) = \text{Cov}(X_n, X_{n+h}) = \gamma_h.$$

Another function used in the analysis of stationary models is the *autocorrelation function* defined by

$$\rho_h = \text{Corr}(X_n, X_{n+h}) = \frac{\gamma_h}{\gamma_0}, \quad h = 0, 1, 2, \ldots$$

Based on the observations $x_1, x_2, \ldots, x_N$, the autocovariance functions is estimated by the *sample* autocovariance function

$$\hat{\gamma}_h = \frac{1}{N} \sum_{n=1}^{N-h} (x_n - \bar{x})(x_{n+h} - \bar{x}), \quad \bar{x} = \frac{1}{N} \sum_{n=1}^{N} x_n.$$

In some work, the denominator N in the formula defining $\hat{\gamma}_h$ is replaced by $N - h$. The autocorrelation function is estimated by the sample autocorrelation

Seismogram signal

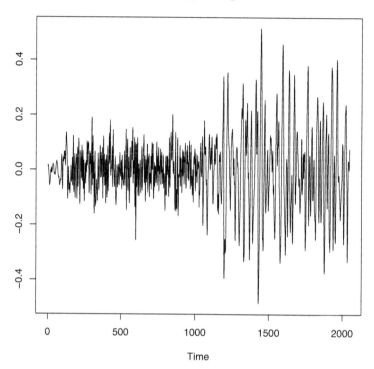

FIGURE 8.2: Part of a seismograph record during an earthquake. The actual observations are the "spike points". Reproduced from Shumway and Stoffer (2011).

function, $\hat{\rho}_h = \hat{\gamma}_h/\hat{\gamma}_0$. Even though $\hat{\rho}(0) = 1$ for every time series, this value is customarily plotted, as in the lower panel of Figure 8.3.

The sample autocorrelation function, and many other more complex tools, are used to determine a suitable stationary model for the transformed time series observations. Predictions of future values are then made using this stationary model, followed by the inverse transformation to obtain predictions of the original data. For example, if a suitable stationary model is the AR(1) model (necessarily with $|\varphi| < 1$), then the *one-step-ahead* prediction is

$$\hat{X}_{n+1} = \hat{\mu} + \hat{\varphi}(X_n - \hat{\mu}).$$

It is obtained by equating the mean zero errors ε_n to zero, and replacing the unknown parameters μ and φ by their estimates.

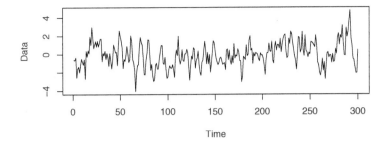

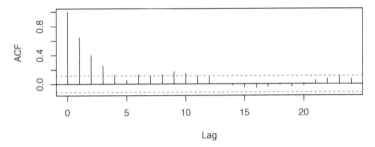

FIGURE 8.3: Top: A realization of an autoregressive process of order 1 with a positive autocorrelation coefficient. Bottom: The corresponding sample autocorrelation function. Only the sample autocorrelations outside the blue dashed lines are statistically significant.

8.2 Functional autoregressive process

We say that a sequence $\{X_n,\ -\infty < n < \infty\}$ of mean zero elements of L^2 follows a functional (mean zero) AR(1), FAR(1), model if

$$X_n = \Phi(X_{n-1}) + \varepsilon_n, \tag{8.1}$$

where $\Phi : L^2 \to L^2$ is an operator transforming a function into another function, and $\{\varepsilon_n,\ -\infty < n < \infty\}$ is a sequence of iid mean zero errors. In applications, the operator Φ is an integral operator defined by

$$\Phi(x)(t) = \int \varphi(t, s)x(s)ds, \quad x \in L^2.$$

Equations (8.1) can thus be rewritten more directly as

$$X_n(t) = \int \varphi(t, s)X_{n-1}(s)ds + \varepsilon_n(t).$$

It is shown in Section 8.8, that there exists a stationary functional sequence $\{X_n\}$ satisfying the above equation if $\iint \varphi^2(t,s)dtds < \infty$, a condition we assume from now on.

A more general definition of the FAR(1) model is sometimes used. Analogously to the scalar case, it allows the ε_n to be a suitably defined white noise in L^2, i.e. to be uncorrelated in an appropriate Hilbert space sense. However, statistical inference requires that the errors be iid, so we will work with this narrower class. To simplify the formulas, we also assume that the mean function is zero. The general FAR(1) process is defined by the equation $X_n - \mu = \Phi(X_{n-1} - \mu) + \varepsilon_n$. All results stated in this chapter are valid for this more general process by replacing X_n by $X_n - \mu$. In applications, the mean function μ is estimated by $\hat{\mu} = \bar{X}_N = N^{-1} \sum_{n=1}^{N} X_n$, and subtracted from all observed functions.

Equation (8.1) defines the functional autoregressive process of order 1. It is easy to define FAR(p) processes, i.e. autoregressive processes of order p by including additional terms of the form $\Phi_k(X_{n-k})$, $1 \leq k \leq p$. In the analysis of functional time series, these processes however play a role smaller than their scalar counterparts.

We now turn to the estimation of the autoregressive operator Φ. It is instructive to consider first the univariate case $X_n = \varphi X_{n-1} + \varepsilon_n$, in which all quantities are scalars. We assume that $|\varphi| < 1$, so that there is a stationary solution such that ε_n is independent of X_{n-1}. Then, multiplying the AR(1) equation by X_{n-1} and taking the expectation, we obtain $\gamma_1 = \varphi \gamma_0$, cf. Section 8.1. The autocovariances γ_h are estimated by the sample autocovariances $\hat{\gamma}_h$, so the usual estimator of φ is $\hat{\varphi} = \hat{\gamma}_1 / \hat{\gamma}_0$. To apply this approach to the functional model, note that by (8.1),

$$E\left[\langle X_n, x \rangle X_{n-1}\right] = E\left[\langle \Phi(X_{n-1}), x \rangle X_{n-1}\right], \quad x \in L^2.$$

Define the lag–1 autocovariance operator by

$$C_1(x) = E[\langle X_n, x \rangle X_{n+1}]$$

and denote with superscript $\cdot^*$ the adjoint operator. Then, $C_1^* = C\Phi^*$ because, by a direct verification, $C_1^* = E\left[\langle X_n, x \rangle X_{n-1}\right]$. This leads to the identity

$$C_1 = \Phi C. \tag{8.2}$$

The above identity is analogous to the scalar case, so we would like to obtain an estimate of Φ by using a finite sample version of the relation $\Phi = C_1 C^{-1}$. The operator C^{-1} cannot, however, be defined on the the whole space L^2. To see it, recall that the inverse operator must satisfy $C^{-1}(C(x)) = x$. Denoting by v_j the eigenfunctions of C, we thus need $v_j = C^{-1}(C(v_j)) = C^{-1}(\lambda_j v_j) = \lambda_j C^{-1}(v_j)$. This implies that $C^{-1}(v_j) = \lambda_j^{-1} v_j$, if $\lambda_j > 0$, an assumption we must make. Since $\{v_j\}$ is an orthonormal basis, every $y \in L^2$ admits representation $y = \sum_{j=1}^{\infty} \langle y, v_j \rangle v_j$. This implies that

$$C^{-1}(y) = \sum_{j=1}^{\infty} \lambda_j^{-1} \langle y, v_j \rangle v_j.$$

However, the above sum converges only if

$$\left\| C^{-1}(y) \right\|^2 = \sum_{j=1}^{\infty} \lambda_j^{-2} \langle y, v_j \rangle^2 < \infty. \tag{8.3}$$

This will not hold for every y, take e.g. $y = \sum_{j=1}^{\infty} \lambda_j v_j$. The above considerations suggest the following solution. Assume that $\lambda_1 \geq \lambda_2 \geq \cdots \geq \lambda_p > 0$. Relation (8.3) clearly holds if $y \in \mathrm{sp}(v_1, v_2, \ldots, v_p)$. We can define C_p^{-1} by $C_p^{-1}(y) = \sum_{j=1}^{p} \lambda_j^{-1} \langle y, v_i \rangle v_i$. The operator C_p^{-1} is defined on the whole of L^2, but it is not a proper inverse, Problem 8.4. It is often called the *pseudo–inverse* and denoted C_p^+.

A sample counterpart of this pseudo–inverse is defined by

$$\widehat{C_p^+}(x) = \sum_{j=1}^{p} \hat{\lambda}_j^{-1} \langle x, \hat{v}_j \rangle \hat{v}_j.$$

The operator $\widehat{C_p^+}$ is defined on the whole of L^2 if $\hat{\lambda}_j > 0$ for $j \leq p$. By judiciously choosing p, we find a balance between retaining the relevant information in the sample, and the instability due to the reciprocals of small eigenvalues $\hat{\lambda}_j$. To derive a computable estimator of Φ, we use a sample version of the relation $\Phi \approx CC_1^{-1}$. Since C_1 is estimated by

$$\widehat{C_1}(x) = \frac{1}{N-1} \sum_{k=1}^{N-1} \langle X_k, x \rangle X_{k+1},$$

we obtain, for any $x \in L^2$,

$$\widehat{C_1}\widehat{C_p^+}(x) = \widehat{C_1}\left(\sum_{j=1}^{p} \hat{\lambda}_j^{-1} \langle x, \hat{v}_j \rangle \hat{v}_j \right)$$

$$= \frac{1}{N-1} \sum_{k=1}^{N-1} \left\langle X_k, \sum_{j=1}^{p} \hat{\lambda}_j^{-1} \langle x, \hat{v}_j \rangle \hat{v}_j \right\rangle X_{k+1}$$

$$= \frac{1}{N-1} \sum_{k=1}^{N-1} \sum_{j=1}^{p} \hat{\lambda}_j^{-1} \langle x, \hat{v}_j \rangle \langle X_k, \hat{v}_j \rangle X_{k+1}.$$

The estimator $\widehat{C_1}\widehat{C_p^+}$ can be used in principle, but typically an additional dimension reduction step is introduced by using the approximation $X_{k+1} \approx \sum_{i=1}^{p} \langle X_{k+1}, \hat{v}_i \rangle \hat{v}_i$. This leads to the estimator

$$\widehat{\Phi}_p(x) = \frac{1}{N-1} \sum_{k=1}^{N-1} \sum_{j=1}^{p} \sum_{i=1}^{p} \hat{\lambda}_j^{-1} \langle x, \hat{v}_j \rangle \langle X_k, \hat{v}_j \rangle \langle X_{k+1}, \hat{v}_i \rangle \hat{v}_i. \tag{8.4}$$

The estimator (8.4) is a kernel operator with the kernel

$$\hat{\varphi}_p(t, s) = \frac{1}{N-1} \sum_{k=1}^{N-1} \sum_{j=1}^{p} \sum_{i=1}^{p} \hat{\lambda}_j^{-1} \langle X_k, \hat{v}_j \rangle \langle X_{k+1}, \hat{v}_i \rangle \hat{v}_j(s) \hat{v}_i(t). \tag{8.5}$$

All quantities at the right–hand side of (8.5) are available as output of the R function `pca.fd`, so this estimator is easy to compute. Section 8.7 shows how the FAR(1) model can be generated and estimated within the package `fda`.

8.3 Forecasting with the Hyndman–Ullah method

The method of forecasting functional time series which we describe in this section has been developed in the context of mortality curves. It is known in actuarial science as the Hyndman–Ullah method. It can however be applied to any functional time series which can be well approximated with a truncated Karhunen–Loéve expansion. Its advantage over methods based on the FAR(1) model is that it assumes neither stationarity nor a specific autocorrelation structure, and can be applied almost automatically using many procedures designed for forecasting of univariate time series. Its optimality has not been investigated theoretically. We explain this method in a general setting, and then illustrate its application to the prediction of mortality curves.

Suppose X_n, $n = 1, 2, \ldots, N$, are observations of a sufficiently smooth functional time series. Consider the approximation

$$X_n^{(J)}(t) = \hat{\mu}(t) + \sum_{j=1}^{J} \hat{\xi}_{n,j} \hat{v}_j(t).$$

The truncation level J can be chosen by the cumulative variance method, but in the context of forecasting, it is more useful to examine the residual curves $e_n^{(J)} = X_n - X_n^{(J)}$. These curves should be a functional white noise, so that they do not contain information useful for prediction. One way of verifying it is to apply a Portmanteau test described in Chapter 7 of Horváth and Kokoszka (2012). A simple exploratory tool is to construct projections $\langle e_n^{(J)}, u \rangle$ for several test functions u and examine the autocorrelation plots of these univariate time series.

Once the truncation level J has been selected, all that remains is to construct forecasts $\hat{\xi}_{N+h|N,j}$ of the future scores $\hat{\xi}_{N+h,j}$. The curve X_{N+h} is then forecasted by

$$X_{N+h|N}^{(J)}(t) = \hat{\mu}(t) + \sum_{j=1}^{J} \hat{\xi}_{N+h|N,j} \hat{v}_j(t). \tag{8.6}$$

The easiest way to predict $\hat{\xi}_{N+h,j}$ is to use only the available scores of the jth EFPC $\hat{v}_j$. In other words, we work separately with each univariate time series $\hat{\xi}_{1,j}, \hat{\xi}_{2,j}, \dots, \hat{\xi}_{N,j}$. Many automated methods of prediction of univariate time series exist, for example the seasonal ARIMA method with the model selected by an information criterion, like the AIC. Most time series textbooks treat prediction, Chapter 3 of Shumway and Stoffer (2011) provides a good account with R code. It should be noted that the sample mean $\hat{\mu}(t) = N^{-1} \sum_{n=1}^{N} X_n(t)$ does not change with the lead time h. The Hyndman–Ullah method assumes that the mean function does not change. If there is some visual evidence that the mean changes, its dynamics should be modeled separately, and the constant $\hat{\mu}$ replaced by an appropriate forecast.

We now illustrate the general approach described above by applying it to mortality curves. We follow the exposition in Booth *et al.* (2014). Certain actuarial definitions are simplified. Denote by $D_n(t)$ the count of deaths in calendar year n of people aged t. We treat t is a continuous variable in some fixed range, e.g. $(0, 110)$; but in practice t is measured in years. Denote by $P_n(t)$ the population of age t in year n. Quantities $D_n(t)$ and $P_n(t)$ are available for specific countries. We will work with the data for the United States. They can be accessed as follows:

```
library(RCurl);   library(demography); library(MortalitySmooth)
usa <- hmd.mx("USA", "username", "password", "USA")
usa1950 <- extract.years(usa, years=19MortalitySmooth50:2010)
```

The `username` and `password` can be obtained from the Human Mortality Database which can be accessed through the help file of the function `hmd.mx`. (Human Mortality Database, University of California, Berkeley (USA), and Max Planck Institute for Demographic Research (Germany). Available at www.mortality.org or www.humanmortality.de, data downloaded May 5, 2015)

The object `usa1950` contains mortality rates for the United States from 1950 to 2010. The mortality rate is defined as $m_n(t) = D_n(t)/P_n(t)$. For any fixed year n, $m_n(t)$ is the proportion of the population of age t that has died. The curve $m_n(t)$ increases rapidly with t, so it is usual to work with the log mortality rate, $\ln m_n(t)$. Except for very large or small t, the curves $\ln m_n(t)$ are smooth. To remove the noisy behavior at the ends and create functional objects, we perform a smoothing step and obtain smooth versions of the curves $\ln m_n$, which we denote by X_n. Our objective is to predict future curves X_{N+h} based on the available curves $X_1, X_2, \dots, X_N$. These predictions are treated

USA: male death rates (2003)

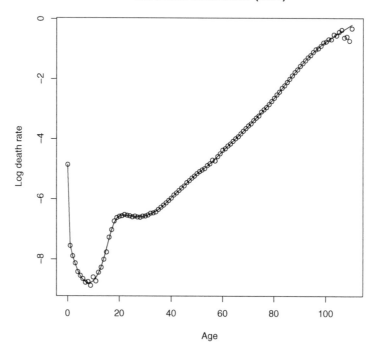

FIGURE 8.4: US male log mortality rates in 2003 with a smooth curve X_n.

as the predictions of the curves $\ln m_{N+h}$. Smoothing can be done in a number of ways; a custom function in the demography package can be used as follows to produce the curve in Figure 8.4:

```
smus <- smooth.demogdata(usa1950)
plot(usa1950, years=2003, series="male", type="p", pch=1)
lines(smus, years=2003, series="male")
```

The implementation of the Hyndman–Ullah method proceeds with the following code, which also produces Figures 8.5 and 8.6

```
fdm.male <- fdm(smus, series="male", order=3)
forecast.fdm.male <- forecast.fdm(fdm.male, h=30)
plot(forecast.fdm.male, plot.type="component")
plot(forecast.fdm.male)
```

The function fdm calculates the EFPC's and the series of scores. The function forecast.fdm computes univariate forecasts. These are shown in Figure 8.5. The confidence bands for the predictions of the univariate series of scores show that a pronounced downward trend is visible only for the first EFPC scores $\hat{\xi}_{n1}$. This trend is reflected in Figure 8.6 which indicates a trend toward a lower mortality rate, $\hat{v}_1(t)$ is positive, with particular gains around the ages of 5–15 and 50–70, which correspond, respectively, to the maximum and the local maximum of the curve $\hat{v}_1(t)$.

The above code is highly customized for data sets in the demography package, but all steps can be implemented using package fda and one of the packages for time series analysis e.g. astsa. In Problem 8.9 we illustrate the application of the package ftsa, in which the Hyndman–Ullah method is also implemented.

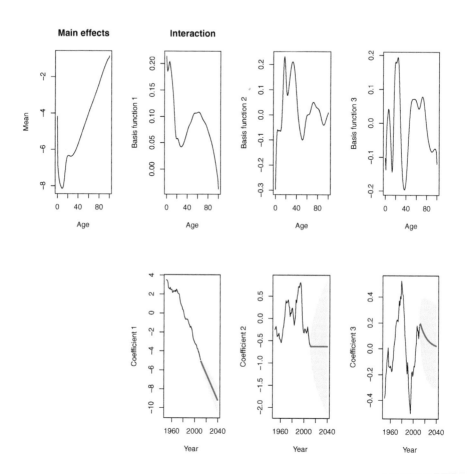

FIGURE 8.5: Prediction analysis of the smoothed US male log mortality rate curves. The top row shows the estimated mean function, $\hat{\mu}$, called the "main effects", and the first three EFPC's $\hat{v}_j, j = 1, 2, 3,,$ called "interactions". The bottom row shows the time series of corresponding scores (black) and their predictions (navy) with confidence bounds (yellow).

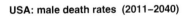

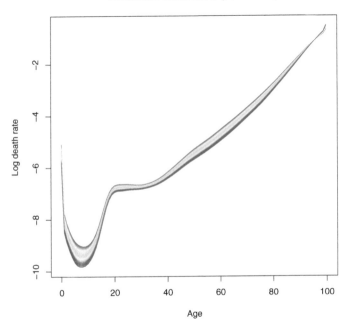

FIGURE 8.6: Predicted US male log mortality rate curves in rainbow code; years close to 2011 are in red, those close to 2040 in violet.

8.4 Forecasting with multivariate predictors

The method of Section 8.3 treats the series of scores $\hat{\xi}_{n,j}$ as univariate time series and constructs predictions $\hat{\xi}_{N+h|N,j}$ separately for every j. A natural extension is to consider the vectors of scores

$$\boldsymbol{\Xi}_n^{(J)} = [\hat{\xi}_{n,1}, \hat{\xi}_{n,2}, \ldots, \hat{\xi}_{n,J}]^T$$

and predict the whole vector $\boldsymbol{\Xi}_{N+h}^{(J)}$. This leads to the following general algorithm:

a) Select J by one of the methods discussed in Section 8.3 and compute the vectors $\boldsymbol{\Xi}_n^{(J)}$, $n = 1, 2, \ldots, N$.

b) Use a multivariate prediction technique to construct a predictor of the vector $\boldsymbol{\Xi}_{N+h}^{(J)}$. Denote this predictor by

$$\widehat{\boldsymbol{\Xi}}_{N+h|N}^{(J)} = [\hat{\xi}_{N+h|N,1}, \hat{\xi}_{N+h|N,2}, \ldots, \hat{\xi}_{N+h|N,J}]^T.$$

c) Predict the function X_{N+h} with (8.6).

We emphasize that even though both the Hyndman–Ullah and the multivariate method of this section use the same functional predictor (8.6), the Hyndman–Ullah method is in fact a special case of the multivariate method. It uses a specific approach to the prediction of the vector $\boldsymbol{\Xi}_{N+h}^{(J)}$; each component is predicted separately. In some cases, this simpler approach may reduce the variability of the forecasts, but if the sample cross–covariances of $\hat{\xi}_{n,j}$ and $\hat{\xi}_{m,k}$ are large, it may be inferior. Recall that, in general, the above covariances are zero for $k \neq j$ only if $n = m$.

In its current implementation, the above algorithm assumes, in addition, that the process $\{X_n\}$ is stationary; the mean function μ is treated as constant and the EFPC's $\hat{v}_j$ are assumed to approximate well the FPC's v_j, which lead to optimal approximations, cf. Chapter 3. The method is implemented in the package ftsa. We illustrate it with the following code which produces Figure 8.7.

```
require(ftsa); require(vars)
x = seq(0, 23.5, by = 0.5)
multi_forecast_sqrt_pm10 = farforecast(ftsm(pm_10_GR_sqrt), h
    = 30, PI=F)
plot(multi_forecast_sqrt_pm10, ylim=c(5.2,7.5), xlab="Hour",
    ylab="Square root of pm10", lwd=2)
oneStep_forecast_sqrt_pm10 = farforecast(ftsm(pm_10_GR_sqrt),
    h=1)
lines(oneStep_forecast_sqrt_pm10, lwd=3, lty=3)
```

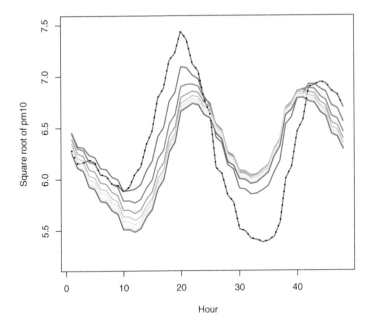

FIGURE 8.7: Predicted pm10 pollution curves; one–step–ahead prediction highlighted with black dots.

The data set we use contains half–hourly measurements of the concentration of particulate matter of certain size in Graz, Austria, October 1, 2010 –March 31, 2011. The square root transformation is applied to stabilize variance. The function `farforecast` computes the predicted curves. The "far" in its name indicates that a vector autoregressive scheme is used to predict the vectors of scores; these predictions are then used to construct the functional forecasts via (8.6). Figure 8.7 shows that these predictions converge to the overall mean function of the functional time series; this is a well–known property of predictions of stationary series.

8.5   Long–run covariance function

In many formulas and procedures of time series analysis, in place of the usual variance or sample variance, a quantity known as the *long–run variance* (LRV), and its estimators, must be used. We explain in this section how an analog of the LRV is defined for functional time series and how it can be estimated. In Section 8.6, we illustrate the application of this concept to testing the stationarity of functional time series. We begin by introducing the LRV of a scalar time series.

Suppose $\{X_n\}$ is a scalar stationary process, cf. Definition 8.1.1. Using the autocovariances $\gamma_h = \mathrm{Cov}(X_n, X_{n+h})$, we define the LRV as

$$\sigma^2 = \sum_{h=-\infty}^{\infty} \gamma_h = \gamma_0 + 2 \sum_{h=1}^{\infty} \gamma_h. \tag{8.7}$$

The motivation for definition (8.7) is as follows. If $X_1, X_2, \ldots, X_N$ are iid with $EX_i = \mu$, $\mathrm{Var}[X_i] = \sigma^2$, then the variance of the sample mean $\bar{X}_N = N^{-1} \sum_{n=1}^{N} X_n$ is given by the usual formula

$$\mathrm{Var}[\bar{X}_N] = N^{-2} \sum_{n=1}^{N} \mathrm{Var}[X_n] = N^{-1}\sigma^2.$$

If the $X_1, X_2, \ldots, X_N$ are a realization of a general stationary process, then the above formula is no longer valid. The variance of the sample mean is calculated as follows:

$$\mathrm{Var}[\bar{X}_N] = N^{-2}\mathrm{Var}\left[\sum_{n=1}^{N} X_n\right] = N^{-2} \sum_{n,m=1}^{N} \mathrm{Cov}(X_n, X_m).$$

Due to stationarity, the covariances depend only on on the differences $h = n - m$. We can rewrite the above formula by grouping covariances with the same difference h:

$$\mathrm{Var}[\bar{X}_N] = N^{-2} \sum_{h=-N+1}^{N-1} (N - |h|)\gamma_h = N^{-1} \sum_{h=-N+1}^{N-1} \left(1 - \frac{|h|}{N}\right)\gamma_h. \tag{8.8}$$

Using the Dominated Convergence Theorem, a frequently used result in probability theory, see e.g. Rudin (1987), it can be shown that if $\sum_{h=-\infty}^{\infty} |\gamma_h| < \infty$, then

$$\lim_{N\to\infty} N\mathrm{Var}[\bar{X}_N] = \sum_{h=-\infty}^{\infty} \gamma_h.$$

The above convergence is easy to justify intuitively: for each h, $h/N \to 0$, and the summation range expands to all integers, as $N \to \infty$. One can also show that under suitable assumptions on the dependence structure of the process $\{X_n\}$,

$$\sqrt{N}(\bar{X}_N - \mu) \overset{d}{\to} N\left(0, \sum_{h=-\infty}^{\infty} \gamma_h\right). \tag{8.9}$$

Result (8.9) should be contrasted with the usual Central Limit Theorem for iid X_n; instead of $\mathrm{Var}[X_1]$, the limit variance in (8.9) is the LRV (8.7). As a consequence, the confidence interval for μ has the form $\bar{X}_N \pm z_{1-\alpha/2}N^{-1/2}\hat{\sigma}$, where $\hat{\sigma}$ is an estimate of the σ defined in (8.7), rather than the usual sample variance.

A common approach to the estimation of the LRV (8.7) is to use an estimator of the form

$$\hat{\sigma}^2 = \hat{\sigma}^2(K, q) = \sum_{h=-N+1}^{N+1} K\left(\frac{h}{q}\right) \hat{\gamma}_h. \tag{8.10}$$

The sample autocovariances $\hat{\gamma}_h$ are defined in Section 8.1. The weight function, or lag window, K has the following properties: $K(x) = 0$ if $|x| > 1$; $K(-x) = K(x)$, $K(0) = 1$. Additional smoothness properties are required to establish large sample properties of such estimators, but we will not discuss them here; the function K must be at least continuous. The parameter q is called the bandwidth. It is a positive integer smaller than N; asymptotic arguments require that $q/N \to 0$ and $q \to \infty$, as $N \to \infty$. A large body of research is dedicated to methods of finding an optimal q, but we cannot discuss these issues in this brief section.

The general form of the estimator (8.10) is motivated by formula (8.8), which is similar to (8.10) with

$$K(x) = 1 - |x|, \quad |x| \le 1, \tag{8.11}$$

and with $q = N$. Note, however, that (8.8) involves population autocovariances γ_h, while (8.10) is defined in terms of the sample autocovariances $\hat{\gamma}_h$. The sampling variability of the $\hat{\gamma}_h$ is the reason why it must be assumed that $q/N \to 0$ to ensure that $\hat{\sigma}^2(K, q)$ converges to σ^2. The condition $K(x) = 0$ if $|x| > 1$ implies that we use only the $\hat{\gamma}_h$ with $|h| \le q$. Function (8.11) is called the *Bartlett window*. Other commonly used windows are listed in many time series textbooks, see e.g. Section 10.4 of Brockwell and Davis (1991).

We are now ready to turn to the case of a stationary functional time series. We will use the following definition:

DEFINITION 8.5.1 A functional time series model $\{X_n, n \in \mathbb{Z}\}$ is stationary if

$$\mu(t) = EX_n(t) \quad \text{and} \quad \gamma_h(t, s) = \mathrm{Cov}(X_n(t), X_{n+h}(s)), \quad h \in \mathbb{Z},$$

do not depend on n.

Definition 8.5.1 assumes that each function X_n is a square integrable element of L^2, i.e. $E \int X_n^2(t)dt < \infty$, cf. Chapter 3. Note also that for functional time series we must carefully distinguish between negative and positive lags h; if X_n is a scalar stationary sequence, then $\gamma_{-h} = \text{Cov}(X_n, X_{n-h}) = \text{Cov}(X_{n-h}, X_n) = \gamma(h)$. However, for a functional stationary sequence,

$$\gamma_{-h}(t, s) = \text{Cov}(X_n(t), X_{n-h}(s)) = \text{Cov}(X_{n-h}(s), X_n(t)) = \gamma_h(s, t).$$

If the X_n are independent, then the autocovariances $\gamma_h(t, s)$ are zero except for $h = 0$. The lag zero autocovariance function $\gamma_0(t, s)$ is equal to the covariance function $c(t, s)$.

By analogy to the LRV of a scalar sequence, cf. (8.7), the *long–run covariance function* (LRCF) of a stationary functional sequence is defined by

$$\sigma(t, s) = \sum_{h=-\infty}^{\infty} \gamma_h(t, s) = \gamma_0(t, s) + 2 \sum_{h=1}^{\infty} \left(\gamma_h(t, s) + \gamma_h(s, t) \right). \qquad (8.12)$$

In order for the infinite sequence in (8.12) to converge in the space of square integrable functions on $[0, 1] \times [0, 1]$, we must assume that

$$\sum_{h=-\infty}^{\infty} \iint \gamma_h^2(t, s)dtds < \infty. \qquad (8.13)$$

As in the scalar case, the LRCF (8.12) is related to the covariances of the sample mean function. Following the derivation presented in the scalar case, one can show that

$$N\text{Cov}(\bar{X}_N(t), \bar{X}_N(s)) = \sum_{h=-N+1}^{N-1} \left(1 - \frac{|h|}{N} \right) \gamma_h(t, s) \to \sum_{h=-\infty}^{\infty} \gamma_h(t, s). \qquad (8.14)$$

Analogously to (8.9), one can also show that

$$\sqrt{N}(\bar{X}_N - \mu) \overset{d}{\to} Z, \qquad (8.15)$$

where Z is a Gaussian random function with $EZ(t) = 0$ and $\text{Cov}(Z(t), Z(s)) = \sigma(t, s)$. Stationarity and condition (8.13) are not sufficient to imply (8.14) and (8.15). Additional assumptions on the dependence structure of the sequence $\{X_n\}$ are needed. References are given in Section 8.9.

We conclude this section by introducing an estimator of the LRCF which is analogous to the estimator (8.10). For a lag window K and a bandwidth $q < N$, we set

$$\hat{\sigma}(t, s) = \sum_{h=-N+1}^{N-1} K \left(\frac{h}{q} \right) \hat{\gamma}_h(t, s) \qquad (8.16)$$

$$= \hat{\gamma}_0(t, s) + \sum_{h=1}^{N-1} K \left(\frac{h}{q} \right) \{\hat{\gamma}_h(t, s) + \hat{\gamma}_h(s, t)\},$$

where

$$\hat{\gamma}_h(t,s) = \frac{1}{N} \sum_{j=1}^{N-h} \left(X_j(t) - \bar{X}_N(t)\right)\left(X_{j+h}(s) - \bar{X}_N(s)\right).$$

Under suitable assumptions on the dependence structure of the FTS $\{X_n\}$, and on the window K, estimator (8.16) converges to $\sigma(t,s)$:

$$\iint \{\hat{\sigma}(t,s) - \sigma(t,s)\}^2 \, dtds \xrightarrow{P} 0, \quad \text{as } N \to \infty.$$

8.6   Testing stationarity of functional time series

As we have seen in previous sections of this chapter, many procedures require that a functional time series be transformed to a stationary series. In this section, we discuss tests which allow us to verify if a given FTS is stationary. We present heuristic derivations, which illustrate several important techniques used in the analysis of FTS's. References to research papers which state precise assumptions and provide the requisite theory are provided in Section 8.9.

We want to test the null hypothesis

$$H_0: \quad X_i(t) = \mu(t) + \eta_i(t),$$

where $\{\eta_i\}$ is a mean zero strictly stationary sequence, cf. Definition 8.8.1. The mean function $\mu(t) = EX_i(t)$ is unknown. The general alternative hypothesis is simply that H_0 is not true. However, two specific alternatives motivate the form of the test statistics. The first one is the change point alternative:

$$H_{A,1}: \quad X_i(t) = \mu(t) + \delta(t)I\{i > k^*\} + \eta_i(t).$$

The *change point* $1 \le k^* < N$ is an unknown integer; the curves up to time k^* have mean function $\mu(t)$, after time k^* their mean function is $\mu(t) + \delta(t)$. The second alternative is the random walk:

$$H_{A,2}: \quad X_i(t) = \mu(t) + \sum_{\ell=1}^{i} u_\ell(t).$$

The sequence $\{u_\ell\}$ is stationary with mean zero, the u_ℓ are often assumed to be even iid.

We emphasize, that each function $t \mapsto X_i(t)$ is typically a realization of a continuous time nonstationary process observed at some discrete points t_j. The stationarity refers to the sequence of functions $X_1, X_2, \ldots, X_N$. This point is illustrated in Figure 8.8. The top panel shows price curves on five consecutive days. Each of these curves can be denoted X_i, and only $N = 5$ curves are shown. In typical applications, N is much larger, from several dozen

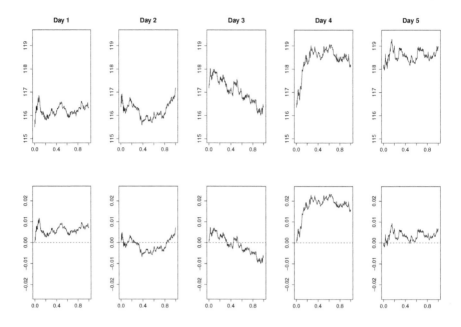

FIGURE 8.8: Top: IBM price curves on five consecutive trading days. Bottom: Cumulative intraday returns on these prices.

to several thousand. The sequence of price curves is in general not stationary. Even for the five displayed curves an upward trend is seen, such a trending or random walk behavior is much more pronounced for longer series. The Bottom panel of Figure 8.8 shows the same curves, but suitably normalized. Even though each curve is a realization of a nonstationary stochastic process, such normalized curves, known as cumulative intraday return curves, form a stationary functional time series.

Test statistics are based on the partial sum process

$$S_N(x,t) = N^{-1/2} \sum_{i=1}^{[Nx]} X_i(t), \quad 0 \le x \le 1,$$

where $[y]$ is the integer part of y, e.g. $[9.57] = 9$. A convenient way to look at the process S_N is to treat it as a stochastic process with index $x \in [0,1]$, and values in the space L^2, i.e. for each x, $S_N(x)$ is a random function. Realizations of the process $\{S_N(x), \ x \in [0,1]\}$ have jumps at points $x_k = k/N$. Observe that

$$S_N\left(\frac{k}{N},t\right) = N^{-1/2} \sum_{i=1}^{k} X_i(t), \quad k = 1, 2, \dots N.$$

The next step in the construction of test statistics is to define the process

$$U_N(x) = S_N(x) - xS_N(1), \quad 0 \le x \le 1.$$

Under H_0, the mean function μ cancels, and

$$U_N\left(\frac{k}{N}\right) = \frac{1}{\sqrt{N}}\left\{\sum_{i=1}^{k} \eta_i - \frac{k}{N}\sum_{i=1}^{N} \eta_i\right\}. \tag{8.17}$$

This cancelation occurs only under H_0. For example, under the change point alternative $H_{A,1}$,

$$U_N\left(\frac{k^*}{N}\right) = \frac{1}{\sqrt{N}}\left\{\sum_{i=1}^{k^*} \eta_i - \frac{k^*}{N}\sum_{i=1}^{N} \eta_i\right\} + \frac{k^*(N-k^*)}{N^{3/2}}\delta.$$

The term involving the jump function δ has a more complicated form for $k \ne k^*$, Problem 8.13. The point is that under $H_{A,1}$, the process U_N consists of two terms: the right–hand side of (8.17) and an additional term whose size is approximately $N^{1/2}\delta$. The same is true under $H_{A,2}$, except that the additional term is random, it involves the partial sums of the u_ℓ. The idea of testing is that under H_0 functionals of the process U_N converge to some distributions, but under various alternatives they diverge in probability to infinity, exceeding the critical values found under H_0 with increasing probability as the sample size N tends to infinity.

Our first test statistic has the form

$$\widehat{T}_N = \int_0^1 \|U_N(x)\|^2\, dx = \int_0^1 \left\{\int U_N^2(x,t)dt\right\} dx.$$

It tells us how large the process U_N is when averaged over all x. If H_0 is true, $\widehat{T}_N$ can be computed using the normalized partial sums $N^{-1/2}\sum_{i=1}^{k} \eta_i$. These partial sums converge to Gaussian random functions, cf. (8.15). The statistic $\widehat{T}_N$ therefore converges to a certain limit distribution. Under $H_{A,1}$ or $H_{A,2}$, the extra terms cause $\widehat{T}_N$ to be so large that it exceeds critical values with probability approaching 1, as $N \to \infty$. The derivation of the limit distribution of $\widehat{T}_N$ requires more background than can be presented in this section, so we merely formulate the final result.

Denote by σ_η the LRCF of the η_i, cf. (8.12). Next, denote by λ_j and φ_j, respectively, the eigenvalues and the eigenfunctions of σ_η, which are defined by

$$\int \sigma_\eta(t, s)\varphi_j(s)ds = \lambda_j\varphi_j(t), \quad 1 \le j < \infty.$$

To specify the limit distribution of $\widehat{T}_N$ we also need to introduce the *Brownian bridge*. It is a real–valued Gaussian stochastic process $\{B(x), x \in [0, 1]\}$ which is discussed in some detail in Section 11.4. Its value at point $x_k = k/N$ can be approximated as

$$B\left(\frac{k}{N}\right) \approx \frac{1}{\sqrt{N}} \left\{ \sum_{i=1}^{k} Z_i - \frac{k}{N} \sum_{i=1}^{N} Z_i \right\}, \tag{8.18}$$

where the Z_i are iid standard normal random variables. The similarity of this expression to (8.17) is obvious. An approximate trajectory of a Brownian bridge over $[0,1]$ is easy to simulate using $Z_1, Z_2, \ldots, Z_N$.

With the above background, we can state that, as $N \to \infty$, the distribution of $\widehat{T}_N$ approaches the distribution of the random variable

$$T := \sum_{j=1}^{\infty} \lambda_j \int_0^1 B_j^2(x)dx,$$

where the B_j are independent Brownian bridges. This approximation is valid if H_0 is true. The random variable T is defined in terms of an infinite sum which involves unknown eigenvalues λ_j and independent random variables $\int_0^1 B_j^2(x)dx$. The distribution of T must therefore be approximated. We explain in some detail how a suitable approximation to the distribution of T is constructed. Similar issues arise in many tests developed for functional data. The reader should keep in mind that the distribution of T is already an asymptotic approximation to the distribution of $\widehat{T}_N$.

There are two levels of the approximation to the distribution of T. First, we replace T by an approximation $T^\star$ which can be computed from the observed functions $X_1, X_2, \ldots, X_N$. In the second step, we generate a large number, say $R = 10E4$, of replications of $T^\star$. The empirical distribution of the R values of $T^\star$ is used in place of the distribution of T. For example, if 382 values $T^\star$ are greater than $\widehat{T}_N$, then the P–value is 0.0382, assuming $R = 10E4$. We therefore focus on the first step, finding a computable approximation to T.

The unknown λ_j must be replaced by their estimates $\hat{\lambda}_j$ which are defined by

$$\int \hat{\sigma}_\eta(t,s)\hat{\varphi}_j(s)ds = \hat{\lambda}_j\hat{\varphi}_j(t), \quad 1 \le j < \infty.$$

The $\hat{\lambda}_j$ are the eigenvalues of an estimator of the LRCF $\sigma_\eta(t,s)$; estimator (8.16) can be used. The number of available $\hat{\lambda}_j$ depends on a specific numerical implementation of the above eigenvalue problem. Assuming that D eigenvalues are computed, the distribution of T is approximated by the distribution of $T^\star = \sum_{j=1}^{D} \hat{\lambda}_j I_j^\star$, where $I_j^\star$ is an approximation to $\int_0^1 B_j^2(x)dx$. A simple way to compute $I_j^\star$ is to use the Riemann sum

$$I^\star = \frac{1}{N} \sum_{k=1}^{N} B^2\left(\frac{k}{N}\right),$$

with $B(k/N)$ defined in (8.18).

The test described above is often referred to as a Monte Carlo test because the critical values have to be found by simulation for each specific data set. With modern computing devices, tests of this form do not pose problems. However, tests which use the same critical values for every data sets are often more convenient to apply. To illustrate the difference in a familiar setting, suppose we observe a sample of iid scalar random variables $X_1, X_2, \ldots, X_N$ which have normal distribution with unknown expectation μ and unknown variance σ^2. If we want to test $H_0 : \mu = 0$, we use the fact that under H_0, $\sqrt{N}\bar{X}_N \sim N(0, \sigma^2)$. The usual approach is to estimate σ^2 by the sample variance $\hat{\sigma}^2$ and approximate the distribution of $\sqrt{N}\bar{X}_N/\hat{\sigma}$ by the standard normal distribution. The test uses the same normal quantiles as critical values.

In the present testing problem, the eigenvalues λ_j collectively play a role analogous to the unknown variance. We would like to normalize the test statistic $\hat{T}_N$ by somehow dividing it by their estimators $\hat{\lambda}_j$ to get a limit distribution which does not depend on the distribution of the data. It turns out that an appropriate test statistic has the form

$$\hat{T}_N^0(d) = \sum_{j=1}^{d} \hat{\lambda}_j^{-1} \int_0^1 \langle U_N(x, \cdot), \hat{\varphi}_j \rangle \, dx$$

$$= \sum_{j=1}^{d} \hat{\lambda}_j^{-1} \int_0^1 \left\{ \int U_N(x, t)\hat{\varphi}_j(t)dt \right\} dx.$$

As $N \to \infty$, the distribution of $\hat{T}_N^0(d)$ approaches the distribution of the random variable

$$T^0(d) = \sum_{j=1}^{d} \int_0^1 B_j^2(x)dx.$$

The distribution of $T^0(d)$ does not depend on any unknown parameters, but it depends on d. Critical values are given in Table 8.1. They were computed using (8.18) with $N = 1,000$ and 10E5 replications. The selection of an "optimal" d is not always obvious. A simple rule is to use the cumulative variance approach: we choose the smallest d for which $\sum_{j=1}^{d} \hat{\lambda}_j > 0.85 \sum_{j=1}^{D} \hat{\lambda}_j$, where D is the maximum number of eigenvalues available from solving numerically the eigenvalue problem. The number d is typically much smaller than D. While D is comparable to N, d is often a single digit number. If d is chosen too large, the statistic $\hat{T}_N^0(d)$ will involve division by small eigenvalues $\hat{\lambda}_j$, and this will increase its variability. On the other hand, if d is too small, the space of functions which are orthogonal to the functions $\hat{\varphi}_j, j \le d$, will be large. The test statistic will not detect departures from stationarity which take place in this large space.

Both tests described above are implemented in the package `ftsa`. We illustrate their application using the functional time series `pm_10_GR_sqrt` introduced in Section 8.4. The call

TABLE 8.1: Critical values of the distribution of $T^0(d)$, $1 \le d \le 18$.

Nominal size	d					
	1	2	3	4	5	6
10%	0.3452	0.6069	0.8426	1.0653	1.2797	1.4852
5%	0.4605	0.7489	1.0014	1.2397	1.4690	1.6847
1%	0.7401	1.0721	1.3521	1.6267	1.8667	2.1259
	7	8	9	10	11	12
10%	1.6908	1.8974	2.0966	2.2886	2.4966	2.6862
5%	1.8956	2.1241	2.3227	2.5268	2.7444	2.9490
1%	2.3423	2.5892	2.8098	3.0339	3.2680	3.4911
	13	14	15	16	17	18
10%	2.8842	3.0669	3.2689	3.4620	3.6507	3.8377
5%	3.1476	3.3363	3.5446	3.7402	3.9491	4.1362
1%	3.7080	3.9040	4.1168	4.3171	4.5547	4.7347

```
T_stationary(pm_10_GR_sqrt$y)
```

is the default application of the test based on the statistic $\widehat{T}_N$. It produces the following output

```
Monte Carlo test of stationarity of a functional time series
null hypothesis: the series is stationary
p-value = 0.082
N (number of functions) = 182
number of MC replications = 1000
```

The P–value of 8.2% indicates that the series can be treated as stationary. Since the P-value is less than 10%, using a larger sample size might reveal some nonstationarity due to seasonal effects. The last line indicates that $R = 10E3$ replications, the default value, were used to approximate the distibution of the limit T. The function `T_stationary` uses the approximation

$$\int_0^1 B^2(x)dxx \approx I^\star(J) = \sum_{j=1}^{J} \frac{Z_j^2}{j^2\pi^2},$$

where the J is an argument with the default value $J = 500$. A reader interested in the derivation of this approximation is referred to Shorack and Wellner (1986), pp. 210–211. By default, it uses the approximation $T^\star = \sum_{j=1}^{d} \hat{\lambda}_j I_j^\star(J)$, where d is chosen to explain 90% of the cumulative variance. This default can be changed with the argument `cumulative_var`.

The argument `pivotal` chooses between the tests based on $\widehat{T}_N$ and $\widehat{T}_N^0(d)$. The default `pivotal = F` chooses $\widehat{T}_N$. The following code illustrates the application of the test based on $\widehat{T}_N^0(d)$.

```
T_stationary(pm_10_GR_sqrt$y, J=100, MC_rep=5000, h=20,
    pivotal=T)
```

It produces the (part) output

```
p-value = 0.1188
N (number of functions) = 182
number of MC replications = 5000
```

The argument h is the kernel bandwidth used in the estimation of the LRVF σ_η, see (8.16). As implemented in the function `T_stationary`, the pivotal test does not use the critical values in Table 8.1, but rather replications of $T^{*0}(d) = \sum_{j=1}^{d} I_j^*(J)$ to approximate the null distribution of $\widehat{T}_N^0(d)$.

8.7 Generation and estimation of the FAR(1) model using package `fda`

This section shows how to generate and estimate the FAR(1) model introduced in Section 8.2 using basic functions available in R and the package `fda`. In addition to enhancing the familiarity with the FAR(1) model, it provides additional practice in using many important function in that package. The package `far`, see Problem 8.6, provides a quicker route, but additional technical background is needed to understand the implementation it uses.

Our first objective is to generate a realization of the FAR(1) process (8.1) of length $N = 200$. We use the kernel $\varphi(t, s) = \alpha st$ with $\alpha = 9/4$ so that $\|\Phi\| = \{\iint \varphi^2(t, s)dtds\}^{1/2} = 3/4$, cf. Example 8.8.1 in Section 8.8. As innovations, we use

$$\varepsilon_n(t) = Z_{n1} \sin(\pi t) + \frac{1}{2} Z_{n2} \cos(2\pi t), \quad t \in [0, 1],$$

where the Z_{n1} and Z_{n2} are independent standard normal random variables. Simulation of any stationary time series typically requires a burn in period which allows the simulated variables to settle into a stationary pattern. We will thus generate 250 functions and discard the first 50. The following code accomplishes our objective.

```
library(fda)
m=100  # each function is observed at m+1 points, including 0
       and 1
burnin=50  # the first 50 functions are a burn in period
N=200  # number of functions to simulate
N1=N+burnin
alpha=9/4
# Create 2 matrices, whose entries are 0s.
# Each column represents one function.
X<- matrix(rep(0, (m+1)*N1),m+1,N1)
```

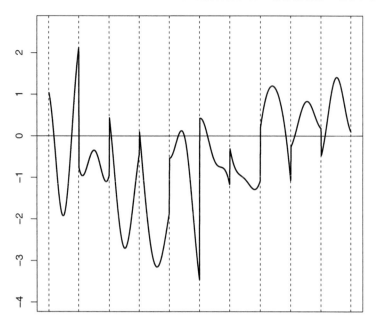

FIGURE 8.9: Ten observations of a simulated FAR(1) process.

```
epsilon<- matrix(rep(0, (m+1)*N1),m+1,N1)
epsilon[,1]<-rnorm(1)*sin(pi*(0:m/m))+0.5*rnorm(1)*cos(2*pi*
   (0:m/m))
# the following loop simulates FAR(1).
for(i in 2:N1){
   epsilon[,i]<-rnorm(1)*sin(pi*(0:m/m))+0.5*rnorm(1)*cos(2*pi*
      (0:m/m))
   X[,i]<-alpha*(1/m)^2*sum((1:m)*X[2:(m+1),i-1])*(0:m/m)+
      epsilon[,i]
                         }
X=X[,-(1:burnin)] # Remove the burn in period functions
```

The following code generates Figure 8.9 which displays the last ten simulated functions:

```
last=10
plot.ts(c(X[,(N-last+1):N]),ylim=c(min(X[,(N-last):N])-0.5,0.5
+max(X[,(N-last):N]))),axes=F,xlab="",ylab="",lwd=2)
axis(2);   axis(1,tick=F,labels=F);   abline(h=0)
abline(v=seq(0,last*(m+1),by=m+1), lty=2);   box()
```

Analogously to a scalar autoregressive process with a positive autoregres-

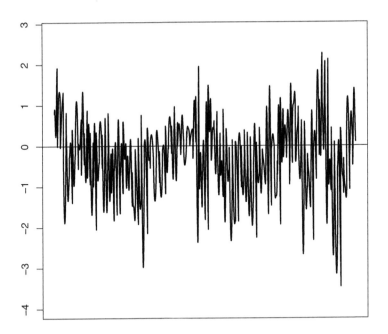

FIGURE 8.10: One hundred functional observations of a simulated FAR(1) process. The last ten functions are those shown in Figure 8.9.

sion coefficient, the functions tend to stay above or below the mean (zero in this case) for several values of n. This pattern of "positive persistence" is further illustrated in Figure 8.10 which shows the last 100 functions of the simulated FAR(1) process.

Our next objective is to represent the functions forming the realization of the FAR(1) process as functional objects. This is accomplished with the following code:

```
basisfd=10 # number of basis functions to represent each
    functional observation
basis=create.bspline.basis(c(0,1), nbasis=basisfd, norder=4)
fdX=Data2fd(argvals=0:m/m, X, basis)
```

With the time series functions available as functional objects, we can perform the functional principal components analysis as follows:

```
p=4 # number of EFPC's
fdXpca=pca.fd(fdX, nharm=p)
eigenvalues=fdXpca$values;   scoresX=fdXpca$scores
# jth column of scoresX contains scores of the jth EFPC
harmonicsX=fdXpca$harmonics   # extract the EFPC's
```

```
varianceprop=fdXpca$varprop  #proportion of variance explained
    by the EFP's
round(varianceprop*100,0)    [1] 71 25  4  0
```

The last output shows the percentage of variance explained by each EFPC. The first component explains 71% of variance, the first two components explain 96% of variance. The estimates together with the true surface $\varphi(t,s)$ are shown in Figure 8.11. The following code was used to produce it:

```
phi=9/4*(0:m/m)%*%t(0:m/m)
# True surface evaluated on a discrete bivariate grid
par(mar=c(1.5, 1.5, 3.5, 0.2), mfrow=c(2,2), oma = c(4, 4, 0.2
    , 0.2))
# 4 panels 2 rows and 2 columns arrangement
axislabelsize=1.5    # controls the size of labels of axes
axisticksize=0.8     # controls the size of ticks of axes
persp((0:m/m),(0:m/m),z=phi,cex.axis = axisticksize,
cex.lab=axislabelsize, xlab="t",ylab="s", zlab=" ", theta=30,
    phi=30,
ticktype="detailed", main="True")

# Next we compute  hat(phi)
# vivj is the matrix whose entries are products of v_j(s_k)*
    v_i(t_l).
# Blocks of vivj of m by m matrices represent products of
# v_j(s)v_i(t) evaluated on the (m+1) by (m+1) grid

for(npc in 1:3){
  vivj=matrix(0,p*(m+1),p*(m+1))
  for(j in 1:npc){
    for(i in 1:npc){
      vivj[1:(m+1)+(m+1)*(i-1),1:(m+1)+(m+1)*(j-1)]
      =eval.fd(evalarg=0:m/m, harmonicsX[i])
      %*%t(eval.fd(evalarg=0:m/m, harmonicsX[j]))
    }
  }
# phip will be the estimated surface.
  phip=matrix(0,m+1,m+1)
  for(k in 1:(N-1)){
    temp=matrix(0,m+1,m+1)
    for(j in 1:npc){
      temp1=matrix(0,m+1,m+1)
      for(i in 1:npc){
        temp1=temp1
        +scoresX[k+1,i]*vivj[1:(m+1)+(m+1)*(i-1),1:(m+1)+(m+1)
    *(j-1)]
      }
      temp=temp+(eigenvalues[j])^(-1)*scoresX[k,j]*temp1
    }
    phip=phip+temp
```

```
    }
    phip=(1/(N-1))*phip

if (npc==1)
persp((0:m/m),(0:m/m), z=phip, cex.axis =axisticksize, cex.lab
    =axislabelsize,
    xlab="t",ylab="s", zlab=" ", theta=30, phi=30, ticktype="
    detailed", main="p=1")
else if (npc==2)
persp((0:m/m),(0:m/m), z=phip, cex.axis = axisticksize,
    cex.lab=axislabelsize,
    xlab="t", ylab="s",zlab=" ",theta=30,phi=30,ticktype="detailed
    ",
main="p=2")
else if (npc==3)
persp((0:m/m),(0:m/m), z=phip, cex.axis = axisticksize,
    cex.lab=axislabelsize,
    xlab="t", ylab="s", zlab=" ", theta=30, phi=30, ticktype="
    detailed",
main="p=3")
}
```

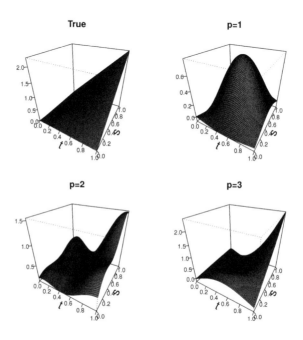

FIGURE 8.11: The surface $\varphi(t, s) = \alpha ts$ together with its estimates obtained using formula (8.5).

8.8 Conditions for the existence of the FAR(1) process

This section requires familiarity with the theory developed in Chapters 10 and 11. We formulate conditions which ensure that the FAR(1) process defined by equations (8.1) exists and is stationary. As noted in Section 8.1, for the scalar AR(1) equations, the required condition is $|\varphi| < 1$. Stationary time series models are typically defined by equations which relate the X_n to past $X_k, k < n$. The first question that must be answered is whether there exists a stationary stochastic process $\{X_n\}$ which satisfies the specified equations. Our goal in this section is to establish sufficient conditions for a stationary FAR(1) process to exist. We will use the strongest form of stationarity known as the *strict* stationarity. The following definition is valid if the X_n take values in any metric space, in particular if they are scalars or functions in L^2. To lighten the notation, we set in this section, $\| \cdot \|_{\mathcal{L}} = \| \cdot \|$.

DEFINITION 8.8.1 A sequence $\{X_n, -\infty < n < \infty\}$ is said to be strictly stationary if for any d and any time points $n_1, n_2, \ldots, n_d$, the distribution of the vector $[X_{n_1+h}, X_{n_2+h}, \ldots, X_{n_d+h}]$ does not depend on h.

THEOREM 8.8.1 *If $\|\Phi\| < 1$, then there is a unique strictly stationary solution to (8.1). This solution is given by*

$$X_n = \sum_{j=0}^{\infty} \Phi^j(\varepsilon_{n-j}). \tag{8.19}$$

The series (8.19) converges almost surely:

$$\left\| X_n - \sum_{j=0}^{m} \Phi^j(\varepsilon_{n-j}) \right\| \overset{a.s.}{\to} 0, \quad \text{as } m \to \infty \tag{8.20}$$

and in the squared mean:

$$E \left\| X_n - \sum_{j=0}^{m} \Phi^j(\varepsilon_{n-j}) \right\|^2 \to 0, \quad \text{as } m \to \infty. \tag{8.21}$$

PROOF: All random functions are defined on the same probability space Ω. We first establish the a.s. convergence of the sequence

$$X_n^{(m)} = \sum_{j=0}^{m} \Phi^j(\varepsilon_{n-j}).$$

The index n is fixed. Set $b = \|\Phi\|$, so that for any $j \geq 0$, $\left\|\Phi^j\right\| \leq \|\Phi\|^j = b^j$.

Observe that

$$E\left(\sum_{j=0}^{\infty}\|\Phi^j\|\|\varepsilon_{n-j}\|\right)^2 \leq \sum_{j,k=0}^{\infty}\|\Phi^j\|\|\Phi^k\|E\|\varepsilon_0\|^2$$

$$\leq E\|\varepsilon_0\|^2\left(\sum_{j=0}^{\infty}b^j\right)^2 < \infty.$$

It follows that $\sum_{j=0}^{\infty}\|\Phi^j\|\|\varepsilon_{n-j}\| < \infty$ a.s. This implies that $\sum_{j=0}^{\infty}\|\Phi^j(\varepsilon_{n-j})\| < \infty$ a.s. so the sequence $X_n^{(m)}$ is Cauchy a.s. Its limit is denoted X_n.

To establish (8.21), we work with the space $L^2(\Omega, L^2([0,1]))$, which is a Hilbert space with the inner product $E\langle X, Y\rangle$, $X, Y \in L^2([0,1])$. Thus, to show that the sequence $X_n^{(m)}$ has a limit in $L^2(\Omega, L^2([0,1]))$, it again suffices to check that it is Cauchy. By Lemma 12.1.1,

$$E\left\|\sum_{j=m}^{m'}\Phi^j(\varepsilon_{n-j})\right\|^2 = \sum_{j=m}^{m'}\sum_{k=m}^{m'}E\langle\Phi^j(\varepsilon_{n-j}), \Phi^k(\varepsilon_{n-k})\rangle$$

$$= \sum_{j=m}^{m'}E\|\Phi^j(\varepsilon_{n-j})\|^2.$$

Note that $E\Phi^j(\varepsilon_{n-j}) = 0$ because the expectation commutes with bounded operators. Therefore,

$$E\left\|\sum_{j=m}^{m'}\Phi^j(\varepsilon_{n-j})\right\|^2 \leq \left(\sum_{j=m}^{m'}\|\Phi^j\|^2\right)E\|\varepsilon_0\|^2 \leq E\|\varepsilon_0\|^2\sum_{j=m}^{m'}b^{2j}.$$

The series (8.19) is strictly stationary and it satisfies equation (8.1), Problem 8.16. Suppose $\{X_n'\}$ is another strictly stationary sequence satisfying (8.1). Then, iterating (8.1), we obtain, for any $m \geq 1$,

$$X_n' = \sum_{j=1}^{m}\Phi^j(\varepsilon_{n-j}) + \Phi^{m+1}(X_{n-m+1}') = X_n^{(m)} + \Phi^{m+1}(X_{n-m+1}').$$

Therefore,

$$E\|X_n' - X_n^{(m)}\| \leq \|\Phi^{m+1}\|E\|X_{n-m+1}'\| \leq b^{m+1}E\|X_0'\|.$$

Thus X_n' is equal a.s. to the limit of $X_n^{(m)}$ i.e. to X_n. ∎

The condition $\|\Phi\| < 1$ can be weakened to the condition that $\|\Phi^j\| < 1$, for some $j \geq 0$, Problems 8.17 and 8.18. In the scalar case both conditions are equivalent. In practice, we assume $\|\Phi\| < 1$, or an even stronger condition, as the following example shows.

EXAMPLE 8.8.1 Consider an integral Hilbert–Schmidt operator on L^2 defined by

$$\Phi(x)(t) = \int \varphi(t,s)x(s)ds, \quad x \in L^2, \tag{8.22}$$

which satisfies

$$\iint \varphi^2(t,s)dt\,ds < 1. \tag{8.23}$$

Recall that the left–hand side of (8.23) is equal to the Hilbert–Schmidt norm $\|\Phi\|_{\mathcal{S}}^2$. Since $\|\Phi\| \leq \|\Phi\|_{\mathcal{S}}$, we see that condition (8.23) is sufficient for the existence of an FAR(1) process whose autoregressive operator is the integral operator (8.22). □

8.9 Further reading and other topics

The theory of functional autoregressive processes is developed in depth in Bosq (2000) who also provides references to several applications. Background presented in Chapters 10 and 11 is needed to understand it. The FAR(1) process is also referred to as the ARH(1) process; the "H" refers to an arbitrary Hilbert space. The method described in Section 8.3 was introduced by Hyndman and Ullah (2007) and elaborated on in Hyndman and Booth (2008) and Hyndman and Shang (2009). These references contain more details, the last one contains a discussion of forecasting methods for functional time series. The method of Section 8.4 was introduced by Aue *et al.* (2015) who provide large sample arguments justifying it, as well as several extensions. Shang (2017) proposes an updating approach to forecasting FTS.

Several chapters in Horváth and Kokoszka (2012) are devoted to functional time series. They cover order determination and change–point detection in the FAR process as well as the theory of general weakly dependent stationary functional processes. The central limit theorem for such processes states that $N^{1/2}(\bar{X}_N - \mu) \xrightarrow{d} G$, in the space L^2, where G is a Gaussian random function. Gaussian random functions are defined in Chapter 11. It follows that $\|\bar{X}_N - \mu\|$ converges in probability to zero, i.e. the functional sample mean is a consistent estimator of the mean function for weakly dependent stationary processes. Estimation of the long–run variance is also studied in greater detail, see also Horváth *et al.* (2013). Theory of stationarity tests for functional time series is developed in Horváth *et al.* (2014). Tests of trend stationarity are proposed in Kokoszka and Young (2016).

An important aspect of time series analysis is the spectral methodology. It is based on the decomposition of a series into sine and cosine functions. Extension of these methods to FTS's are complex, but they provide some additional insights. The relevant theory was developed by Panaretos and Tavakoli (2013b, 2013a). Using spectral methods, Hörmann *et al.* (2015) introduced orthonormal functions, which they call dynamic functional principal components. These basis functions lead to optimal expansions for dependent stationary processes (EFPC's introduced in Chapter 3 are optimal if the observed functions are iid).

8.10 Chapter 8 problems

8.1 It can be shown that if $|\varphi| < 1$, then the scalar mean zero AR(1) process $X_n = \varphi X_{n-1} + \varepsilon_n$ admits the expansion

$$X_n = \sum_{j=0}^{\infty} \varphi^j \varepsilon_{n-j},$$

where the series converges in mean square, e.g. $\lim_{m\to\infty} E|X_n - \sum_{j=0}^{m} \varphi^j \varepsilon_{n-j}|^2 = 0$. Using the above expansion, find the autocovariance and the autocorrelation functions.

8.2 The least squares estimate $\hat{\varphi}$ in the scalar AR(1) process $X_n = \varphi X_{n-1} + \varepsilon_n$ is defined as the the value of φ which minimizes $S_N(\varphi) = \sum_{n=2}^{N}(x_n - \varphi x_{n-1})^2$. Find $\hat{\varphi}$.

8.3 The scalar AR(1) process $X_n = \varphi X_{n-1} + \varepsilon_n$ with $\varphi = 1$ and $X_0 = 0$ is called the random walk. Show that this process is not stationary.

8.4 Recall the partial inverse operator C_p^{-1} defined in Section 8.2. Find functions x and y such that $C_p^{-1}(C(x)) \neq x$ and $C(C_p^{-1}(y)) \neq y$.

8.5 This problem elaborates on the functional principal component analysis of the simulated FAR(1) process considered in Section 8.7. It assumes that the code displayed in that section has been entered.

Use the following code to display the percentage of variance and cumulative variance explained by the first four EFPC's:

```
par(mtrow=c(1,2))
plot(round(varianceprop*100,0),type="b",
xlab="Eigenfunction",ylab="Percentage of explained variance",
    axes=F)
axis(2);   axis(1,at=1:4);   box()
plot(cumsum(round(varianceprop*100,0)), ylim=c(80,100), type="
    b",
xlab="Eigenfunction",ylab="Cumulative percentage of variance",
    axes=F)
axis(2);   axis(1,at=1:4);  box()
```

Using the following code, display the first four EFPC's:

```
par(mfrow=c(2,2))
plot.fd(harmonicsX[1], ylab= "", main=expression(hat(italic(v))
    [1]))
plot.fd(harmonicsX[2], ylab= "", main=expression(hat(italic(v))
    [2]))
plot.fd(harmonicsX[3], ylab= "", main=expression(hat(italic(v))
    [3]))
plot.fd(harmonicsX[4], ylab= "", main=expression(hat(italic(v))
    [4]))
```

Comment on the shape of $\hat{v}_4$.

8.6 The package `far` contains many functions for simulating and estimating functional autoregressive processes. The following code generates a sequence of $N = 100$ observations following a FAR(1) process and graphs the first five:

```
require(far)
far1 <- simul.far(m=64,n=100)
plot(select.fdata(far1,date=1:5), whole=TRUE, separator=TRUE )
```

Plot all one hundred functions without separating them to obtain a plot similar to Figure 8.10.

8.7 Using the help file of the function `hmd.mx` in the `Demography` package, cf. Section 8.3, chose a country that is of interest to you. Construct male log mortality rate curves and their smooth versions for years 1950 through 2010. Plot, on one graph, these curves for 1970 and 2010. Comment on the changes in the mortality pattern. Using the graphs, estimate 2010 mortality rates $m_n(t) = \exp(X_n(t)$ at ages $t = 10$ and $t = 60$ for the country you have chosen, and compare them to the corresponding rates for the United States. Display the predicted log mortality curves for years 2011 and 2040, and compare these predictions with those for the United States.

8.8 Using the function `hmd.mx` in the `demography` package, cf. Section 8.3, construct log mortality rate curves for years 1900–1945 for the total male population of France (use FRATNP). Display the curves and their smooths for years 1900 and 1945 and comment on the difference. Do you see the effect of World War II? Display predictions for the next 30 years using the method of Section 8.3. Next, display the predictions which use a robust method, use the call `fdm(XXXXX.smooth, series="male", method="rapca")`. The argument `rapca` causes the outliers to be taken into account. Comment on the differences between the two sets of predicted curves.

8.9 This problem illustrates the application of the R package `ftsa` to implement the forecasting method explained in Section 8.3. It is based on an example presented in Shang (2013). The grey lines in Figure 8.12 show Australian fertility rate curves observed from 1921 to 2006. The value $X_n(t)$ is the count of live births in year n per 1,000 females of age $t \in [15, 49]$. The top

Forecasted fertility rates (2007–2026)

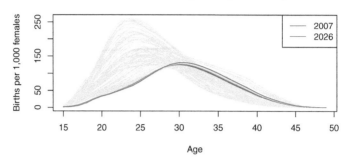

Recursive Forecasts

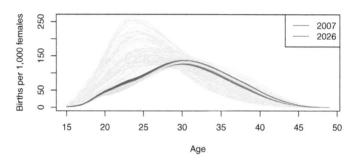

FIGURE 8.12: Predictions of fertility rates in Australia

panel shows the predictions obtained using the "exact method" and the bottom panel those obtained with the "recursive method". In the exact method, the forecasts of $X_{N+1}, X_{N+2}, \ldots, X_{N+h}$ are computed all at once using the data $X_1, X_2, \ldots, X_N$. In the recursive method, the one step ahead prediction $\hat{X}_{N+1|N}$ is obtained and treated as a real observation to predict X_{N+2}. This process is continued until the prediction of X_{N+h} is obtained.

Reproduce Figure 8.12 using the following code:

```
library("ftsa"); par(mfrow=c(2,1))
# Plot the historical data in gray
plot(Australiasmoothfertility, col = gray(0.8), xlab = "Age",
ylab = "Births per 1,000 females",
main = "Forecasted fertility rates (2007-2026)")
# Plot the forecasts in rainbow color
plot(forecast(ftsm(Australiasmoothfertility, order = 2), h =
    20), add = TRUE)
legend("topright", c("2007", "2026"), col = c("red", "blue"),
    lty = 1)
```

```
# Repeat the above steps but use recursive forecasts
plot(Australiasmoothfertility, col=gray(0.8), xlab="Age", ylab =
    "Births per 1,000 females", main = "Recursive Forecasts")
plot(ftsmiterativeforecasts(Australiasmoothfertility,
    components = 2, iteration = 20), add = TRUE)
legend("topright", c("2007", "2026"), col = c("red", "blue"),
    lty = 1)
```

What conclusions can you draw from these predictions? Comment on the difference between the predictions obtained using the two methods.

8.10 This problem compares the Hyndman–Ullah method described in Section 8.3 to the multivariate method described in Section 8.4. It uses the fertility data set introduced in Problem 8.9. Use the following code to plot the one–step–ahead predicted curve, with confidence bands, using the multivariate method, and then using the Hyndman–Ullah method:

```
library("ftsa"); library("vars")
aus_farforecast = farforecast(ftsm(Australiafertility), h=20,
    PI=TRUE)
# plot one-step-ahead predicted curve using the multivariate
    method
plot(extract(aus_farforecast$point_fore, timeorder=1), ylim=c
    (0,140),
xlab="Age", lwd=2)
lines(extract(aus_farforecast$PI_lb, timeorder=1), col=2, lty
    =2)
lines(extract(aus_farforecast$PI_ub, timeorder=1), col=2, lty
    =2)
#plot predicted curve using the Hyndman-Ullah method
plot(forecast(ftsm(Australiasmoothfertility, order = 2), h =
    1), add = TRUE,
lty=3, lwd=2)
```

Comment on the differences in the two predicted curves and explain which prediction is more appropriate.

8.11 Suppose $\{X_i\}$ is a scalar time series defined by $X_i = \mu + Z_i + \theta Z_{i-1}$, where the Z_i are iid with mean zero and variance σ_Z^2. Compute the LRV σ^2 of the series $\{X_i\}$. For what values of θ is σ^2 greater than the variance $\mathrm{Var}[X_i]$?

8.12 Show that the processes $\{X_i\}$ defined by $H_{A,1}$ and $H_{A,2}$ of Section 8.6 are not stationary in the sense of Definition 8.5.1. Display the formula for $\mathrm{Cov}(X_i(t), X_i(s))$ under $H_{A,2}$. Consider the case of a general stationary sequence $\{u_\ell\}$ and then specialize to the iid u_ℓ.

8.13 Consider the process $\{U_N(x), 0 \le x \le 1\}$ defined in Section 8.6. Assuming $H_{A,1}$ holds, compute $U_N(k/N)$ for $k = 1, 2, \ldots N$.

8.14 Apply the stationarity test of Section 8.6 to the Elnino series. Look up the definitions of this series in the help files of the package `ftsa`. Use the test based on the statistic $\widehat{T}_N$, first with with default values, then using 5,000 replications and d so large that it explains 99% of the cumulative variance. Display the output and state the conclusions.

8.15 Definition 8.1.1 can be extended to L^2–valued functions as follows: a sequence $\{X_n\}$ is stationary if the mean functions $\mu_n = EX_n$ and the covariance operators $C_n(h)$ defined by

$$C_n(h)(x) = E[\langle X_n - \mu_n, x \rangle (X_{h+h} - \mu_{n+h})]$$

do not depend on n. Show that if $\{X_n\}$ is strictly stationary in the sense of Definition 8.8.1 and $E\|X_n\|^2 < \infty$ for each n, then it is stationary in the above sense (also referred to as weakly stationary).

8.16 Show that the series (8.19) is strictly stationary, and it satisfies equation (8.1).

8.17 Show that for any $\Phi \in \mathcal{L}$, the following two conditions are equivalent:

C0: There exists an integer j_0 such that $\|\Phi^{j_0}\| < 1$.

C1: There exist $a > 0$ and $0 < b < 1$ such that for every $j \geq 0$, $\|\Phi^j\| \leq ab^j$.

8.18 Using problem 8.17 show that in Theorem 8.8.1 the assumption $\|\Phi\| < 1$ can be weakened to the condition that $\|\Phi^j\| < 1$ for some $j \geq 0$.

9

Spatial functional data and models

This chapter provides an introduction to the analysis of functions indexed by spatial locations. While there are other types of spatial functional data, this chapter considers data which take a form of a collection of curves, $X(\mathbf{s}_k), k = 1, 2, \ldots, K$, where the $\mathbf{s}_k$ are spatial locations. Each $t \mapsto X(\mathbf{s}_k; t)$ is a curve, a function. For example, the $\mathbf{s}_k$ can be locations of meteorological stations, and $X(\mathbf{s}_k, t)$ can be some measurement, for example temperature or wind speed at time t. In traditional spatial statistics, the observations are scalars, for example temperatures at the same time. The most common problem of spatial statistic is to predict a value at a location at which no measurement is available using the available measurements. Prediction in the context of spatial data is called *kriging*. In our temperature example, if we consider t to be day, it would mean that we want to predict the value $X(\mathbf{s}_0, t_0)$ for some location $\mathbf{s}_0$ using the values $X(\mathbf{s}_1, t_0), X(\mathbf{s}_2, t_0), \ldots, X(\mathbf{s}_K, t_0)$. For a large spatial domain, like a continent, it is clear that $X(\mathbf{s}_0, t_0)$ depends also on the values $X(\mathbf{s}_k, t)$ for $t \neq t_0$. For example, high temperatures early in the spring at some locations may predict high summer time temperatures at different locations. Similar arguments may apply to wind speeds, strong winds at certain times and certain locations may predict strong winds at later times at different locations. It is therefore important to present fundamentals of prediction of the whole curve $X(\mathbf{s}_0, \cdot)$ from a collection of curves $X(\mathbf{s}_1, \cdot), X(\mathbf{s}_2, \cdot), \ldots, X(\mathbf{s}_K, \cdot)$. This is the objective of this chapter.

In Section 9.1, we present the fundamental concepts of spatial statistics of scalar observations. Section 9.2 introduces the framework of this chapter by defining stationary and isotropic random fields of functions. Functional kriging and the estimation of the mean function of spatially indexed curves are considered, respectively, in Sections 9.3 and 9.4. Implementation of these procedures in the R package `geofd` is presented in Section 9.5. The chapter concludes with Section 9.6 which discusses some extensions and provides references to a few more advanced topics.

9.1   Fundamental concepts of spatial statistics

There are many excellent monographs and textbooks on spatial statistics. The monograph of Cressie (1993) provides a comprehensive account of the subject. The more recent monograph of Cressie and Wikle (2011) focuses on spatio–temporal data emphasizing Bayesian methods, which are also treated in detail by Banerjee *et al.* (2004). The book of Schabenberger and Gotway (2005) is a modern textbook with end–of–chapter problems. Initial chapters of Wackernagel (2003) and Sherman (2011) also provide accessible introductions to the subject of spatial statistics. The collection of Gelfand *et al.* (2010) presents recent developments in the field. This section summarizes the main concepts of spatial statistics. In the following sections of this chapter, we will see how they are extended to functional data. In this section, we assume that all data are scalars.

Geostatistical spatial data

A sample of *geostatistical* spatial data has the form

$$\{X(\mathbf{s}_k),\ \mathbf{s}_k \in \mathbf{S},\ k = 1, 2, \ldots, N\}.$$

The spatial domain $\mathbf{S}$ is typically a subset of the two–dimensional plane or sphere. The points $\mathbf{s}_k$ at which observations are available are generally scattered over $\mathbf{S}$ in some irregular manner. In some applications, e.g. satellite pictures, these points form a grid.

   The observed value $X(\mathbf{s}_k)$ is viewed as a realization of a random variable, so $\{X(\mathbf{s}),\ \mathbf{s} \in \mathbf{S}\}$ is a scalar random field. Just as in time series analysis, stationary random fields play a fundamental role in modeling spatial data. To define arbitrary shifts, we must assume that $\mathbf{S}$ is either the whole Euclidean space $\mathbb{R}^d$, or the whole sphere. The random field $\{X(\mathbf{s}),\ \mathbf{s} \in \mathbf{S}\}$ is then strictly stationary if

$$\{X(\mathbf{s}_1 + \mathbf{h}), X(\mathbf{s}_2 + \mathbf{h}), \ldots, X(\mathbf{s}_m + \mathbf{h})\} \stackrel{d}{=} \{X(\mathbf{s}_1), X(\mathbf{s}_2), \ldots, X(\mathbf{s}_m)\} \quad (9.1)$$

for any points $\mathbf{s}_1, \mathbf{s}_2, \ldots, \mathbf{s}_m \in \mathbf{S}$ and any shift $\mathbf{h}$. If we assume only that the means $EX(\mathbf{s})$ and the covariances $\mathrm{Cov}(X(\mathbf{s}), X(\mathbf{s} + \mathbf{h}))$ do not depend on $\mathbf{s}$, then the field is called second–order stationary. For such a field, we define the covariance function

$$C(\mathbf{h}) = \mathrm{Cov}(X(\mathbf{s}), X(\mathbf{s} + \mathbf{h})).$$

   In spatial statistics, the concept of *intrinsic* stationarity is useful. The field $\{X(\mathbf{s}),\ \mathbf{s} \in \mathbf{S}\}$ is said to be intrinsically stationary if $\mathrm{Var}[X(\mathbf{s}+\mathbf{h}) - X(\mathbf{s})]$ does not depend on $\mathbf{s}$. A second–order stationary field is intrinsically stationary, cf. (9.2). The converse is not true, Problem 9.4. If $\{X(\mathbf{s}),\ \mathbf{s} \in \mathbf{S}\}$ is intrinsically stationary, we define the *semivariogram* by

$$\gamma(\mathbf{h}) = \frac{1}{2}\mathrm{Var}[X(\mathbf{s} + \mathbf{h}) - X(\mathbf{s})].$$

(The variogram is defined as $2\gamma(\cdot)$.) The semivariogram of a second order stationary field with the covariance function $C(\cdot)$ is given by

$$\gamma(\mathbf{h}) = C(\mathbf{0}) - C(\mathbf{h}). \qquad (9.2)$$

If $C(\mathbf{h})$ depends only on the length h of $\mathbf{h}$, we say that the random field is isotropic. The covariance function of an isotropic random field is typically parametrized as

$$C(h) = \sigma^2 \phi(h), \quad h \geq 0, \ \phi(0) = 1.$$

The function ϕ is then called the correlation function. It quantifies the strength of linear dependence between observations distance h apart. The following correlation functions are frequently used. The powered exponential correlation function is defined by

$$\phi(h) = \exp\left\{ -\left(\frac{h}{\rho}\right)^p \right\}, \quad \rho > 0, \ 0 < p \leq 2.$$

If $p = 1$, this correlation function is called exponential, if $p = 2$, it is called Gaussian. A very general family of correlation functions is the so–called Matérn class. The Matérn class correlation functions are defined as

$$\phi(h) = \frac{2^{1-\nu}}{\Gamma(\nu)} \left(\frac{h}{\rho}\right)^\nu K_\nu(h/\rho), \quad \rho > 0, \ \nu > 0,$$

where K_ν is is a special function called the modified Bessel function, see Stein (1999). The function K_ν decays monotonically and approximately exponentially fast; numerical approximations show that $K_\nu(s)$ practically vanishes if $s > \nu$.

It follows from Problem 9.2 that every correlation function must be non-negative definite in the sense that $\sum_{i,j=1}^m a_i a_j \phi(d(\mathbf{s}_i, \mathbf{s}_j)) \geq 0$, where $d(\mathbf{s}_i, \mathbf{s}_j)$ is the distance between s_i and s_j. Such a nonnegative definite function ϕ is called *valid*. It is difficult to verify directly that any specific correlation function is valid. Theory of spatial Fourier transforms must be used, see Stein (1999).

Estimation of the covariance function proceeds through the estimation of the semivariogram. Very often isotropy is assumed, and we explain the procedure in this case. First, an empirical variogram is computed at several available lags $h > 0$. There are several versions of the empirical variogram. The simplest estimator is given by

$$\hat{\gamma}(d) = \frac{1}{|N(d)|} \sum_{N(d)} (X(\mathbf{s}_k) - X(\mathbf{s}_\ell))^2, \qquad (9.3)$$

where $N(d)$ is the set of pairs $(\mathbf{s}_k, \mathbf{s}_\ell)$ approximately distance d apart, and $|N(d)|$ is the count of pairs in $N(d)$. The function $\hat{\gamma}$ is noisy. It is approximated by a function $C(0) - C(d)$, cf. (9.2), where C is a valid covariance function.

For example, if we wish to fit an exponential model, we must find σ and ρ such that

$$\hat{\gamma}(d) \approx \sigma^2 \left(1 - \exp\{-d/\rho\}\right).$$

The details of this process and its variations are described in spatial statistics textbooks. Section 9.5 illustrates this process by suitable graphs in the context of functional data.

Kriging

In spatial spatistics, the term *kriging* refers to linear prediction. Using the observed values $X(\mathbf{s}_k)$, we want to predict the value $X(\mathbf{s})$ at a location $\mathbf{s}$ which is not one of the $\mathbf{s}_k, 1 \leq k \leq N$, at which data are available. The procedure is named after D. G. Krige (1919-2013), a South African mining engineer. We describe it in the simplest case which assumes that the data are observations of a second order stationary random field. Such a field can be represented as

$$X(\mathbf{s}) = \mu + e(\mathbf{s}), \quad \mu = EX(\mathbf{s}), \quad Ee(\mathbf{s}) = 0.$$

We want to find weights w_k such that the predictor

$$\widehat{X}(\mathbf{s}) = \mu + \sum_{k=1}^{N} w_k(X(\mathbf{s}_k) - \mu) \tag{9.4}$$

minimizes the expected squared error of prediction $E\left(\widehat{X}(\mathbf{s}) - X(\mathbf{s})\right)^2$. It can be shown, Problem 9.5, that

$$\sigma^2(\mathbf{s}) := E\left(\widehat{X}(\mathbf{s}) - X(\mathbf{s})\right)^2 = C(\mathbf{0}) + \sum_{k,j=1}^{N} w_k w_j C(\mathbf{s}_k - \mathbf{s}_j) - 2\sum_{k=1}^{N} w_k C(\mathbf{s}_k - \mathbf{s}).$$
$$\tag{9.5}$$

Minimizing $\sigma^2(\mathbf{s}) = \sigma^2(\mathbf{s}; w_1, w_2, \ldots, w_N)$ with respect to the weights $w_1, w_2, \ldots, w_N$ shows that they must satisfy the system of N linear equations

$$\sum_{j=1}^{N} C(\mathbf{s}_k - \mathbf{s}_j)w_j = C(\mathbf{s}_k - \mathbf{s}), \quad k = 1, 2, \ldots N, \tag{9.6}$$

Problem 9.6. If the matrix $[C(\mathbf{s}_k - \mathbf{s}_j), 1 \leq j, k \leq N]$ is nonsingular, these kriging weights are unique. In practice, the unknown covariances must be replaced by their variogram–based estimates.

Kriging the mean

If the observations $X_1, X_2, \ldots, X_N$ form a simple random sample, i.e. are iid, with common mean μ, then the most natural estimator of μ is the sample mean $\bar{X}_N = N^{-1}\sum_{k=1}^{N} X_k$. Intuitively, in case of iid observations, there is no difference between them, and each contains the same information about the unknown mean μ. This intuition has been formalized in classical estimation theory discussed in many textbooks, see e.g. Chapter 7 of Casella and Berger (2002). If the observations are taken at spatial locations, then, in most cases, observations at close–by locations are similar and those at locations far apart may be very different. This intuition is captured by the spatial covariances which decay with distance. For this reason, if $X(\mathbf{s}_1), X(\mathbf{s}_2), \ldots, X(\mathbf{s}_N)$ are observations of a second-order stationary random field, the mean μ is estimated by the weighted average

$$\hat{\mu} = \sum_{k=1}^{N} w_k X(\mathbf{s}_k).$$

If a location $\mathbf{s}_\ell$ is close to another location $\mathbf{s}_k$, then adding $X(\mathbf{s}_\ell)$ to the sample does not increase much the information about the mean already available in $X(\mathbf{s}_k)$ because $X(\mathbf{s}_\ell)$ is close to $X(\mathbf{s}_k)$. Both $X(\mathbf{s}_k)$ and $X(\mathbf{s}_\ell)$ should thus get smaller weights than values at locations further away. We now explain how the weights are computed assuming $\{X(\mathbf{s})\}$ is second order stationary. The first requirement is that $\hat{\mu}$ be an unbiased estimator, i.e. $E\hat{\mu} = \mu$. This implies that $\sum_{k=1}^{N} w_k = 1$. Then, the criterion that determines the w_k is that $E(\hat{\mu} - \mu)^2$ be minimum subject to $\sum_{k=1}^{N} w_k = 1$. It can be shown that the weights can be found by solving the system of $N + 1$ equations:

$$\sum_{j=1}^{N} C(\mathbf{s}_k - \mathbf{s}_j)w_j - r = 0, \quad k = 1, 2, \ldots, N, \quad \sum_{j=1}^{N} w_j = 1, \qquad (9.7)$$

where r is an additional unknown. Details of the argument leading the above equations are explained in Problem 9.7.

Other spatial data structures

Our focus in this chapter will be on geostatistical functional data, so the introduction presented above focused on geostatistical scalar data. Other data structures however naturally occur in spatial statistics. *Regional data* are measured not at a point, but over a region. For example, the count of shooting deaths in a county in a given year is not associated with a point in space but with the whole region, a county. *Lattice data* are available at a regular grid. For example, a satellite picture of a region consists of values at pixels on a rectangular grid. Each pixel can represent, e.g., the intensity of vegetation over a region it captures. In this sense, lattice data can be considered as regional data, and both terms are sometimes used interchangeably. *Spatial point process* data are concerned with locations at which events occur. For example, the points at which lightning strikes occur during a storm form a pattern. To describe such patterns, point processes are used. If a value is associated with each point, e.g. if the intensity of a discharge can be determined, then point process data can be, in principle, viewed as geostatistical data. However, in the analysis of geostatistical data, the locations are treated as fixed; for point process data, they are considered random.

9.2   Functional spatial fields

We focus on geostatistical functional data. Just as in the case of scalar data of this type, the model for functional geostatistical data is the random field $\{X(\mathbf{s}), \mathbf{s} \in \mathbf{S}\}$, with the difference that now each $X(\mathbf{s})$ is a random function. We assume that each $X(\mathbf{s})$ is an element of $L^2 = L^2([0,1])$. The value of $X(\mathbf{s})$ at time t is denoted by $X(\mathbf{s};t)$. We assume that each $X(\mathbf{s})$ is square integrable, i.e.

$$E \, \|X(\mathbf{s})\|^2 = \int EX^2(\mathbf{s};t)dt < \infty.$$

If we assume that $\mathbf{S}$ is the whole Euclidean space, we can define strict stationarity by condition (9.1). The equality of distributions is now in the product space $(L^2)^m$ rather than in $\mathbb{R}^m$. If this is the case, then $E \, \|X(\mathbf{s})\|^2$ does not depend on $\mathbf{s}$. Square integrability and strict stationarity imply that the mean function

$$\mu(t) = EX(\mathbf{s};t)$$

and the covariances

$$C(\mathbf{h};t,u) = \mathrm{Cov}(X(\mathbf{s};t), X(\mathbf{s}+\mathbf{h};u))$$

exist and do not depend on $\mathbf{s}$. When working with functional geostatistical data, it is important to keep in mind that the function $\mathbf{h} \mapsto C(\mathbf{h};t,u)$ is nonnegative definite, see (9.18) in Problem 9.2, only if $t = u$. In that case, $C(\cdot;t) := C(\cdot;t,t)$ is the spatial covariance function of the scalar field $\{X(\mathbf{s};t)\}$.

We will also assume that the covariances are isotropic, i.e. $C(\mathbf{h};t,u) = C(h;t,u)$ depends only on $h = \|\mathbf{h}\|$. The assumptions listed above can be weakened, but to keep the exposition simple, *in the remainder of this chapter, unless stated otherwise, we assume that the field $\{X(\mathbf{s})\}$ is square integrable, strictly stationary and isotropic.*

9.3   Functional kriging

We observe functions $X(\mathbf{s}_k)$ at spatial locations $\mathbf{s}_1, \mathbf{s}_2, \ldots, \mathbf{s}_N$. Suppose $\mathbf{s}$ is a different location, and we want to predict the function $X(\mathbf{s})$ using the functions $X(\mathbf{s}_1), X(\mathbf{s}_2), \ldots, X(\mathbf{s}_N)$. This is the kriging problem introduced for scalar data in Section 9.1. In the functional setting, the available data can be represented as a matrix of scalars

$$[X(\mathbf{s}_k; t_j), \ k = 1, 2, \ldots, N, \ j = 1, 2, \ldots, J],$$

assuming all functions are observed at the same points t_j. In some situations, the points t_j may depend on the location $\mathbf{s}_k$, so the scalar data points are $X(\mathbf{s}_k, t_{kj})$. There are many approaches to computing a predictor $\widehat{X}(\mathbf{s})$ of the unobserved curve $X(\mathbf{s})$. We discuss one of them, with additional references given in Section 9.6.

Taking a functional point of view, we treat each curve $X(\mathbf{s}_k)$ as an indivisible data object. We want to determine the weights w_k for the predictor of the form (9.4), but we keep in mind that the $X(\mathbf{s}_k)$ are now functions. In the least squares criterion, we must replace $(\widehat{X}(\mathbf{s}) - X(\mathbf{s}))^2$ by the squared L^2 norm. We thus want to find weights $w_1, w_2, \ldots, w_N$ which minimize

$$E \left\| \widehat{X}(\mathbf{s}) - X(\mathbf{s}) \right\|^2 = E \int \left(\widehat{X}(\mathbf{s}; t) - X(\mathbf{s}; t) \right)^2 dt.$$

Observe that

$$\left\| \widehat{X}(\mathbf{s}) - X(\mathbf{s}) \right\|^2 = \left\| \mu + \sum_{k=1}^{N} w_k (X(\mathbf{s}_k) - \mu) - X(\mathbf{s}) \right\|^2$$

$$= \left\langle \sum_{k=1}^{N} w_k (X(\mathbf{s}_k) - \mu) - (X(\mathbf{s}) - \mu), \sum_{\ell=1}^{N} w_\ell (X(\mathbf{s}_\ell) - \mu) - (X(\mathbf{s}) - \mu) \right\rangle$$

$$= \sum_{k,\ell=1}^{N} w_k w_\ell \langle X(\mathbf{s}_k) - \mu, X(\mathbf{s}_\ell) - \mu \rangle$$

$$- 2 \sum_{k=1}^{N} w_k \langle X(\mathbf{s}_k) - \mu, X(\mathbf{s}) - \mu \rangle + \langle X(\mathbf{s}) - \mu, X(\mathbf{s}) - \mu \rangle.$$

Introduce functional spatial covariances

$$C(\mathbf{s}, \mathbf{s}') = E \left[\langle X(\mathbf{s}) - \mu, X(\mathbf{s}') - \mu \rangle \right]. \tag{9.8}$$

Using this notation, we see that

$$E \left\| \widehat{X}(\mathbf{s}) - X(\mathbf{s}) \right\|^2 = \sum_{k,\ell=1}^{N} w_k w_\ell C(\mathbf{s}_k, \mathbf{s}_\ell) - 2 \sum_{k=1}^{N} w_k C(\mathbf{s}_k, \mathbf{s}) + C(\mathbf{s}, \mathbf{s}).$$

Finding the kriging weights thus formally reduces to minimizing (9.5), just as for scalar observations, but using the functional covariances (9.8). Therefore, the weights w_k are found by solving the system of equations

$$\sum_{\ell=1}^{N} C(\mathbf{s}_k, \mathbf{s}_\ell) w_\ell = C(\mathbf{s}_k, \mathbf{s}), \quad k = 1, 2, \ldots N.$$

The argument leading to the above system is the same as in the scalar case, cf. (9.6) and Problem 9.6. It only requires computing the partial derivatives and equating them to zero.

To implement this approach in practice, we must estimate the covariances $C(\mathbf{s}_k, \mathbf{s}_\ell)$ and $C(\mathbf{s}_k, \mathbf{s})$. We now explain how this can be done. Using the assumptions of stationarity and isotropy, we see that

$$C(\mathbf{s}, \mathbf{s}') = E \int \left(X(\mathbf{s}; t) - \mu(t) \right) \left(X(\mathbf{s}'; t) - \mu(t) \right) dt$$

$$= \int C(\|\mathbf{s} - \mathbf{s}'\| ; t) dt,$$

where

$$C(h; t) = \mathrm{Cov}(X(\mathbf{s} + \mathbf{h}; t), X(\mathbf{s}; t)), \quad h = \|\mathbf{h}\|,$$

is the spatial covariance function of the scalar field $\{X(\cdot; t)\}$ obtained by fixing t. For each t, the covariance function $h \mapsto C(h; t)$ can be estimated by any method that is applicable to covariance estimation for scalar spatial data. As noted in Section 9.1, this is typically done by estimating the semivariogram.

The kriging predictor is often expressed in a different way. We want to find a predictor $X^\star(\mathbf{s})$ of the form $X^\star(\mathbf{s}) = \sum_{k=1}^N \lambda_k X(\mathbf{s}_k)$, subject to the condition

$$\sum_{k=1}^N \lambda_k = 1. \tag{9.9}$$

By contrast, equations (9.6) do not imply that $\sum_{j=1}^N w_j = 1$. In many applications, the weights w_j are very close to the weights λ_j, and the predicted function $\widehat{X}(\mathbf{s})$ and $X^\star(\mathbf{s})$ are also very close. This is illustrated in Problem 9.12.

9.4   Mean function estimation

We now explain how to estimate the mean function of spatially indexed curves $X(\mathbf{s}_k), 1 \le k \le N$. Estimation of the mean function is needed to implement the kriging method described in Section 9.3. As explained in Section 9.1, even for scalar data, a proper approach is to use a weighted average rather than a simple average $N^{-1} \sum_{k=1}^N X(\mathbf{s}_k; t)$. In the functional context, a weighted average takes the form

$$\hat{\mu}(t) = \sum_{k=1}^N w_k X(\mathbf{s}_k; t), \tag{9.10}$$

and the problem becomes how to compute the optimal weights w_k. As in Section 9.3, in (9.10) we treat the curves $X(\mathbf{s}_k)$ as indivisible statistical objects, and so we apply one weight to the whole curve. One could consider weights

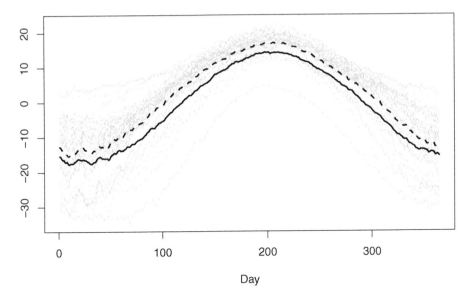

FIGURE 9.1: Gray lines: annual temperature curves at 35 locations shown in Figure 9.2 (averaged over several decades). Dashed line: simple average of these curves. Continuous black line: mean function estimated by functional kriging.

$w_k(t)$, defined as functions, but the simple approach we describe in this section is easy to implement, and produces estimates which significantly improve on the simple average. Figure 9.1 illustrates the difference between the simple average ($w_k = 1/N$) and the weighted average we are about to derive, using annual temperature curves in Canada. It is helpful to refer to Figure 9.2 which shows the locations at which the curves are available. There are plenty of locations in the southern part of Canada, especially in the densely populated South–East of the country, and very few in the North. If we postulate the model $X(\mathbf{s}_k; t) = \mu(t) + \varepsilon(\mathbf{s}_k; t)$, then μ should represent the typical annual temperature profile for the whole country. The simple average will mostly reflect temperature curves in the region where there are many stations, i.e. mostly in the South. To obtain a more informative kriged mean function, stations far in the North must receive larger weights. Some stations in the south will receive even slightly negative weights. In this section, we present the derivation of these weights, Problem 9.13 shows how to implement this procedure in R.

We define the optimal weights as those that minimize

$$E \left\| \sum_{k=1}^{N} w_k X(\mathbf{s}_k) - \mu \right\|^2 = \int \left\{ \sum_{k=1}^{N} w_k X(\mathbf{s}_k; t) - \mu(t) \right\}^2 dt \qquad (9.11)$$

subject to the condition $\sum_{k=1}^{n} w_k = 1$ which ensures that $E\hat{\mu} = \mu$. As in the scalar case, the method of the Lagrange multiplier, Problem 9.7, shows that the weights w_k must satisfy the following system of $N+1$ linear equations:

$$\sum_{k=1}^{N} w_k = 1, \quad \sum_{k=1}^{N} w_k C(\mathbf{s}_k, \mathbf{s}_n) - r = 0, \quad n = 1, 2, \ldots N, \qquad (9.12)$$

in which $w_1, w_2, \ldots, w_N, r$ are $N+1$ unknowns, Problem 9.9. The functional covariances $C(\mathbf{s}_k, \mathbf{s}_n)$ can be estimated as explained in Section 9.3. Another approach, based on *Functional variogram*, is explained in Problem 9.10.

9.5 Implementation in the R package geofd

Methods for spatial functional data discussed in this chapter, as well as several other methods, are implemented in the R package geofd. In this section, we provide an example of the code which implements the kriging method derived in Section 9.3. It includes the code for the estimation of the functional mean. In addition to the package fda, the implementation we present requires two more packages: fda.usc and maps. For illustration, we use the well–known Canadian weather data set available in the package fda. The assumptions of stationarity and isotropy may not hold for this data set, but the predictions are still fairly good, and our objective is merely to provide guidance on how to use the relevant packages.

The initial lines of the code are:

```
library(fda); library(fda.usc); library(geofd); library(maps)

data("CanadianWeather")

dailyAv <- CanadianWeather$dailyAv
dim(dailyAv)
Temperature    <- dailyAv[,,1]
Precipitation <- dailyAv[,,2]
log10Precip    <- dailyAv[,,3]

place <- CanadianWeather$place
coordinates <- CanadianWeather$coordinates
coordinates <- coordinates[,2:1]
coordinates[,1] <- - coordinates[,1]
```

FIGURE 9.2: The 35 Canadian weather stations, Calgary marked with a red square.

```
geo.dist.35 <- dist(coordinates)
Day <- 1:365
n<-dim(Temperature)[2];   nt <- dim(Temperature)[1]
```

Next, we produce a map of Canada and mark the 35 locations at which data are available. After that, "Calgary" is removed from sample locations and treated as a location at which functions are to be predicted. It is marked with a red square in Figure 9.2.

```
map('world',ylim=c(42,78),xlim = c(-170,-40))
map('world',region="Canada",add=TRUE)
points(coordinates,col='black',pch=19)
i.0 <- which(place=="Calgary")
coord.0 <- coordinates[i.0,]
points(coord.0[1],coord.0[2],  pch= 15)
```

In the next step, using the Fourier basis, we define the functional object Tempe.34 which contains the temperature curves for the remaining 34 stations. The code below also plots these temperature curves.

```
Tempe.34 <- Temperature[,-i.0]
coord.34 <- coordinates[-i.0,]
K <- min(99, max(49, 1 + 4 * round(sqrt(nt))))
fourier.basis <- create.fourier.basis(rangeval=range(Day),
    nbasis=K)
temp.fd.Fb.34 <- Data2fd(argvals=Day, y=Tempe.34, basisobj=
    fourier.basis)
```

```
temp.fd.Fb.0 <- Data2fd(argvals=Day, y=Temperature[,i.0],
    basisobj=fourier.basis)
plot(temp.fd.Fb.34, col="grey",
    xlab="Day", ylab="Temperature (degrees C)",
    main="Average daily temperatures")
lines(temp.fd.Fb.0, lwd=2)
```

The following code fits the exponential model to the empirical variogram and plots Figure 9.3. Binning is used to reduce the chance variability of the variogram cloud; each point in the variogram cloud is based on only one pair of points $(\mathbf{s}_k, \mathbf{s}_\ell)$. We group these pairs into bins that contain points of roughly the same distance $d \approx \|\mathbf{s}_k - \mathbf{s}_\ell\|$, and then compute the empirical variogram (9.3) with $X(\mathbf{s}_k) - X(\mathbf{s}_\ell)$ replaced by $\|X(\mathbf{s}_k) - X(\mathbf{s}_\ell)\|$. Such a variogram is called the *trace variogram* in the package geofd.

```
# computing L2 norms between functions, using Fourier basis
    expansions
L2norm.Fb.34 <- dist(t(temp.fd.Fb.34$coefs))^2
# Calculating the empiricial trace bin variogram
emp.trace.vari.34 <- trace.variog(coords=coord.34, L2norm=
    as.matrix(L2norm.Fb.34), bin=TRUE)
# fitting an exponential vriogram
sigma2.0 <- quantile(emp.trace.vari.34$v, 0.75)
phi.0    <- quantile(emp.trace.vari.34$Eu.d, 0.75)
fit.vari.34 <- variofit(emp.trace.vari.34, ini.cov.pars=c(
    sigma2.0, phi.0), cov.model="exponential")

plot(as.dist(emp.trace.vari.34$Eu.d),L2norm.Fb.34, col="grey",
    xlab="Geographical distances", ylab="L2 distances",
    main="Empirical variogram")
points(emp.trace.vari.34$u,emp.trace.vari.34$v, col="black",pch
    =19)
lines(fit.vari.34, col="black",lwd=2)
legend("topleft", c("Variogram cloud",  "Binned variogram",  "
    Fitted variogram"),
        col=c(8,1,1),  lwd=c(-1,-1,2),  pch=c(1,19,-1) )
```

Now we compute the kriging weights w_ℓ. First, using the exponential variogram model, we find the estimates of the covariances $C(\mathbf{s}_k, \mathbf{s}_\ell)$ and $C(\mathbf{s}_k, \mathbf{s})$, where $\mathbf{s}$ is the location of Calgary and the $\mathbf{s}_k$ are the remaining 34 locations. Since we use the functional version of model (9.4), we also need to find an estimate of the mean function μ. This is done in after finding the weights w_ℓ.

```
# computing the covariances
hat.C.34 <- cov.spatial(emp.trace.vari.34$Eu.d,
                        cov.model= fit.vari.34$cov.model,
                        cov.pars=fit.vari.34$cov.pars)
geo.dist.0.34 <- as.matrix(geo.dist.35)[-i.0,i.0]
hat.C.0 <- cov.spatial(geo.dist.0.34,
                        cov.model= fit.vari.34$cov.model,
```

Empirical variogram

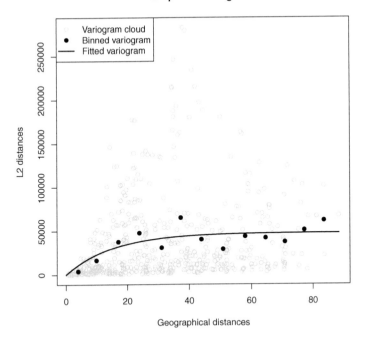

FIGURE 9.3: Estimation of the functional variogram.

```
                          cov.pars=fit.vari.34$cov.pars)
# computing the weights for the mean of 34 stations
inv.hat.C.34 <- solve(hat.C.34)
v.34 <- apply(inv.hat.C.34,1,sum)
w.m.34 <- v.34/sum(v.34)
# computing the weights for kriging
w0.k <- solve(hat.C.34,hat.C.0)
w.k <- w0.k + w.m.34*(1-sum(w0.k))
# computing the weighted functional mean of 34 stations
mean.temp.Fb.34 <- mean(temp.fd.Fb.34)
w.mean.temp.Fb.34 <- mean.temp.Fb.34
w.mean.temp.Fb.34$coefs <- apply(temp.fd.Fb.34$coefs,1,
weighted.mean,w=w.m.34)
temp.fd.Fb.34.ctrd <-temp.fd.Fb.34
temp.fd.Fb.34.ctrd$coefs <- temp.fd.Fb.34.ctrd$coefs
- matrix(w.mean.temp.Fb.34$coefs,ncol=34,nrow=77,byrow=FALSE)
```

The following code is used to display the kriging weights just computed.
They are shown in Figure 9.4 as a function of the distance to Calgary. The
gray line indicates the weight of 1/34 which would be used without spatial
kriging. We see that stations far away from Calgary receive weights practically

Kriging weights

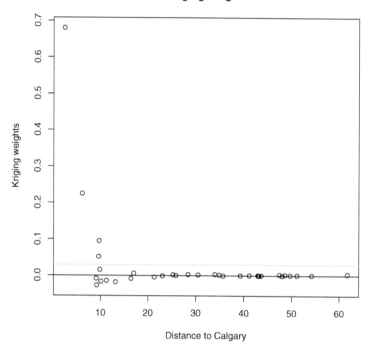

FIGURE 9.4: Kriging weights for the prediction of the temperature curve at Calgary. The gray line is equal to $1/34$.

equal to zero; the nearby locations receive larger weights, including slightly negative weights in some cases. The station closest to Calgary, Edmonton, receives by far the largest weight, almost 0.7. A different way to display these weights, using a map of Canada, is presented in Problem 9.11.

```
plot(geo.dist.0.34,w.k, xlab="Distance to Calgary", ylab="
   Kriging weights",
      main="Kriging weights")
abline(h=1/34, col="gray",lwd=2); abline(h=0,col="black")
```

The last part of the code computes the kriged temperature function at Calgary and compares it in Figure 9.5 to the true temperature function.

```
kriging.i.0 <- w.mean.temp.Fb.34
kriging.i.0$coefs <- kriging.i.0$coefs +
  apply(temp.fd.Fb.34.ctrd$coefs,1,weighted.mean,w=w.k)
plot(temp.fd.Fb.34, col=8, xlab= "Day")
lines(temp.fd.Fb.0,lwd=2)
lines(kriging.i.0,lty=2,lwd=2)
legend("bottomright", c("True temperature function",
          "Kriged function"),    lwd=c(2,2), lty=c(1,2))
```

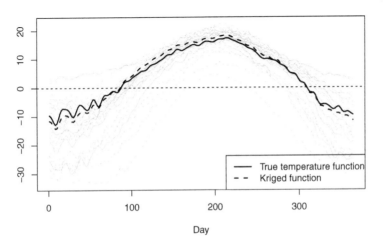

FIGURE 9.5: True and predicted (kriged) temperature functions at Calgary. The remaining 34 functions used for kriging are shown in grey.

9.6   Other topics and further reading

In addition to the method outlined in Section 9.3, Delicado *et al.* (2010) give references to several other methods of kriging functional data, and review research on regional and point process functional data. A relatively simple kriging method they discuss, which gives predictions as good as more complex methods can be summarized as follows. The predictor has the form

$$\widehat{X}(\mathbf{s};t) = \sum_{k=1}^{N} w_k(t) X(\mathbf{s}_k;t)$$

where the unknown functions $w_k(\cdot)$ satisfy $\sum_{k=1}^{N} w_k(t) = 1$, for each t. The observed functions $X(\mathbf{s}_k)$ and the unknown weight functions $w_k(\cdot)$ are expanded using some basis system:

$$X(\mathbf{s}_k;t) = \sum_{m=1}^{M} c_{km} B_m(t), \quad w_k(t) = \sum_{m=1}^{M} b_{km} B_m(t).$$

In particular, one can use the EFPC's $\hat{v}_m$ as the B_m. The coefficient matrix $\mathbf{B} = [b_{km}, \; 1 \le k \le N, 1 \le m \le M]$ is also found by minimizing $E\|\widehat{X}(\mathbf{s}) - X(\mathbf{s})\|^2$. This approach is similar to the method of co–kriging used for multivariate functional data, e.g. Wackernagel (2003). It is more computationally expensive than the method explained in Section 9.3, but it gives more accurate predictions in the example considered by Delicado *et al.* (2010). We have assumed in this chapter that $EX(\mathbf{s};t) = \mu(t)$ does not depend on the location $\mathbf{s}$. Caballero *et al.* (2013) and Menafoglio *et al.* (2013) show how to perform kriging under a more flexible assumption that $EX(\mathbf{s};t) = \sum_{\ell=1}^{L} \beta_\ell(t) f_\ell(\mathbf{s})$. In Section 9.3 we described a method of the estimation of the mean functions of spatially distributed curves. Other methods, as well as the estimation of the FPC's are presented in Chapter 17 of Horváth and Kokoszka (2012). The problem of testing the equality of two mean functions, each defined for functions observed over a different spatial region, is considered in Gromenko and Kokoszka (2012).

In the remainder of this section, we describe in some detail several problems involving inference for spatially indexed functional data which require applications of more advanced statistical methodology and mathematical derivations. We therefore do not describe all details, but provide enough information to give the reader an idea what a problem is and how it is solved. All problems in the remainder of this section lead to specific significance tests for functions, or time series of functions, indexed by spatial locations.

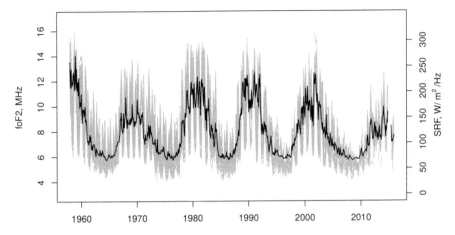

FIGURE 9.6: Gray lines represent ionosonde measurements obtained at observatories located in mid-latitude northern hemisphere, with the scale on the left-hand side. The black line represents the observed solar radio flux with the scale on the right-hand side.

Global cooling in the ionosphere

The first test is motivated by an interesting and extensively studied problem of space physics. The account presented here is based on Gromenko and Kokoszka (2013) and Gromenko *et al.* (2016). We first describe the space physics problem, and then explain the idea of the test. A chief difficulty is that the functions are observed with sometimes huge gaps. An analogous problem for iid functions is studied by Kraus (2015).

Increased concentration of greenhouse gases in the upper atmosphere is associated with global warming in the lower troposphere (the atmosphere roughly below 10 km). Roble and Dickinson (1989) suggested that the increasing amounts of these radiatively active gases, mostly CO_2 and CH_4, would lead to a global cooling in the ionosphere (atmosphere roughly 300 km above the Earth's surface). Rishbeth (1990) pointed out that this would result in a thermal contraction of the ionosphere. The height of the ionosphere can be approximately computed using data from a radar–type instrument called the ionosonde. Relevant measurements have been made for many decades by globally distributed ionosondes. In principle, these observations could be used to quantitatively test the hypothesis of Roble and Dickinson (1989). The difficulty in testing the contraction hypothesis comes from several sources. The height of the ionosphere depends on magnetic coordinates, the season, long term changes in the strength and direction of the internal magnetic field, and, most importantly, on the solar cycle; more solar radiation leads to greater ionization. This is illustrated in Figure 9.6. Another difficulty stems from the

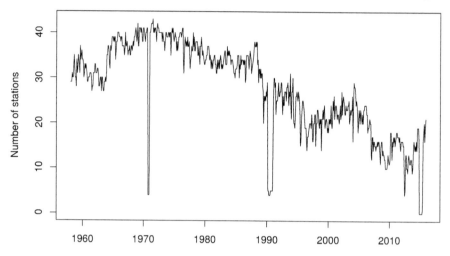

FIGURE 9.7: Number of available stations in the mid–latitude northern hemisphere.

fact that ionosonde records are not complete. Most observation stations do not operate continuously for many decades. They start and end operation at different times, some of them are out of service for many years, or even decades. In the mid-latitude northern hemisphere, there are 81 ionosonde stations, but at any given time, data from no more than 40 are available, as shown in Figure 9.7. This means that the estimation methods designed for fully observed curves cannot be used, as they require complete records to compute various integrals. More complex methods that work for incomplete records are needed.

Let $Y(\mathbf{s}_k; \tau_i)$ be the original record at location $\mathbf{s}_k$, measured from 1958 to 2015, possibly with long gaps. The set of all locations is $\{\mathbf{s}_k, 1 \le k \le K\}$, and the set of time points at which measurements may be available is $\{\tau_i, 1 \le i \le T\}$; in Gromenko *et al.* (2016) these are months from January 1958 to December 2015. The following spatio–temporal model is postulated:

$$Y(\mathbf{s}; \tau) = \mu(\mathbf{s}; \tau) + \varepsilon(\mathbf{s}; \tau) + \theta(\mathbf{s}; \tau), \tag{9.13}$$

where $\mathbf{s}$ is a generic location in a region of interest, and τ is continuous time. A simple form of the mean function relevant to the space physics problem is

$$\mu(\mathbf{s}; \tau) = \beta_1 + \beta_2 \tau + \beta_3 \operatorname{SRF}(\tau) + \beta_4 M(\mathbf{s}; \tau),$$

where $\operatorname{SRF}(\tau)$ is the solar radio flux, cf. Fig. 9.6, and $M(\mathbf{s}; \tau)$ is a suitable function computed from the coordinates of the internal magnetic field. The interest lies in the estimation of the mean function, and testing if it contains a linear trend, i.e. testing $H_0 : \beta_2 = 0$. The function $\mu(\cdot, \cdot)$ is treated as an unknown deterministic functional parameter. The second term, $\varepsilon(\mathbf{s}; \tau)$, describes the spatio–temporal variability away from the mean function. Stochastic modeling of this term is needed to develop inferential procedures. The term $\theta(\mathbf{s}; \tau)$ represents a random error, which can be associated with measurement error. The details of the estimation and testing procedures are too complex to describe here. The conclusion is that β_2 is significantly negative, confirming the hypothesis of global ionospheric contraction. The software to perform the test is available and can be used to test for the presence of global trends in other data of this type, for example in near-surface temperatures.

Detection of change in the mean function

We now turn to the problem of testing for a change in the mean function of functional time series observed at a number of spatial locations. The presentation is based on Gromenko *et al.* (2016). General theory for change point detection for scalar data is presented in Csörgő and Horváth (1997), several extensions to functional data are presented in Horváth and Kokoszka (2012). In the spatio–temporal setting we consider here, the data are assumed to follow the model

$$X_n(\mathbf{s}, t) = \mu_n(\mathbf{s}; t) + \varepsilon_n(\mathbf{s}; t).$$

In most applications, n denotes year, $\mathbf{s}$ and t time within a year. For a fixed $\mathbf{s}$, $\{X_n(\mathbf{s}), n = 1, 2, \ldots\}$ is a time series of functions, one function per year. While we assume that the curves are observed densely in time, they are only observed at a finite number of spatial locations $\{\mathbf{s}_k : k = 1, \ldots, K\}$. The goal is to determine if the mean functions, μ_n, are the same across n, or if there is a change in the mean at some unknown time point (year). In other words, we aim to evaluate the null hypothesis

$$H_0 : \mu_1 = \mu_2 = \cdots = \mu_N \qquad \text{against} \qquad H_A : \mu_1 = \ldots \mu_{n^*} \neq \mu_{n^*+1} = \cdots = \mu_N.$$

The alternative of a sudden change at a single year n^* is to be viewed as a mathematical approximation. As with all testing problems, the null hypothesis of no change is tested, and the alternative specifies the violations of the null hypothesis which will be detected with the highest power. The tests described below, will also detect different violations of H_0, including gradual changes taking place over a few years.

To carry out the tests, we assume that the covariance structure of the ε_n is separable, a test of this assumption is described below. Separability implies that

$$\text{Cov}(\varepsilon_n(\mathbf{s}; t), \varepsilon_n(\mathbf{s}'; t')) = v(t, t')u(\mathbf{s}, \mathbf{s}').$$

The decomposition above is unique only up to a constant, so we assume, wlog, that $\int v(t, t)\, dt = 1$. Gromenko *et al.* (2016) do not require a specific procedure for estimating $v(t, t')$ and $\mathbf{U} = \{u(\mathbf{s}_k, \mathbf{s}_{k'})\}$, instead formulating their tests assuming arbitrary consistent estimators. Examples of how to estimate these quantities are given in their appendix, as well as in Aston *et al.* (2016) and Constantinou *et al.* (2015). The first test involves using a temporal FPCA, normalizing by the corresponding eigenvalues, and then pooling across space. The tests statistics is

$$\hat{\Lambda}_1 = \frac{1}{N^2} \sum_{k=1}^{K} \hat{w}_k \sum_{i=1}^{p} \hat{\lambda}_i^{-1} \sum_{r=1}^{N} \left\langle \sum_{n=1}^{r} X_n(\mathbf{s}_k) - \frac{r}{N} \sum_{n=1}^{N} X_n(\mathbf{s}_k), \hat{v}_i \right\rangle^2.$$

The weights, $\hat{w}_k$, can be chosen however the user prefers, but specific recommendations are given. It has now become well recognized in the FDA community that dividing by estimated eigenvalues can cause stability problems with certain data. Thus a second test statistic was also proposed which omits normalizing by the $\hat{\lambda}_i$:

$$\hat{\Lambda}_2 = \frac{1}{N^2} \sum_{k=1}^{K} \hat{w}_k \sum_{i=1}^{p} \sum_{r=1}^{N} \left\langle \sum_{n=1}^{r} X_n(\mathbf{s}_k) - \frac{r}{N} \sum_{n=1}^{N} X_n(\mathbf{s}_k), \hat{v}_i \right\rangle^2.$$

For p large, $\hat{\Lambda}_2$ is essentially an approximation of the third and final test statistic

$$\hat{\Lambda}_2^\infty = \frac{1}{N^2} \sum_{k=1}^{K} \hat{w}_k \sum_{i=1}^{p} \hat{\lambda}_i^{-1} \sum_{r=1}^{N} \left\| \sum_{n=1}^{r} X_n(\mathbf{s}_k) - \frac{r}{N} \sum_{n=1}^{N} X_n(\mathbf{s}_k) \right\|^2.$$

Each statistic has a slightly different limiting distribution. In particular, under H_0

$$\hat{\Lambda}_1 \xrightarrow{d} \sum_{k=1}^{K} w(k) \sum_{i=1}^{d} \int B_{ik}^2(t) \, dt$$

$$\hat{\Lambda}_2 \xrightarrow{d} \sum_{k=1}^{K} w(k) \sum_{i=1}^{d} \lambda_i \int B_{ik}^2(t) \, dt$$

$$\hat{\Lambda}_2^\infty \xrightarrow{d} \sum_{k=1}^{K} w(k) \sum_{i=1}^{\infty} \lambda_i \int B_{ik}^2(t) \, dt,$$

where $B_{ik}(t)$ are Brownian bridges which satisfy

$$\text{Cov}(B_{ik}(t), B_{i'k'}(t')) = 1_{i=i'} \min\{t, t'\}(1 - \min\{t, t'\})\sigma(\mathbf{s}_k, \mathbf{s}_{k'}).$$

These asymptotics can be used for choosing rejection regions with proper Type 1 error. To do so, one can use Monte Carlo, though Gromenko *et al.* (2016) also provide a normal approximation which works well when considering a relatively large number of spatial points. These tests can be implemented using the scpt package in R, which is available from http://www.personal.psu.edu/ mlr36/codes.html and includes an example code for implementing the methodology.

   If a change point is detected, its location can be identified as the year

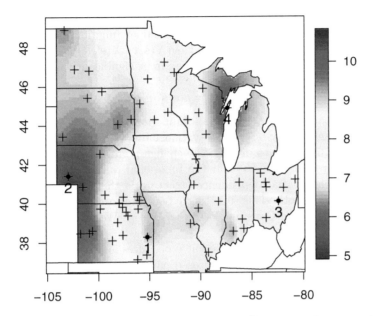

FIGURE 9.8: The spatial field showing the L^2 distance between the mean log–precipitation before and after 1966. There is an increase in precipitation throughout the year in the area around location 4, decrease in the first half of the year in the area around location 1. Locations close to 2 and 3 do not show a large change nor a consistent pattern.

when a test statistics attains maximum. Gromenko *et al.* (2016) applied the above test procedures (with suitable finite sample calibration) to test if the precipitation pattern over the midwest US states has changed. All tests led to the same conclusion: one change point in second half of the sixties. The pattern of the change is shown in Fig. 9.8. The biggest changes are in the area around Michigan Lake and south–west of the lake. To visualize the spatial distribution of change, the authors used the spatial field

$$\hat{\phi}(\mathbf{s}) = \|\hat{\mu}_1(\mathbf{s}) - \hat{\mu}_2(\mathbf{s})\|,$$

where

$$\hat{\mu}_1(\mathbf{s};t) = \hat{r}^{-1} \sum_{n=1}^{\hat{r}} X_n(\mathbf{s};t), \quad \hat{\mu}_2(\mathbf{s};t) = (N - \hat{r})^{-1} \sum_{n=\hat{r}+1}^{N} X_n(\mathbf{s};t).$$

They performed kriging with the exponential covariance model to obtain the heat map shown in Fig. 9.8. The application of the significance tests confirms that the heat map shows a statistically significant change over a region, not a variation in the magnitude of change which may be due to chance.

Separability tests

The second order structure of a random field of functions is described by the covariance function $\sigma(\mathbf{s}, \mathbf{s}'; t, t') = \mathrm{Cov}(X(\mathbf{s}, t), X(\mathbf{s}', t'))$. Theoretical and computational aspects of most procedures can be significantly simplified if one can assume that

$$\sigma(\mathbf{s}, \mathbf{s}'; t, t') = u(\mathbf{s}, \mathbf{s}')v(t, t'), \tag{9.14}$$

that is, if the spatio–temporal covariance function factors into the product of a purely spatial and purely temporal covariance functions. The above condition is, for example, required to develop the change point methodology described above. If (9.14) holds, we say that the functional random field is *separable*. In particular, the spatial dependence structure is the same for any time t. Tests whose null hypothesis is (9.14) have been proposed by Liu *et al.* (2014), Aston *et al.* (2016) and Constantinou *et al.* (2015). We discuss the latter two approaches here.

   We begin by discussing two different procedures for estimating $v(t, t')$ and $\mathbf{U} = \{u(\mathbf{s}_k, \mathbf{s}_{k'})\}$. Aston *et al.* (2016) used partial trace operators to estimate these quantities, while Constantinou *et al.* (2015) combined bases expansions with a "flip-flop" approach, which basically iterates between estimating v and $\mathbf{U}$. Partial trace operators are used to eliminate either the temporal or spatial component of the covariance, in particular, define the estimated covariance function as

$$\hat{\sigma}(\mathbf{s}_k, \mathbf{s}'_k, t, t') = \frac{1}{N} \sum_{n=1}^{N} X_n(\mathbf{s}_k, t) X_n(\mathbf{s}_{k'}, t').$$

(In the above formula, and in the remainder of this section, we assume that the mean function is zero.) Then the spatial covariance is estimated using

$$\hat{\mathbf{U}}_{kk'} = \int \hat{\sigma}(\mathbf{s}_k, \mathbf{s}'_k, t, t) \, dt,$$

and the temporal covariance is estimated using

$$\hat{v}(t, t') = \frac{\sum_{k=1}^{K} \hat{\sigma}(\mathbf{s}_k, \mathbf{s}_k, t, t')}{\sum_k \int \hat{\sigma}(\mathbf{s}_k, \mathbf{s}_k, t, t) \, dt}.$$

The normalization above ensures that the trace of $v(t, t')$ is normalized to be 1, though other normalizations are possible. One thing to note is that the full covariance $\hat{\sigma}$ does not actually need to be estimated, as one can move to estimating the $\hat{\mathbf{U}}$ and $\hat{v}$ directly.

   The second approach to estimating v and $\mathbf{U}$ comes form Constantinou *et al.* (2015) and begins with basis expansions. In particular, one approximates each temporal curve using J basis functions:

$$X_n(\mathbf{s}_k, t) \approx \sum_{j=1}^{J} \xi_{jn}(\mathbf{s}_k) e_j(t).$$

The basis $e_j(t)$ can either be deterministic or data driven, such as FPCA. Testing separability of X_n is then translated into testing separability of $\xi_{jn}(\mathbf{s}_k)$, so that the null becomes

$$H_0 : \mathrm{Cov}(\xi_{jn}(\mathbf{s}_k), \xi_{j'n}(\mathbf{s}_{k'}) = \mathbf{V}_{jj'}\mathbf{U}_{kk'}.$$

However, this is now a multivariate problem and we can utilize multivariate methods for estimating $\mathbf{V}$ and $\mathbf{U}$, in particular, the maximum likelihood estimators, assuming the data are Gaussian, satisfy

$$\hat{\mathbf{V}} = \frac{1}{N}\sum_{n=1}^{N}\mathbf{\Xi}_n\hat{\mathbf{U}}^{-1}\mathbf{\Xi}_n^{\top} \qquad \hat{\mathbf{U}} = \frac{1}{N}\sum_{n=1}^{N}\mathbf{\Xi}_n^{\top}\hat{\mathbf{V}}^{-1}\mathbf{\Xi}_n,$$

where $\mathbf{\Xi}_n = \{\xi_{jn}(\mathbf{s}_k)\}$ is a $J \times K$ matrix. To compute these estimators, one chooses an initial value, for example $\hat{\mathbf{U}}_0 = \mathbf{I}$, and then iterates between the two estimates, hence the name "flip-flop method".

With these estimators in hand, one can now carry out a number of different tests. To construct these tests first define

$$\mathbf{\Sigma} := \mathrm{E}\,\mathrm{vec}(\mathbf{\Xi}_n)\,\mathrm{vec}(\mathbf{\Xi}_n)^{\top},$$

where vec is the vectorization operator, which stacks the columns of a matrix into a vector. Under H_0 one can show that

$$\mathbf{\Sigma} = \mathbf{V} \otimes \mathbf{U},$$

where $\otimes$ is the Kronecker product. Constantinou *et al.* (2015) consider three different approaches

$$T_W = N(\hat{\mathbf{V}} \otimes \hat{\mathbf{U}} - \widehat{\mathbf{\Sigma}})^{\top}\widehat{\mathbf{W}}^{+}(\hat{\mathbf{V}} \otimes \hat{\mathbf{U}} - \widehat{\mathbf{\Sigma}}),$$
$$T_L = N(J\log\det(\hat{\mathbf{U}}) + K\log\det(\hat{\mathbf{V}}) - \log\det(\widehat{\mathbf{\Sigma}})),$$
$$T_F = N\|\hat{\mathbf{V}} \otimes \hat{\mathbf{U}} - \widehat{\mathbf{\Sigma}}\|_F^2.$$

The first test uses the Moore–Penrose generalized inverse of the estimated covariance matrix of $\hat{\mathbf{V}} \otimes \hat{\mathbf{U}} - \widehat{\mathbf{\Sigma}}$, $\widehat{\mathbf{W}}^{+}$, to normalize the difference and create a Wald type test statistic. The form for $\widehat{\mathbf{W}}$ is complex, but can be found in Constantinou *et al.* (2015). The generalized inverse is necessary since the symmetry of the matrices and the non-uniqueness of $\mathbf{U}$ and $\mathbf{V}$ create a number of linear constraints. This turns out to be a fairly unstable test statistic in terms of Type 1 error, and works well only for very small J and K. The second test is derived from the likelihood ratio and had previously been explored in the multivariate literature in Lu and Zimmerman (2005); Mitchell *et al.* (2006). This test is also quite unstable, but only when using a chi-squared distribution as an asymptotic approximation. Mitchell *et al.* (2006) provided a Monte-Carlo method which provides a much more stable (in terms of type 1 error) test. The last test forgoes normalizing the test statistic and instead uses an asymptotic distribution given by a weighted sum of chi-squares. This test is very stable and, in the settings considered by Constantinou *et al.* (2015) exhibited excellent power. Lastly, we note that Aston *et al.* (2016) provided a testing procedure that focussed on testing the separability in the directions dictated by the eigenfunctions. They utilized a bootstrap approach instead of asymptotic distributions, and their individual tests can be pooled to construct a global test for evaluating separability overall.

The tests described above assume that the functions $X_n(\mathbf{s}, \cdot)$ are independent across n. Extension of the tests of Constantinou *et al.* (2015) to functional time series are are developed in Constantinou *et al.* (2016).

Spatio–temporal extremes

In this section, we summarize the work of French *et al.* (2016) which deals with the computation of probabilities of heat waves. The raw data are spatially–indexed time series of daily temperature measurements, $X(\mathbf{s}_k, j)$, denoting the temperature at a spatial location $\mathbf{s}_k$, $k = 1, \cdots, K$, on day j. As argued above, due to the natural annual climate cycle, for each site, we partition $\{X(\mathbf{s}_i, j)\}$ into years, and view the resulting 365-dimensional vectors as samples from a functional time series:

$$X_n(\mathbf{s}_k; \cdot) = \{X_n(\mathbf{s}_k; t_i), \ i \in \{1, 2, \ldots 365\}\}. \tag{9.15}$$

Here, $t \mapsto X_n(\mathbf{s}_k; t)$ is the temperature curve at site $\mathbf{s}_k$ for year n, viewed as a function of time t in days. In contrast to the setting of previous sections, the data used by French *et al.* (2016) are not historical records, but data generated by a computer climate model. These artificial data are of much higher quality than historical records; there are no gaps, and the daily records are available at 16,100 locations forming a grid covering much of North America. It is, at this point, not clear how to extend the methodology of French *et al.* (2016) to historical records. The advantage of using computer model data is that they are predicted future temperatures (French *et al.* (2016) work with the period 2041-2070), which are more relevant to the prediction of future heat waves. On the other hand, these data do depend on a model, and the poor quality, geostatistical historical records are the real data.

French *et al.* (2016) propose many functionals that can quantify a heat wave, but here we focus on one specific approach that explains the general idea. A heat wave is characterized by its spatial and temporal extents and by its intensity. The intensity is typically quantified by a threshold. Public health concerns call for a fixed threshold, like 105^0F. However, in climate studies of large spatial regions, with many climatic zones, such a fixed threshold is not appropriate. Also the variability of temperatures depends greatly on the geographical location, with coastal locations exhibiting much smaller variability than locations far away from large bodies of water. It is therefore reasonable to work with standardized temperatures

$$Z_n(\mathbf{s}_k, t_i) = \frac{X_n(\mathbf{s}_k, t_i) - \overline{X}(\mathbf{s}_k, t_i)}{\mathrm{SD}(\mathbf{s}_k, t_i)}, \tag{9.16}$$

where

$$\overline{X}(\mathbf{s}_k, t_i) = \frac{1}{N} \sum_{n=1}^{N} X_n(\mathbf{s}_k, t_i) \text{ and } \mathrm{SD}^2(\mathbf{s}_k, t_i) = \frac{1}{N-1} \sum_{n=1}^{N} (X_n(\mathbf{s}_k, t_i) - \overline{X}(\mathbf{s}_k, t_i))^2. \tag{9.17}$$

If the $Z_n(\mathbf{s}_k, t_i)$ exceed a fixed threshold z, e.g. $z = 2$, for a number of neighboring locations and over a period of time, then we have observed a heat wave (the $Z_n(\mathbf{s}_k, t_i)$ are practically normal). The severity of a heat wave increases with the size of the region, the duration and the threshold z that is exceeded. Suppose the $X_n(\mathbf{s}_k, t_i)$ are maximum daily temperatures, and set

$$Z_n^{\star}(\mathbf{s}_k, t_j) = \frac{1}{\ell} \sum_{t_j - \ell < t_i \leq t_j} Z_n(\mathbf{s}_k, t_i).$$

This is the average maximum temperature over the ℓ days preceding day t_j. Next, define

$$Z_n^{\star}(t_j) = \min_{1 \leq k \leq K} Z_n^{\star}(\mathbf{s}_k, t_j).$$

If $Z_n^{\star}(t_j) > z$, then the average maximum temperature over ℓ days over K (neighboring) locations exceeds, z; this corresponds to a heat wave defined by this specific functional. We are interested in the probability of a heat wave in any given year. We assume that this probability does not depend on year n. We thus want to compute, for some relevant $z > 0$,

$$p(z) = P(\exists \, j \; : \; Z_n(t_j) > z) = P\left(\max_{1 \leq j \leq J} Z_n^{\star}(t_j) > z \right) = P(M_J > z),$$

where $J = 365$, and

$$M_J \stackrel{d}{=} M_{J,n} := \max_{1 \leq j \leq J} Z_n^{\star}(t_j).$$

The concatenated sequence $Z^{\star}(t_j)$ is stationary and weakly dependent, so (see e.g. Beirlant *et al.* (2006), Chapter 10), there are sequences a_J and b_J such that

$$\lim_{J \to \infty} P\left(\frac{M_J - b_J}{a_J} \leq z \right) \to H(z),$$

where H is a univariate Generalized Extreme Value distribution function. The function H depends on three parameters, which can be estimated, together with the constants a_J and b_J, using now standard R implementations.

Figure 9.9 shows examples of regions corresponding to 50, 150 and 450 neighboring locations. Figure 9.10 shows a map of the probability of a heat wave for $d = 50$ for three durations ℓ, with (a) corresponding to $\ell = 3$, (b)

FIGURE 9.9: A map of the neighborhood structures for different locations using 50, 150, and 450 nearest neighbors. Each × marks a neighborhood centroid and the sequences of grey shading mark the extents of the increasing neighborhood sizes.

to $\ell = 10$, and (c) to $\ell = 30$. When $\ell = 3$, there is a surprisingly high probability of localized heat waves over the Labrador Peninsula. Such short heat spells may occur with probability approaching 50%, that is on average every second year. While our EVT approximation may break down for such high probabilities, it is nevertheless obvious that that part of Canada will see heat spells much more frequently than in the past. Generally, we see that the area around the Hudson Bay will experience an increased frequency of hot spells lasting a few days. There is a noticeable drop in the probability of such a heat wave around the Rocky Mountain range. The probability is also very low along the Eastern seaboard of the United States. Increasing the duration to $\ell = 10$ days, dramatically reduces the probability of a heat wave of the corresponding magnitude. The reader will note the different probability scale. Many parts of Canada once again show an increased probability of a heat wave of this magnitude, as well as parts of Iowa and Illinois, certain regions in Texas, and, most visibly, the Pacific Ocean off the Southern California coast. Increasing the duration to approximately 1 month ($\ell = 30$), causes the probability of a heat wave to drop even further; generally, throughout North America, heat waves of this magnitude will occur with probability of less than 1%, i.e. once per one hundred years, on average. Over the Canadian plains and the Canadian Rockies, this probability increases only slightly to about 1.5%. There are two patches, in Arizona and Southern Texas, with probabilities elevated to 2-3%.

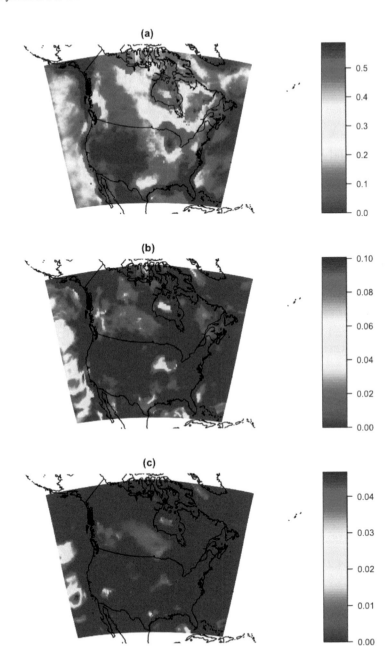

FIGURE 9.10: Probability of a heat wave with amplitude more than 2 standard deviations above the mean for spatial extent $d = 50$ and durations of (a) $\ell = 3$, (b) $\ell = 10$, and (c) $\ell = 30$.

9.7   Chapter 9 problems

9.1 [Elementary properties of the covariance function] Suppose $\{X(\mathbf{s}), \mathbf{s} \in \mathbb{R}^d\}$ is a second order stationary scalar random field and C is its covariance function defined in Section 9.1. Verify the following properties: 1) $C(\mathbf{0}) \geq 0$; 2) $C(-\mathbf{h}) = C(\mathbf{h})$; 3) $|C(\mathbf{h})| \leq C(\mathbf{0})$.

9.2 [Valid covariance function] Suppose $\{X(\mathbf{s}), \mathbf{s} \in \mathbb{R}^d\}$ and C are the same as in Problem 9.1. Show that for any m, any locations $\mathbf{s}_1, \mathbf{s}_2, \ldots, \mathbf{s}_m$ and any real numbers $a_1, a_2, \ldots, a_m$,

$$\sum_{i,j=1}^{m} a_i a_j C(\mathbf{s}_i - \mathbf{s}_j) \geq 0. \tag{9.18}$$

We have referred to a function C with the above property as nonnegative definite. In spatial statistics, a function C satisfying (9.18) is called *valid*. The covariance function is thus always valid, yet the term *a valid covariance function* is often used. This is because very often some preliminary estimates or models lead to expressions for C which do not satisfy (9.18). In such situation, some further regularization is needed to obtain a valid covariance function.

9.3 [Semivariogram] Suppose $\{X(\mathbf{s}), \mathbf{s} \in \mathbb{R}^d\}$ and C are the same as in Problem 9.1. Verify relation (9.2).

9.4 Suppose $\varepsilon_i, i \geq 1$, are iid random variables with mean zero and variance σ^2. The random walk is defined as $X(j) = \sum_{i=1}^{j} \varepsilon_i$, $j \geq 1$. It can be viewed as random field defined on the grid $\{1, 2, 3, \ldots\}$. For such a domain, one can consider only nonnegative shifts h, so X is stationary if for every $h \geq 0$, $\mathrm{Cov}(X(j), X(j+h))$ does not depend on j, and intrinsically stationary if $\mathrm{Var}[X(j) - X(j+h)]$ does not depend on j. (In both cases the mean function cannot depend on j.) Show that the random walk is intrinsically stationary, but not stationary.

9.5 Show that predictor (9.4) satisfies $E[\widehat{X}(\mathbf{s}) - X(\mathbf{s})] = 0$ and verify relation (9.5).

9.6 Show that the weights $w_1, w_2, \ldots, w_K$ minimizing (9.5) satisfy equations (9.6).

9.7 [Equations for kriging the mean] Suppose $\{X(\mathbf{s})\}$ is second order stationary. To derive equations (9.7), we must find $w_1, w_2, \ldots, w_N$ which minimize $E(\sum_{k=1}^{N} w_k X(\mathbf{s}_k) - \mu)^2$ subject to the constraint $\sum_{k=1}^{N} w_k = 1$. Problems of this type are solved using the method of Lagrange multipliers, which is explained in many multivariate calculus textbooks, see e.g. Chapter 7 of Apostol

(1957). If there is only one constraining equation, this method can be formulated as follows:

If a function $f(x_1, x_2, \ldots, x_n)$ has an extremum subject to the constraint $g(x_1, x_2, \ldots, x_n) = 0$, then at any point $\mathbf{x}_0$ at which an extremum is reached, there exists λ such that

$$\frac{\partial f}{\partial x_\ell}(\mathbf{x}_0) = \lambda \frac{\partial g}{\partial x_\ell}(\mathbf{x}_0), \quad \ell = 1, 2, \ldots n.$$

To find potential extremal points $\mathbf{x}_0$, we thus set up a system consisting of the n equations above and the equation $g(x_1, x_2, \ldots, x_n) = 0$.

Use the above method to derive equations (9.7).

9.8 This problem illustrates the application of equations (9.7) in a very simple setting. Consider a second order stationary random field $\{X(s), s \in \mathbb{R}\}$, the domain is the real line. Suppose the covariance function is given by $C(h) = \exp(-h)$. Observations are available at three locations: $s_1 = 0, s_2 = 1, s_3 = 10$, and their values are $X(s_1) = 1, X(s_2) = 2, X(s_3) = 3$. Find the weights w_1, w_2, w_3 and compute the estimated mean $\hat{\mu}$. Provide intuition explaining why $\hat{\mu}$ is greater than the sample mean $(1 + 2 + 3)/3 = 2$.

9.9 Show that weights w_k which minimize (9.11) satisfy equations (9.12). *Hint:* work through Problem 9.7 first.

9.10 Define the functional semivariogram by

$$\gamma(\mathbf{s}_k, \mathbf{s}_\ell) = \frac{1}{2} E \|X(\mathbf{s}_k) - X(\mathbf{s}_\ell)\|^2.$$

Show that under the assumptions of Section 9.2 (stationarity and isotropy),

$$\gamma_f(h) = C_f(0) - C_f(h), \tag{9.19}$$

where

$$h = \|\mathbf{s}_k - \mathbf{s}_\ell\|, \quad \gamma_f(h) = \gamma(\mathbf{s}_k, \mathbf{s}_\ell), \quad C_f(h) = C(\mathbf{s}_k, \mathbf{s}_\ell),$$

with the functional spatial covariances $C(\mathbf{s}_k, \mathbf{s}_\ell)$ defined by (9.8).

The above relations imply that the covariances $C(\mathbf{s}_k, \mathbf{s}_\ell)$ can be estimated as follows. First we compute an empirical variogram $\hat{\gamma}_f$, using e.g. (9.3), with $(X(\mathbf{s}_k) - X(\mathbf{s}_\ell))^2$ replaced by $\|X(\mathbf{s}_k) - X(\mathbf{s}_\ell)\|^2$. Then we fit a parametric covariance model to $\hat{\gamma}_f$ using relation (9.19). Finally, we set $C(\mathbf{s}_k, \mathbf{s}_\ell) = C_f(h)$ with $h = \|\mathbf{s}_k - \mathbf{s}_\ell\|$.

9.11 Use the code below to produce a map of Canada with the kriging weights displayed in Figure 3.4 indicated by size and color. Study the display and the code and explain what the different colors and sizes indicate.

```
map ('world',ylim=c (42,78),xlim = c (-180,-40))
map ('world',region="Canada",add=TRUE,fill=TRUE,col="
    lightyellow")
points (coord.34,col=3-sign (w.k),pch=19, cex=0.75+4*abs (w.k)/
    max (abs (w.k)))
points (coord.34,col='black',pch=19,cex=.5)
points (coord.0[1],coord.0[2],pch=19,col=3,cex=1.5)
title (main="Kriging weights")
```

9.12 The function okfd in the package geofd implements a version of kriging discussed around equation (9.9), i.e. it does not require estimation of the
mean function. We illustrate its application continuing the example presented
in Section 9.5.

Use the following code to compute and display the kriging weights λ_k for
the prediction of the temperature curve in Calgary:

```
lambda.okfd <- okfd.res$functional.kriging.weights
lambda.okfd <- lambda.okfd[-length (lambda.okfd)]
plot (geo.dist.0.34,lambda.okfd,
      xlab="Distance to Calgary",
      ylab="Kriging weights",
      main="Kriging weights by 'okfd' function")
abline (h=1/34,col="gray",lwd=2); abline (h=0,col="blue")
```

Compare the weights you have obtained to those in Figure 9.4.

Modify the code presented in Problem 9.11 and display the weights λ_k on
a map of Canada. Mark Canada in light blue.

Modify the code presented in Section 9.5 and plot the okfd kriged Calgary
curve in red together with the true curve in black.

Use the following code to compare the kriged curve obtained in Section 9.5
with the kriged curve obtained using function okfd and comment.

```
plot (okfd.res$datafd, lty=1,col=8, xlab="Day",
      ylab="Temperature (degrees C)", main="Predictions at
    Calgary")
lines (kriging.i.0, col=4, lwd=2)
lines (okfd.res$argvals,okfd.res$krig.new.data,col="red",lwd=2)
legend ("bottomright", "Kriged function at Calgary", "Kriged
    function at Calgary using 'okfd' function"), lwd=c (2,2,2),
    col=c (1,4,2))
```

9.13 This problem presents the steps required to computed the weighted
average of temperature curves shown in Figure 9.1. It is assumed that the code
presented in Section 9.5 has been entered. We will work with raw temperature
curves, without any smoothing.

Use the following code to estimate and plot the variogram:

```
temp.fd <- fdata (t (Temperature),argvals=Day, names=list (main="
    Canadian Weather. Temperature", xlab="Day",ylab="
    Temperature"))
```

```
# computing L2 norms between functions, using fda.usc library
L2norm.raw <- as.dist(metric.lp(temp.fd,lp=2))^2
# Calculating the empirical trace bin variogram
emp.trace.vari <- trace.variog(coords=coordinates,
L2norm=as.matrix(L2norm.raw), bin=TRUE)
# fitting an exponential variogram
sigma2.0 <- quantile(emp.trace.vari$v, 0.75)
phi.0     <- quantile(emp.trace.vari$Eu.d, 0.75)
fit.vari <- variofit(emp.trace.vari, ini.cov.pars=c(sigma2.0,
    phi.0), cov.model="exponential")

plot(geo.dist,L2norm.raw,col="grey")
points(emp.trace.vari$u,emp.trace.vari$v,col="red",pch=19)
lines(fit.vari,col="blue",lwd=2)
```

Use the following code to compute the weights and plot them as a function of the station index.

```
hat.C <- cov.spatial(emp.trace.vari$Eu.d, cov.model= fit.vari$
    cov.model,
                     cov.pars=fit.vari$cov.pars)
inv.hat.C <- solve(hat.C)
v <- apply(inv.hat.C,1,sum); w <- v/sum(v)

plot(w)
abline(h=1/35,col="gray",lwd=2); abline(h=0,col="blue")
```

The following code represents these weights in a map of Canada:

```
map('world',ylim=c(42,78),xlim = c(-180,-40))
map('world',region="Canada",add=TRUE,fill=TRUE,col="
    lightyellow")
points(coordinates,col=3-sign(w),pch=19,cex=1+abs(w)*35)
points(coordinates,col='black',pch=19,cex=.5)
```

Now use the following code to reproduce Figure 9.1

```
mean.temp <- apply(t(Temperature),2,mean)
w.mean.temp <- w %*% t(Temperature)
plot(temp.fd,col=8, main=" ")
lines(Day,mean.temp, lty= 2, lwd=2)
lines(Day,w.mean.temp,lwd=2)
```

10

Elements of Hilbert space theory

In Chapter 3, we described the main elements of the framework of random functions which is commonly used in FDA. In this Chapter, and in Chapter 11, we present the concepts introduced in Chapter 3 in much greater detail. These two chapters are recommended to readers who would like to gain a fairly solid grounding in the mathematical theory underlying FDA, but who do not wish to study extensive, and even more detailed, mathematical monographs

Most FDA tools make heavy use of Hilbert space theory, and thus it is important to gain comfort using some tools from functional analysis, a field of mathematics that studies objects in abstract vector spaces, rather than in the Euclidean space. We do not intend (nor could we) cover all of the material found in a traditional functional analysis course. Instead, we present basic definitions and facts of Hilbert space theory which have been extensively used in functional data analysis. A reader with some familiarity with multivariate statistics should be able to quickly pick up the core ideas in this chapter. The problems of this section are simple exercises intended to help the reader become familiar with these definitions and theorems by using them. The textbook of Debnath and Mikusinski (2005) is an accessible exposition that presents a broader picture and applications to other areas of mathematics and science. The monograph of Akhiezier and Glazman (1993) is an extensive survey of the classical theory. Chapter 4 of Rudin (1987) provides a concise and rigorous introduction to the most important ideas. There are dozens of other textbooks on Hilbert spaces; the monographs of Hsing and Eubank (2015) is an exposition taylored specifically to the needs of researchers working in the field of FDA.

10.1 Hilbert space

A vector space is defined over a field of scalars. In this book, the scalars are either the real or the complex numbers. A vector space is a set $\mathcal{V}$ whose elements are called vectors, and in which two operations are defined: addition and scalar multiplication. The following are the axioms of a vector space.

AXIOM 10.1.1 *To every pair of vectors* $\mathbf{x}$ *and* $\mathbf{y}$, *there corresponds a vector*

$\mathbf{x} + \mathbf{y}$, *and*
$$\mathbf{x} + \mathbf{y} = \mathbf{y} + \mathbf{x}, \qquad \mathbf{x} + (\mathbf{y} + \mathbf{z}) = (\mathbf{x} + \mathbf{y}) + \mathbf{z}$$

The space $\mathcal{V}$ *contains a unique vector* $\mathbf{0}$ *such that* $\mathbf{x} + \mathbf{0} = \mathbf{x}$ *for every* $\mathbf{x} \in \mathcal{V}$. *For each* $\mathbf{x} \in \mathcal{V}$, *there is a unique vector* $-\mathbf{x}$ *such that* $\mathbf{x} + (-\mathbf{x}) = \mathbf{0}$.

AXIOM 10.1.2 *To every pair* $(a, \mathbf{x})$, *where* a *is a scalar and* $\mathbf{x}$ *is a vector, there corresponds a vector* $a\mathbf{x}$, *and*
$$1\mathbf{x} = \mathbf{x}, \qquad a(b\mathbf{x}) = (ab)\mathbf{x};$$

$$a(\mathbf{x} + \mathbf{y}) = a\mathbf{x} + a\mathbf{y}, \qquad (a + b)\mathbf{x} = a\mathbf{x} + b\mathbf{x}.$$

All the usual intuitive properties follow from these axioms. For example, if 0 is the scalar zero, then $0\mathbf{x} = \mathbf{0}$.

A set $\mathcal{V}_1 \subset \mathcal{V}$ is called a *linear subspace* , or sometimes a *vector subspace* of $\mathcal{V}$ if it is a vector space with respect to the operations defined on $\mathcal{V}$. Clearly, $\mathcal{V}_1 \subset \mathcal{V}$ is a subspace if and only if for any scalars a, b and $\mathbf{x}, \mathbf{y} \in \mathcal{V}_1$, $a\mathbf{x} + b\mathbf{y} \in \mathcal{V}_1$.

We say that a set of vectors $e_1, \dots, e_d$ of $\mathcal{V}$ are *linearly independent* if
$$a_1 \mathbf{e}_1 + a_2 \mathbf{e}_2 + \dots + a_d \mathbf{e}_d = 0 \Rightarrow a_1 = \dots = a_d = 0.$$

Furthermore, if every element of $\mathcal{V}$ can be expressed as a linear combination of $\mathbf{e}_1, \dots, \mathbf{e}_d$, then we call that set of vectors a basis of $\mathcal{V}$ and say that $\mathcal{V}$ has dimension d.

In the following, we drop the bold face to denote vectors, as it will readily follow from the context which objects are vectors and which are scalars. We will reserve the bold face for finite dimensional vectors (and matrices), as is usual in statistics. This book is primarily concerned with infinite dimensional vector spaces, i.e. those that do not have a finite basis.

DEFINITION 10.1.1 A vector space, $\mathcal{V}$, is called an *inner product space* if for each pair of vectors, (x, y), there is a scalar $\langle x, y \rangle$, called the inner product of x and y, which satisfies, for all $(x, y) \in \mathcal{V} \times \mathcal{V}$,

a) $\langle x, y \rangle = \overline{\langle y, x \rangle}$;

b) $\langle a_1 x_1 + a_2 x_2, y \rangle = a_1 \langle x_1, y \rangle + a_2 \langle x_2, y \rangle$;

c) $\langle x, x \rangle \geq 0$ with equality only for $x = 0$.

The bar in $\overline{\langle y, x \rangle}$ denote the complex conjugate. An inner product in a vector space over the field of real numbers is symmetric: $\langle x, y \rangle = \langle y, x \rangle$.

In an inner product space, one can define the inner product norm by
$$\|x\| = \sqrt{\langle x, x \rangle}.$$

The following proposition follows from the definitions stated above

PROPOSITION 10.1.1 *The inner product defines a norm,* $\| \cdot \|$, *on the inner product space which has the following properties:*

a) $\|ax\| = |a|\|x\|$;

b) $|\langle x, y \rangle| \leq \|x\|\|y\|$ *(Cauchy–Schwarz inequality)*;

c) $\|x + y\| \leq \|x\| + \|y\|$ *(triangle inequality)*;

d) $d(x, y) = \|x - y\|$ *is a metric (distance)* , *i.e.* $d(x, y) = d(y, x) > 0$, *if* $x \neq y$, $d(x, x) = 0$, *and* $d(x, y) \leq d(x, z) + d(z, y)$.

The last part of Proposition 10.1.1 implies that every inner product space is a metric space. In fact, every inner product space is a normed space. A normed space is a vector space with the norm which satisfies conditions 1 and 3 in Proposition 10.1.1 and $\|x\| = 0$ if and only if $x = 0$. Every normed space is a metric space with the metric defined by $d(x, y) = \|x - y\|$.

Recall that a sequence $\{x_n\}$ of points in a metric space converges to a point x if $d(x_n, x) \to 0$, as $n \to \infty$. A sequence $\{x_n\}$ is said to be a Cauchy (or a fundamental) sequence if $d(x_n, x_m) \to 0$, as $n, m \to \infty$. Since $d(x_n, x_m) \leq d(x_n, x) + d(x, x_m)$, every convergent sequence is a Cauchy sequence. The converse statement is not true in general. This leads to the following definition.

DEFINITION 10.1.2 A metric space in which every Cauchy sequence has a limit is called *complete*.

Under Euclidean distance, the open interval $(0, 1)$ is not a complete metric space, but the closed interval $[0, 1]$ is. The Euclidean space $\mathbb{R}^d$ is complete, see e.g. Chapter 3 of Rudin (1976) for a proof.

DEFINITION 10.1.3 A complete inner product space is called a *Hilbert space*.

EXAMPLE 10.1.1 The space ℓ^2 is defined as the collection of sequences $x = \{x_1, x_2, \ldots\}$ such that $\sum_{i=1}^{\infty} |x_i|^2 < \infty$. It is a vector space with the operation defined componentwise, i.e.

$$x + y = \{x_1 + y_1, x_2 + y_2, \ldots\}; \quad ax = \{ax_1, ax_2, \ldots\}.$$

The inner product is defined by

$$\langle x, y \rangle = \sum_{i=1}^{\infty} x_i \bar{y}_i. \tag{10.1}$$

Problem 10.3 asks for a verification that ℓ^2 with the operations so defined is a Hilbert space. The Cauchy–Schwarz inequality takes the form

$$\left| \sum_{i=1}^{\infty} x_i \bar{y}_i \right| \leq \left(\sum_{i=1}^{\infty} |x_i|^2 \right)^{1/2} \left(\sum_{i=1}^{\infty} |y_i|^2 \right)^{1/2}.$$

□

In this book, we will extensively work with the space of square integrable *real* functions defined on an interval. Without any loss of generality we can assume that this interval is $[0, 1]$ We therefore state the following definition.

DEFINITION 10.1.4 *The space L^2 is the collection of Lebesgue measurable real–valued functions x defined on the interval $[0, 1]$ such that*

$$\int_0^1 x^2(t)dt < \infty.$$

The vector space operations are defined via

$$(x + y)(t) = x(t) + y(t), \quad (ax)(t) = ax(t).$$

The inner product is defined by

$$\langle x, y \rangle = \int_0^1 x(t)y(t)dt.$$

The L^2 space of complex–valued functions is defined with obvious modifications. The inequalities $(a + b)^2 \le 2a^2 + 2b^2$ and $2|ab| \le a^2 + b^2$, which hold for any real numbers a and b show that vector operations and the inner product are well–defined. The axioms of the vector space and of the inner product space are then easy to verify. One can also show that the space L^2 is complete. This is more technical, an interested reader is referred. e.g., to Section 10 of Akhiezier and Glazman (1993). Measurable functions and abstract $L^p(\mu)$ spaces are studied, e.g., in Rudin (1987). We summarize this discussion in the following theorem.

THEOREM 10.1.1 *The space L^2 of Definition 10.1.4 is a Hilbert space.*

Throughout the book, if the limits of integration are not indicated, it is assumed that the integral is over the entire domain on which the functions are define (typically $[0, 1]$). The Cauchy–Schwarz inequality in the space L^2 thus takes the form

$$\left| \int x(t)y(t)dt \right| \le \left(\int x^2(t)dt \right)^{1/2} \left(\int y^2(t)dt \right)^{1/2}.$$

A sequence $\{x_n\}$ converges to a function x in L^2 if $\int [x_n(t) - x(t)]^2 dt \to 0$. We now review several of the most commonly encounter Hilbert spaces for functional data.

EXAMPLE 10.1.2 While $L^2[0, 1]$ is the most commonly countered space, there are many examples which move beyond the $[0, 1]$ interval. For example $L^2(\mathcal{D})$ where $\mathcal{D}$ is a compact subset of $\mathbb{R}^d$. For spatial processes, one commonly uses $d = 2$, while for applications such as fMRI one can take $d = 3$. For space time processes one may take $L^2[\mathcal{T} \times \mathcal{D}]$ where the first argument represents time and second space. □

EXAMPLE 10.1.3 Given that FDA is often applied to smooth process, it may also be of interest to incorporate derivatives. This leads to the Sobolev spaces, $H^K(\mathcal{T})$, which consist of all functions over $\mathcal{T}$ which are square integrable and whose derivatives up to order K are also square integrable. Focussing on square integrable functions ensures the resulting space is a Hilbert space. The inner product is then given by

$$\langle x, y \rangle_{K,2} = \sum_{k=0}^{K} \int x^{(k)}(t) y^{(k)}(t) \; dt,$$

where $x^{(k)}(t)$ and $y^{(k)}(t)$ are the kth order derivatives of x and y, respectively, with respect to t. □

EXAMPLE 10.1.4 Multivariate functional data are commonly encountered, e.g. when examining panels of functional time series. Given two Hilbert space, $\mathcal{H}_1$ and $\mathcal{H}_2$, the cartesian product $\mathcal{H}_1 \times \mathcal{H}_2$ is also a Hilbert space. If $(x_1, x_2) \in \mathcal{H}_1 \times \mathcal{H}_2$ then one defines the inner product norm on the product space as

$$\|(x_1, x_2)\|_{\mathcal{H}_1 \times \mathcal{H}_2}^2 = \|x_1\|_{\mathcal{H}_1}^2 + \|x_2\|_{\mathcal{H}_2}^2.$$

Given that the cartesian product between two Hilbert spaces is a Hilbert space, one can readily generalize beyond one product. □

EXAMPLE 10.1.5 The space $\mathcal{C}[0, 1]$ denotes the set of all continuous functions over the interval $[0, 1]$. Commonly, this space is equipped with the "sup norm" $\|x\| = \sup_{t \in [0,1]} |x(t)|$. Under the sup norm, $\mathcal{C}[0, 1]$ is a complete normed vector space (i.e. a Banach space), but it is not an inner product space. Alternatively, one can equip $\mathcal{C}[0, 1]$ with the L^2 norm defined in Definition 10.1.4, however, under the L^2 norm, $\mathcal{C}[0, 1]$ is not complete. Any metric space can be completed by expanding the space so that all Cauchy sequences have an element to which they converge. The completion of $\mathcal{C}[0, 1]$ with respect to the L^2 norm is $L^2[0, 1]$. □

10.2 Projections and orthonormal sets

Functional data, typically a collection of functions $x_n \in L^2$ defined on a common interval, are viewed as infinite dimensional objects. A common technique is to project these functions onto finite dimensional subspaces to facilitate statistical calculations. In this section, we therefore review in some detail the relevant Hilbert space theory background. Proofs of Theorems 10.2.1 and 10.2.2 can be found in many textbooks, e.g. in Rudin (1987).

THEOREM 10.2.1 *Suppose $\mathcal{G}$ is a closed subspace of a Hilbert space $\mathcal{H}$. For any $x \in \mathcal{H}$ define*

$$\delta = \inf_{z \in \mathcal{G}} \|x - z\| \, .$$

Then there exists a unique $y \in \mathcal{G}$ such that $\|x - y\| = \delta$.

Recall that a set is closed if it contains all of its limit points, i.e, if for any convergent sequence in $\mathcal{G}$, say x_n, we have $x := \lim x_n \in \mathcal{G}$ as well, then we say $\mathcal{G}$ is closed.

The vector y is called the projection of x onto $\mathcal{G}$. It is the point of $\mathcal{G}$ which is nearest to x, but another interpretation is based on the concept of orthogonality. We say that x is orthogonal to y if $\langle x, y \rangle = 0$. If $\mathcal{G}$ is a subspace, then the *orthogonal complement*, $\mathcal{G}^{\perp}$, denotes the set of all $y \in \mathcal{H}$ which are orthogonal to every vector in $\mathcal{G}$. Recall that a mapping $L : \mathcal{H} \to \mathcal{H}$ is linear if for any scalars a, b and vectors x, y,

$$L(ax + by) = aL(x) + bL(y). \tag{10.2}$$

THEOREM 10.2.2 *Suppose $\mathcal{G}$ is a closed subspace of a Hilbert space $\mathcal{H}$.*

a) *Every $x \in \mathcal{H}$ has a unique decomposition $x = P(x) + Q(x)$, where $P(x) \in \mathcal{G}$ and $Q(x) \in \mathcal{G}^{\perp}$.*

b) *$P(x)$ and $Q(x)$ are the nearest points to x in $\mathcal{G}$ and $\mathcal{G}^{\perp}$, respectively.*

c) *The mappings $P : \mathcal{H} \to \mathcal{G}$ and $Q : \mathcal{H} \to \mathcal{G}^{\perp}$ are linear.*

d) *$\|x\|^2 = \|P(x)\|^2 + \|Q(x)\|^2$.*

The linear operator P is called the projection onto $\mathcal{G}$. The following corollary is often used.

COROLLARY 10.2.1 *The projection $P(x)$ is orthogonal to $x - P(x)$.*

Theorem 10.2.2 leads to the following important result, stated as Theorem 10.2.3 and known as the Riesz representation theorem. It is easy to see that for every $y \in \mathcal{H}$, the function $x \mapsto \langle x, y \rangle$ is linear and continuous on $\mathcal{H}$. In fact, there are no other continuous linear functions.

THEOREM 10.2.3 *If $L : \mathcal{H} \to R$ is linear and continuous, then there is a unique $y \in \mathcal{H}$ such that $L(x) = \langle x, y \rangle$.*

PROOF: If $L(x) = 0$ for all x, take $y = 0$. For L which is not identically 0, define

$$\mathcal{G} = \{x : L(x) = 0\} \, .$$

By the linearity of L, $\mathcal{G}$ is a subspace, and by the continuity of L it is closed. We can thus use Theorem 10.2.2. We will show that there is $y \in \mathcal{G}^{\perp}$ such that $L(x) = \langle x, y \rangle$.

Take $z \in \mathcal{G}^{\perp}$ with $\|z\| = 1$ and observe that

$$L(x) = L(x) \langle z, z \rangle = \langle L(x)z, z \rangle = \langle L(z)x + u_x, z \rangle ,$$

where

$$u_x = L(x)z - L(z)x.$$

Note that $u_x \in \mathcal{G}$. Since $z \in \mathcal{G}^{\perp}$, $\langle u_x, z \rangle = 0$. Therefore,

$$L(x) = \langle L(z)x, z \rangle = \left\langle x, \overline{L(z)}z \right\rangle .$$

Thus, we take $y = \overline{L(z)}z$.

To establish uniqueness, suppose $\langle x, y \rangle = \langle x, y' \rangle$ for every x. Taking $x = y - y'$, this implies $\|y - y'\| = 0$, so $y = y'$. $\blacksquare$

As noted at the beginning of this section, in FDA we often work with projections on finite dimensional subspaces. Let $\mathcal{G}$ be a d–dimensional subspace of $\mathcal{H}$ with a basis $e_1, e_2, \ldots, e_d$. Every vector in $\mathcal{G}$ is a unique linear combination of $e_1, e_2, \ldots, e_d$. In particular, for every $x \in \mathcal{H}$,

$$P(x) = a_1 e_1 + a_2 e_2 + \ldots, a_d e_d.$$

By Theorem 10.2.2, $x - P(x)$ is orthogonal to $\mathcal{G}$. Therefore, for each e_j, $\langle x - P(x), e_j \rangle = 0$. Consequently,

$$\langle e_1, e_j \rangle a_1 + \langle e_2, e_j \rangle a_2 + \ldots \langle e_d, e_j \rangle a_d = \langle x, e_j \rangle , \quad j = 1, 2, \ldots, d.$$

This is a system of d linear equations in the unknowns $a_1, a_2, \ldots, a_d$ which can be uniquely determined if we know the inner products $\langle e_i, e_j \rangle$ and $\langle x, e_j \rangle$. The above system simplifies significantly if $\langle e_i, e_j \rangle = 0$, whenever $i \neq j$. If, in addition, $\langle e_j, e_j \rangle = 1$, then $a_j = \langle x, e_j \rangle$. This motivates the following definitions.

DEFINITION 10.2.1 Let A be an arbitrary index set. We say that $\{e_\alpha, \alpha \in A\}$ is an *orthonormal system* if

$$\langle e_\alpha, e_\beta \rangle = \begin{cases} 1 & \text{if } \alpha = \beta, \\ 0 & \text{if } \alpha \neq \beta. \end{cases}$$

DEFINITION 10.2.2 A Hilbert space $\mathcal{H}$ is said to be *separable* if there is a countable orthonormal system $\{e_1, e_2, \ldots\}$ such that every $x \in \mathcal{H}$ admits the expansion

$$x = \sum_{j=1}^{\infty} a_j e_j.$$

Such a system is said to be *complete* (in $\mathcal{H}$).

DEFINITION 10.2.3 A complete orthonormal system in a separable Hilbert space is called an *orthonormal basis*.

A weaker (topological) definition of separability is often used which states that a separable topological space is one which contains a countably dense subset, and then the property stated in Definition 10.2.2 is proven as an equivalent characterization for Hilbert spaces, see e.g. Theorem 3 in Akhiezier and Glazman (1993). Notice that a set $\{e_j, j \geq 1\}$ is an orthonormal basis if there is no nonzero element of $\mathcal{H}$ which is orthogonal to every e_j.

The expansion in Definition 10.2.2 is assumed to converge in the norm of $\mathcal{H}$, i.e.

$$\lim_{J \to \infty} \left\| x - \sum_{j=1}^{J} a_j e_j \right\| = 0. \tag{10.3}$$

By Theorem 10.2.2,

$$\|x\|^2 = \left\| \sum_{j=1}^{J} a_j e_j \right\|^2 + \left\| x - \sum_{j=1}^{J} a_j e_j \right\|^2.$$

The orthonormality of the e_j implies that $\left\| \sum_{j=1}^{J} a_j e_j \right\|^2 = \sum_{j=1}^{J} |a_j|^2$. Together with (10.3), this implies that $\|x\|^2 = \sum_{j=1}^{\infty} |a_j|^2$. Finally, observe that $a_j = \langle x, e_j \rangle$. We summarize these results in the following theorem:

THEOREM 10.2.4 *[Parseval's Theorem] Suppose $\mathcal{H}$ is a separable Hilbert space and $\{e_j\}$ is a complete orthonormal system. Then, for any $x \in \mathcal{H}$,*

$$x = \sum_{j=1}^{\infty} \langle x, e_j \rangle e_j$$

and

$$\|x\|^2 = \sum_{j=1}^{\infty} |\langle x, e_j \rangle|^2. \tag{10.4}$$

Identity (10.4) is known as Parseval's equality.

We conclude this section by giving examples of orthonormal bases which are often used, the bases consisting of trigonometric functions. The most commonly used basis of this type, often referred to as *the Fourier basis* is

$$\begin{aligned}
e_1(t) &= 1; \\
e_2(t) &= \sqrt{2} \sin(2\pi t); \\
e_3(t) &= \sqrt{2} \cos(2\pi t); \\
e_4(t) &= \sqrt{2} \sin(2 \cdot 2\pi t); \\
e_5(t) &= \sqrt{2} \cos(2 \cdot 2\pi t); \\
&\vdots
\end{aligned} \tag{10.5}$$

In applications, we typically project functions on the subspace spanned by the constant function and the functions $\sin(k \cdot 2\pi t), \cos(k \cdot 2\pi t), \ 1 \leq k \leq K$, so we use $2K + 1$ Fourier basis functions to construct approximations.

A real function in $L^2([0,1])$ can be expanded using only cosine functions and the constant function. The functions

$$e_0(t) = 1, \quad e_k(t) = \sqrt{2}\cos(\pi k t), \quad k = 1, 2, \ldots, \quad t \in [0,1], \tag{10.6}$$

form an orthonormal basis. Another orthonormal basis consists of sine functions:

$$e_k(t) = \sqrt{2}\sin(\pi k t), \quad k = 1, 2, \ldots, \quad t \in [0,1], \tag{10.7}$$

Note that the trigonometric functions in (10.5) have arguments that are multiples of $2\pi t$ and those in (10.6) and (10.7) have arguments which are multiples of πt. It is easy to verify that these three systems are orthonormal. The proof that they are complete is more delicate, an interested reader is referred to Section 2.4 of Hsing and Eubank (2015)

Observe that the members of the basis (10.7) are zero at both ends of the interval $[0,1]$, so one may wonder is all functions in L^2 can indeed be expanded using only sine functions. For example, how can we expand the function $x(t) = t$? The point to keep in mind is that we do not claim that a uniform approximation is possible; we only claim the convergence in the L^2 norm, i.e., in the case of the basis (10.7),

$$\lim_{K \to \infty} \int_0^1 \left(x(t) - \sqrt{2}\sum_{k=1}^{K} b_k \sin(\pi k t) \right)^2 dt = 0,$$

where $b_k = \sqrt{2} \int_0^1 x(t)\sin(\pi k t)dt$. Having said that, expanding the function $x(t) = t$ using the sine basis would not be optimal, see Fig. 10.1, reproduced from Hunter and Nachtergaele (2001). Finding optimal basis expansions for a set of observed functions is an important issue in FDA which will be given due attention.

10.3 Linear operators

A function $L : \mathcal{H} \to \mathcal{H}$ is called a linear operator, or a linear transformation, if for any scalars a, b and any $x, y \in \mathcal{H}$ relation (10.2) holds. We will call a linear operator simply an operator. An operator L is bounded if there is a constant K such that for any $x \in \mathcal{H}$, $\|L(x)\| \leq K \|x\|$. The smallest such K is called the operator norm of L and denoted $\|L\|_{\mathcal{L}}$. It is clear that

$$\|L\|_{\mathcal{L}} = \sup_{\|x\|=1} \|L(x)\|.$$

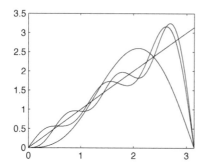 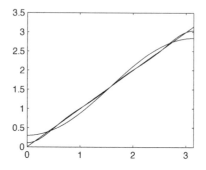

FIGURE 10.1: Approximation of the function $x(t) = t, t \in [0, \pi]$, using sine (left) and cosine (right) functions.

In general, computing the operator norm is not easy. One can show that an operator is bounded if and only if it is continuous, see Problem 10.7. Hence, bounded operators are often called continuous operators.

Bounded operators form a normed vector space with the operations defined pointwise, i.e.

$$(L_1 + L_2)(x) = L_1(x) + L_2(x), \quad (aL)(x) = aL(x).$$

With the norm $\|\cdot\|_{\mathcal{L}}$, this vector space is a Banach space, see Problem 10.8. Since it is more convenient to work in a Hilbert space than in a Banach space, the class of Hilbert–Schmidt operators is often used.

Hilbert–Schmidt operators are defined on a *separable* Hilbert space, cf. Definition 10.2.2. In the remainder of this section, we assume that $\mathcal{H}$ is a separable Hilbert space over the field of *real* numbers. The following version of Parseval's equality is useful to define Hilbert–Schmidt operators:

$$\langle x, y \rangle = \sum_{j=1}^{\infty} \langle x, e_j \rangle \langle e_j, y \rangle, \tag{10.8}$$

where $\{e_j, j \geq 1\}$ is any orthonormal basis in $\mathcal{H}$. Equality (10.8) follows from Theorem 10.2.4:

$$\langle x, y \rangle = \left\langle \sum_{j=1}^{\infty} \langle x, e_j \rangle e_j \, , \, \sum_{j'=1}^{\infty} \langle y, e_{j'} \rangle e_{j'} \right\rangle$$

$$= \sum_{j,j'=1}^{\infty} \langle x, e_j \rangle \langle y, e_{j'} \rangle \langle e_j, e_{j'} \rangle$$

$$= \sum_{j=1}^{\infty} \langle x, e_j \rangle \langle e_j, y \rangle,$$

where in the last equality we used the fact that the e_j are orthormal and the inner product is symmetric.

A bounded linear operator Ψ is said to be Hilbert–Schmidt if for an orthonormal basis $\{e_j, j \geq 1\}$

$$\|\Psi\|_{\mathcal{S}}^2 = \sum_{j=1}^{\infty} \|\Psi(e_j)\|^2 < \infty.$$

The above relation defines the Hilbert–Schmidt norm $\|\cdot\|_{\mathcal{S}}$. Using (10.8), it is easy to check that this definition is correct, i.e. it does not depend on the choice of the basis, see Problem 10.10. We denote the vector space of Hilbert–Schmidt operators by $\mathcal{S}$. The space $\mathcal{S}$ is a separable Hilbert space with the inner product

$$\langle \Psi, \Phi \rangle_{\mathcal{S}} = \sum_{j=1}^{\infty} \langle \Psi(e_j), \Phi(e_j) \rangle. \tag{10.9}$$

If $\{e_j, j \geq 1\}$ is an orthonormal basis in $\mathcal{H}$, then the operators $e_j \otimes e_i$ defined by

$$(e_j \otimes e_i)(x) = \langle e_i, x \rangle e_j$$

form an orthonormal basis in $\mathcal{S}$, see Problem 10.11. The object $e_j \otimes e_i$ is called a tensor and will be discussed in greater detail in Section 10.5.

EXAMPLE 10.3.1 [Integral operator] Consider $\mathcal{H} = L^2 = L^2([0,1])$, and define an operator Ψ by

$$\Psi(x)(t) = \int \psi(t,s)x(s)ds,$$

where $\psi(\cdot, \cdot)$ is a bivariate function, called the *kernel*, which satisfies

$$\iint \psi^2(t,s)dtds < \infty.$$

We first show that the operator Ψ is bounded. It is convenient to introduce the functions ψ_t defined by $\psi_t(s) = \psi(t,s)$. By Fubini's theorem, $\psi_t \in L^2$ for almost all t. For such t, the Cauchy–Schwarz inequality implies that for any $x \in L^2$,

$$\left\{ \int \psi_t(s)x(s)ds \right\}^2 \leq \int \psi_t^2(s)ds \int x^2(s)ds.$$

Therefore,

$$\|\Psi(x)\|^2 = \int \left\{ \int \psi(t,s)x(s)ds \right\}^2 dt$$
$$\leq \int \left\{ \int \psi^2(t,s)ds \int x^2(s)ds \right\} dt$$
$$= \left(\iint \psi^2(t,s)dtds \right) \|x\|^2.$$

We see that

$$\|\Psi\|_{\mathcal{L}} \leq \left(\iint \psi^2(t,s) dt ds \right)^{1/2}.$$

Now we will verify that $\Psi \in \mathcal{S}$. For any orthonormal basis $\{e_j, j \geq 1\}$, by Theorem 10.2.4,

$$\sum_{j=1}^{\infty} \|\Psi(e_j)\|^2 = \int \sum_{j=1}^{\infty} \left\{ \int \psi(t,s) e_j(s) ds \right\}^2 dt$$

$$= \int \sum_{j=1}^{\infty} \langle \psi_t, e_j \rangle^2 \, dt$$

$$= \int \|\psi_t\|^2 \, dt$$

$$= \iint \psi^2(t,s) dt ds.$$

The above calculation shows that

$$\|\Psi\|_{\mathcal{S}} = \left(\iint \psi^2(t,s) dt ds \right)^{1/2}. \tag{10.10}$$

$\square$

We now consider a simple example of an operator which is bounded, but not Hilbert-Schmidt.

EXAMPLE 10.3.2 Let $\mathcal{H}$ be an infinite dimensional separable Hilbert space. The identity operator $\Psi(x) = x$, is bounded, but not Hilbert-Schmidt. Clearly, the operator norm of Ψ is 1 since

$$\sup_{\|x\|=1} \|\Psi(x)\| = \sup_{\|x\|=1} \|x\| = 1.$$

However, the Hilbert-Schmidt norm is infinite since

$$\|\Psi\|_{\mathcal{S}}^2 = \sum_{j=1}^{\infty} \|\Psi(e_j)\|^2 = \sum_{j=1}^{\infty} \|e_j\|^2 = \sum_{j=1}^{\infty} 1 = \infty.$$

The same arguments show that if $\mathcal{H}$ is finite dimensional, then the identity will be Hilbert-Schmidt. This equivalence holds more generally, as all bounded operators over a finite dimensional Hilbert space are also Hilbert-Schmidt. $\square$

The integral operator in Example 10.3.1 satisfies $\|\Psi\|_{\mathcal{L}} \leq \|\Psi\|_{\mathcal{S}}$. This relation holds in general, as stated in Theorem 10.3.1.

THEOREM 10.3.1 *If $\Psi \in \mathcal{S}$, then $\Psi \in \mathcal{L}$ and $\|\Psi\|_{\mathcal{L}} \leq \|\Psi\|_{\mathcal{S}}$.*

Perhaps the easiest way to prove Theorem 10.3.1 is by introducing the concept of the adjoint operator.

DEFINITION 10.3.1 Let $L : \mathcal{H} \to \mathcal{H}$ be a bounded operator. The operator L^* defined by

$$\langle L(x), y \rangle = \langle x, L^*(y) \rangle$$

is called the *adjoint operator* of L.

Problem 10.12 asks for the verification that L^* is bounded and

$$\|L^*\|_{\mathcal{L}} = \|L\|_{\mathcal{L}} .$$

To prove Theorem 10.3.1, it thus suffices to verify that

$$\|\Psi^*(x)\|^2 \leq \left(\sum_{j=1}^{\infty} \|\Psi(e_j)\|^2 \right) \|x\|^2 .$$

The above relation follows from Parseval's equality and Cauchy–Schwarz inequality:

$$\|\Psi^*(x)\|^2 = \sum_{j=1}^{\infty} \langle \Psi^*(x), e_j \rangle^2 = \sum_{j=1}^{\infty} \langle x, \Psi(e_j) \rangle^2 \leq \|x\|^2 \sum_{j=1}^{\infty} \|\Psi(e_j)\|^2 .$$

10.4 Basics of spectral theory

Many concepts and results discussed in this Section do not require that assumption that $\mathcal{H}$ is a Hilbert space; the general framework of a vector space is often sufficient. Motivated by applications to FDA, we assume that the vector space is defined over the filed of real numbers, but all results of this section remain true in the complex case, with minimal adjustments.

Suppose L is an operator on a vector space. A number λ is called an *eigenvalue* of L if there is a nonzero vector x such that

$$L(x) = \lambda x.$$

Every such vector x is called an *eigenvector* of L corresponding to λ. If the operator acts on a function space, x is often called an *eigenfunction*.

If x is an eigenvector corresponding to λ, then ax is clearly also an eigenvector corresponding to λ. In fact, the set of all eigenvectors corresponding to λ forms a linear subspace, as can be verified using the definition of the eigenvector. This space is called the eigenvalue space of λ; its dimension is called the multiplicity of λ. An eigenvalue of multiplicity greater than one is called *multiple.* In the remainder of this section, we state several simple theorems which will be used throughout the book.

We say that T is an invertible operator if there is an operator T^{-1} defined on the whole vector space such that $T(T^{-1}(x)) = T^{-1}(T(x)) = x$. If T^{-1} exists, it is unique.

THEOREM 10.4.1 *Let T be an invertible operator on a vector space, and let L be any operator. The operators L and TLT^{-1} have the same eigenvalues.*

PROOF: Suppose λ is an eigenvalue of L and x a corresponding eigenvector. Since T is invertible, $T(x) \neq 0$. Observe that

$$TLT^{-1}(T(x)) = TL(x) = T(\lambda x) = \lambda T(x),$$

so λ is an eigenvalue of TLT^{-1} (with a corresponding eigenvector $T(x)$). Since $L = T^{-1}(TLT^{-1})T$, the same argument shows that every eigenvalue of TLT^{-1} is also an eigenvalue of L.

∎

DEFINITION 10.4.1 An operator L over an inner product space is called *nonnegative–definite* if for every vector x,

$$\langle L(x), x \rangle \geq 0.$$

It is *positive–definite*, if the equality can hold only for $x = 0$.

Nongetive–definite operators are also called: nonnegative, positive, positive–definite, positive semidefinite. Positive–definite operators are also called: strictly positive and strictly positive-definite.

THEOREM 10.4.2 *All eigenvalues of a nonnegative–definite operator are non-negative. All eigenvalues of a positive–definite operator are positive.*

PROOF: Suppose L is a nonnegative definite operator, λ its eigenvalue and x a corresponding eigenvector. Then

$$\lambda \|x\|^2 = \lambda \langle x, x \rangle = \langle \lambda x, x \rangle = \langle L(x), x \rangle \geq 0.$$

Thus $\lambda \geq 0$. As an eigenvector, x is nonzero, so $\|x\| > 0$. If L is positive definite, then $\lambda \|x\|^2 > 0$, so $\lambda > 0$.

∎

DEFINITION 10.4.2 An operator L over an inner product space is called *symmetric* if for any x, y,

$$\langle L(x), y \rangle = \langle x, L(y) \rangle.$$

In the context of a complex Hilbert space, symmetric operators are called self–adjoint, cf. Definition 10.3.1.

THEOREM 10.4.3 *Eigenvectors corresponding to distinct eigenvalues of a symmetric operator are orthogonal.*

PROOF: Let $\lambda_1 \neq \lambda_2$ be eigenvalues with eigenvectors x_1 and x_2, respectively. Observe that

$$\lambda_1 \langle x_1, x_2 \rangle = \langle L(x_1), x_2 \rangle = \langle x_1, L(x_2) \rangle = \lambda_2 \langle x_1, x_2 \rangle.$$

It follows that $\langle x_1, x_2 \rangle = 0$. ∎

We now state an important result known as the Hilbert–Schmidt theorem.

THEOREM 10.4.4 *Suppose* $\Psi \in S$ *is symmetric. Then, there is an orthonormal system* $\{e_j, j \geq 1\}$ *consisting of eigenvectors of* Ψ *corresponding to nonzero eigenvalues* λ_j *such that every* $x \in \mathcal{H}$ *has a unique representation*

$$x = \sum_{j=1}^{\infty} a_j e_j + v,$$

where v *satisfies,* $\Psi(v) = 0$.

The proof of Theorem 10.4.4 is more complex than the other proofs presented in this section, and is not given here. An interested reader is refereed to Debnath and Mikusinski (2005), or any other textbook covering operators in a Hilbert space. The following corollary is known as the spectral theorem or Mercer's theorem.

COROLLARY 10.4.1 *Suppose* $\Psi \in S$ *is symmetric. Then, there is an orthonormal basis* $\{v_j, j \geq 1\}$ *such that for every* $x \in \mathcal{H}$

$$\Psi(x) = \sum_{j=1}^{\infty} \lambda_j \langle x, v_j \rangle v_j,$$

where λ_j *is an eigenvalue corresponding to the eigenfunction* v_j *(of* Ψ*).*

PROOF: Set $V = \{x \in \mathcal{H} : \Psi(x) = 0\}$. The subspace V is a separable Hilbert space; let $\{v_k^*, k \geq 1\}$ be an orthonormal basis in V. The required basis $\{v_j, j \geq 1\}$ is the union of $\{e_j, j \geq 1\}$ and $\{v_k^*, k \geq 1\}$. The result follows because $a_j = \langle x, e_j \rangle$ and each v_k^* corresponds to the zero eigenvalue. ∎

As we will see in Chapters 11 and 12, Corollary 10.4.1 plays a fundamental role in the analysis of variance of random functions. We conclude this section with a theorem which leads to such results. Suppose $\Psi \in S$ is symmetric and nonnegative definite. Suppose we want to find a vector $x \in \mathcal{H}$ with $\|x\| = 1$ such that $\langle \Psi(x), x \rangle$ is maximum. Since Ψ is nonnegative definite, all eigenvalues λ_j are nonnegative. We arrange them in nonincreasing order, so that $\lambda_1 \geq \lambda_2 \geq \dots$. To ensure a unique solution to the problem, assume that $\lambda_1 > \lambda_2$. Then, by Theorem 10.4.3, v_1 is orthogonal to all remaining eigenvectors. We must maximize

$$\langle \Psi(x), x \rangle = \left\langle \sum_{j=1}^{\infty} \lambda_j \langle x, v_j \rangle, x \right\rangle = \sum_{j=1}^{\infty} \lambda_j \langle x, v_j \rangle^2$$

subject to $\sum_{j=1}^{\infty} \langle x, v_j \rangle^2 = 1$. By our assumptions on the eigenvalues, the maximum is attained if we take $x = v_1$ (or $x = -v_1$). We have thus shown that $\langle \Psi(x), x \rangle$ is maximum if $x = v_1$, the eigenvector corresponding to the largest eigenvalue, and the maximum is equal to λ_1, the largest eigenvalue.

In FDA, it is useful to solve a more general problem which can be formulated as follows: we want to find x with unit norm which maximizes $\langle \Psi(x), x \rangle$, but which is orthogonal to v_1. We will see that $x = v_2$. Next we want to maximize $\langle \Psi(x), x \rangle$ over a subspace orthogonal to v_1 and v_2, etc. To obtain a unique solution for p iterations, we assume that

$$\lambda_1 > \lambda_2 > \ldots \lambda_p > \lambda_{p+1}. \tag{10.11}$$

THEOREM 10.4.5 *Suppose Ψ is a symmetric, nonnegative definite Hilbert–Schmidt operator with eigenfunctions v_j and eigenvalues λ_j satisfying (10.11). Then, for $i \leq p$*

$$\sup \{ \langle \Psi(x), x \rangle : \ \|x\| = 1, \ \langle x, v_j \rangle = 0, \ 1 \leq j \leq i - 1 \} = \lambda_i,$$

and the supremum is achieved if $x = v_i$. The maximizing function is unique up to a sign.

PROOF: The theorem follows from the identities

$$\langle \Psi(x), x \rangle = \sum_{j=i}^{\infty} \lambda_j \langle x, v_j \rangle^2, \quad \sum_{j=i}^{\infty} \langle x, v_j \rangle^2 = 1.$$

∎

Note that the above implies that for a symmetric nonnegative definite linear operator with eigenvalues ordered as $\lambda_1 \geq \lambda_2 \geq \ldots$ we have the relations

$$\|\Psi\|_{\mathcal{S}}^2 = \sum_{i=1}^{\infty} \lambda_i^2 \quad \text{and} \quad \|\Psi\|_{\mathcal{L}} = \lambda_1.$$

This allows us to gain a bit more understanding as to the difference between a bounded operator and one that is Hilbert-Schmidt; a bounded operator has bounded eigenvalues, while a Hilbert-Schmidt operator has eigenvalues which go to zero fast enough to sum.

10.5   Tensors

Later, when examining covariance operators, it is convenient to work with tensor products and spaces. In finite dimensional spaces, use of tensors can be avoided by stacking matrices in the right way. Such an approach does not work in infinite dimensional spaces and thus use of tensors cannot be avoided if one wants to work with general Hilbert spaces. The methods presented here work well for real separable Hilbert spaces. For more general Banach spaces or vectors spaces, the definitions become more complex, and we avoid them. For a more complete treatment we refer to one of the many texts now available on tensor methods and multilinear algebra such as Hackbusch (2012).

While tensors are often thought of as generalizations of matrices, this does not translate well to infinite dimensional spaces. Instead, it is easier to think of tensors as multilinear mappings. We begin with an example in Euclidian space to illustrate this point. First, consider a matrix $\mathbf{A} \in \mathbb{R}^{N \times M}$. A matrix can be viewed in at least three different ways: 1) as a traditional matrix, 2) as a linear transformation from $\mathbb{R}^N \to \mathbb{R}^M$ (or the reverse), or 3) a bilinear functional $\mathbb{R}^N \times \mathbb{R}^M \to \mathbb{R}$. The first is, of course, how we often think of matrices. The second is obtained by multiplying a matrix from the left or right by a vector. One learns in linear algebra that all linear transformations in Euclidean space are of such a form. The last view is maybe less common, but arises often in the study of quadratic forms. Consider two Euclidean spaces $\mathcal{H}_1 = \mathbb{R}^N$ and $\mathcal{H}_2 = \mathbb{R}^M$. Forming the tensor product between to vectors, $x_1 \in \mathbb{R}^N$ and $x_2 \in \mathbb{R}^M$, leads to a matrix:

$$x_1 \otimes x_2 := x_1 x_2^\top.$$

The space $\mathbb{R}^N \otimes \mathbb{R}^M$ is then defined as the space of all finite linear combinations of the above elements. In other words, if $\mathbf{A} \in \mathbb{R}^N \otimes \mathbb{R}^M \to \mathbb{R}$, then there exists J elements $x_{1j} \in \mathbb{R}^N$ and $x_{2j} \in \mathbb{R}^M$ such that

$$\mathbf{A} = \sum_{j=1}^{J} x_{1j} \otimes x_{2j} = \sum_{j=1}^{J} x_{1j} x_{2j}^\top.$$

Since any $N \times M$ matrix can be expressed in such a way, we have that $\mathbb{R}^N \otimes \mathbb{R}^M = \mathbb{R}^{N \times M}$. We will now specify an alternative view, that while equivalent, generalizes more readily to other spaces. Any $N \times M$ matrix, $\mathbf{A}$, leads to a bilinear mapping $\mathbf{A} : \mathbb{R}^N \times \mathbb{R}^M \to \mathbb{R}$ in form of

$$\mathbf{A}(y_1, y_2) := y_1^\top \mathbf{A} y_2.$$

Therefore the space $\mathbb{R}^N \otimes \mathbb{R}^M$ can also be identified as the set of all bilinear maps from $\mathbb{R}^N \times \mathbb{R}^M \to \mathbb{R}$. The object $x_1 \otimes x_2$ can also be identified as a bilinear mapping

$$(x_1 \otimes x_2)(y_1, y_2) = y_1^\top x_1 x_2^\top y_2 = (y_1^\top x_1)(y_2^\top x_2).$$

It is in this way we view tensors. Tensors are multilinear maps from the cartesian product of the spaces involved to a field of scalars. As we will mention later on, if $\{e_{1j}\}$ and $\{e_{2k}\}$ are the standard bases for $\mathbb{R}^N$ and $\mathbb{R}^M$ respectively, then $\{e_{1j} \otimes e_{2k}\}$ is a basis of $\mathbb{R}^N \otimes \mathbb{R}^M$. In this way, one uses this basis to get back to traditional matrices:

$$\mathbf{A}(e_{1j}^\top, e_{2k}) = e_{1j}^\top \mathbf{A} e_{2k} = \mathbf{A}(j, k).$$

We are now in a position to define the tensor product between elements of Hilbert spaces.

DEFINITION 10.5.1 Let $x_1 \in \mathcal{H}_1$ and $x_2 \in \mathcal{H}_2$ be two elements of two real Hilbert spaces. Then the tensor product, $x_1 \otimes x_2 : \mathcal{H}_1 \times \mathcal{H}_2 \to \mathbb{R}$, is a bilinear map defined as

$$(x_1 \otimes x_2)(y_1, y_2) = \langle x_1, y_1 \rangle_{\mathcal{H}_1} \langle x_2, y_2 \rangle_{\mathcal{H}_2},$$

for any $(y_1, y_2) \in \mathcal{H}_1 \times \mathcal{H}_2$.

The tensor space $\mathcal{H}_1 \otimes \mathcal{H}_2$ can be viewed as the space of all tensors which is also complete. To construct it, first define the set of all finite linear combinations of tensors

$$\mathcal{A} = \left\{ \sum_{j=1}^J x_{1j} \otimes x_{2j} : x_{1j} \in \mathcal{H}_1, x_{2j} \in \mathcal{H}_2, 1 \leq J < \infty \right\}.$$

By definition, this is a vector space. One can define an inner product on $\mathcal{A}$ as

$$\left\langle \sum_{j=1}^J x_{1j} \otimes x_{2j}, \sum_{k=1}^K y_{1k} \otimes y_{2k} \right\rangle_{\mathcal{A}} = \sum_{j=1}^J \sum_{k=1}^K \langle x_{1j}, y_{1k} \rangle_{\mathcal{H}_1} \langle x_{2j}, y_{2k} \rangle_{\mathcal{H}_2}.$$

This inner product leads to an inner product norm. We thus define $\mathcal{H}_1 \otimes \mathcal{H}_2$ as the completion of $\mathcal{A}$ with respect to the above inner product norm (note that if the spaces are finite dimensional, then $\mathcal{A}$ is already complete). Since $\mathcal{H}_1 \otimes \mathcal{H}_2$ is also a Hilbert space, we can now define the tensor product of any order since our original definition only requires that $\mathcal{H}_1$ and $\mathcal{H}_2$ be Hilbert spaces.

Below are several properties which follow from our definition.

THEOREM 10.5.1 *Let $x_1, y_1 \in \mathcal{H}_1$ and $x_2, y_2 \in \mathcal{H}_2$ be elements of two Hilbert spaces. Then one has the following*

a) $(x_1 + y_1) \otimes x_2 = x_1 \otimes x_2 + y_1 \otimes x_2$

b) $x_1 \otimes (x_2 + y_2) = x_1 \otimes x_2 + x_1 \otimes y_2$

c) $c(x_1 \otimes x_2) = (cx_1) \otimes x_2 = x_1 \otimes (cx_2)$

THEOREM 10.5.2 *If $\{e_{1j}\}$ and $\{e_{2k}\}$ are orthonormal bases of $\mathcal{H}_1$ and $\mathcal{H}_2$ respectively, then $\{e_{1j} \otimes e_{2k}\}$ is a basis of $\mathcal{H}_1 \otimes \mathcal{H}_2$.*

EXAMPLE 10.5.1 A very relevant example for our applications occurs when $\mathcal{H}_1 = \mathcal{H}_2 = L^2(\mathcal{T})$:

$$(x_1 \otimes x_2)(y_1, y_2) = \int \int x_1(t) x_2(s) y_1(t) y_2(s) \, dt \, ds.$$

This implies that $x_1 \otimes x_2$ has a square integrable kernel $x_1(t) x_2(s)$. We therefore have that $L^2(\mathcal{T}) \otimes L^2(\mathcal{T})$ is equivalent to $L^2(\mathcal{T} \times \mathcal{T})$ (one may even write an equality if the interpretation is clear). Tensors in this case are merely square integrable bivariate functions. This will be convenient later on as covariance functions denote within curve covariance at different time points. □

Our description of tensors is not the only one, and there are other equivalent approaches. We mention one alternative which fits in naturally with the covariance operators we will discuss in subsequent chapters. Once again, consider two real Hilbert spaces $\mathcal{H}_1$ and $\mathcal{H}_2$. For each pair of elements of $(x_1, x_2) \in \mathcal{H}_1 \times \mathcal{H}_2$, one can define an operator $L : \mathcal{H}_1 \to \mathcal{H}_2$ in the following way:

$$L_{x_1, x_2}(y_1) = \langle x_1, y_1 \rangle_{\mathcal{H}_1} x_2,$$

where $y_1 \in \mathcal{H}_1$. Notice that the above operator induces a bilinear form which is equivalent to $x_1 \otimes x_2$,

$$\langle L_{x_1, x_2}(y_1), y_2 \rangle_{\mathcal{H}_2} = \langle x_1, y_1 \rangle_{\mathcal{H}_1} \langle x_2, y_2 \rangle_{\mathcal{H}_2}.$$

Thus the operator approach is sometimes interpreted as the definition of a tensor. An advantage is that it fits in well with covariance operators which are defined as $E[\langle X_1, x \rangle X_2]$ (assuming the random elements are centered). The approaches are completely equivalent mathematically, thus it is not uncommon to move between the two depending on which is more convenient. The downside of this is that it can create quite a bit of ambiguity. For example, when $\mathcal{H}_1 = \mathcal{H}_2 = L^2(\mathcal{T})$ one may interpret $x \otimes y$ as a bilinear functional, an operator, or a bivariate function. For this reason, we will often use the following conventions:

- Operator: $(x_1 \otimes y_1)(y_2) = \langle y_1, y_2 \rangle x_1$,

- Functional: $\langle x_1 \otimes y_1, x_2 \otimes y_2 \rangle = \langle x_1, x_2 \rangle \langle y_1, y_2 \rangle$,

- General Form: $x_1 \otimes x_2$.

Extending the above definitions to the tensor product of multiple spaces is straightforward as one would start with the form

$$(x_1 \otimes \cdots \otimes x_p)(y_1, \ldots, y_p) = \langle x_1, y_1 \rangle \ldots \langle x_p, y_p \rangle.$$

Note that is allows us to take tensor products of tensor spaces as well. However, the operator view of a tensor becomes a bit confusing as there are multiple interpretations.

10.6 Chapter 10 problems

10.1 Prove Proposition 10.1.1.

10.2 Show that in any inner product space, the function $y \mapsto \langle x, y \rangle$ is continuous. (x is an arbitrary vector.)

10.3 Verify that the space ℓ^2 defined in Example 10.1.1 is a Hilbert space.

10.4 Suppose $\mathcal{G}$ is a subspace of Hilbert space $\mathcal{H}$. Show that $\mathcal{G}^{\perp}$ is a closed subspace of $\mathcal{H}$.

10.5 Show that the functions

$$1, \quad \sqrt{2}\sin(\pi k t), \quad \sqrt{2}\cos(\pi k t), \quad k = 1, 2, \ldots, \quad t \in [0, 1],$$

do not form an orthonormal basis in real $L^2 = L^2([0, 1])$.

10.6 Suppose $\{e_j, j \geq 1\}$ is a complete orthonormal system in a Hilbert space. Show that if $\{f_j, j \geq 1\}$ is an orthonormal sequence satisfying

$$\sum_{j=1}^{\infty} \|e_j - f_j\|^2 < 1,$$

then $\{f_j, j \geq 1\}$ is also complete.

10.7 Suppose $L : \mathcal{H} \to \mathcal{H}$ is a linear operator. Prove the equivalence of the following conditions:

(i) L is bounded;

(ii) L is continuous;

(iii) L is continuous at a point.

10.8 A vector space is said to be a *normed space* if to each vector x there is associated a nonnegative number $\|x\|$ such that the following conditions hold:

(i) $\|x + y\| \leq \|x\| + \|y\|$;

(ii) $\|ax\| = |a| \|x\|$ (a is a scalar);

(iii) $\|x\| = 0$ implies $x = 0$.

A *Banach* space is a normed space which is complete in the distance defined by the norm.

Denote by $\mathcal{L}$ the vector space of bounded operators on a Hilbert space $\mathcal{H}$. Show that $\mathcal{L}$ is a Banach space.

10.9 Recall the definition of a Banach space stated in Problem 10.8. Let l^∞ be the space of bounded real sequences $x = \{x_1, x_2, \dots\}$ such that for any $x \in l^\infty$,

$$\|x\|_\infty := \sup_{k \in \mathbb{N}} |x_k| < \infty.$$

Prove that $\|\cdot\|_\infty$ is a norm and l^∞ is a Banach space. (The vector space operations are defined pointwise as in Example 10.1.1.)

10.10 Suppose $\{e_j, j \geq 1\}$ and $\{f_i, i \geq 1\}$ are orthonormal bases in $\mathcal{H}$. Show that for any Hilbert–Schmidt operators Ψ, Φ

$$\sum_{i=1}^\infty \langle \Psi(f_i), \Phi(f_i) \rangle = \sum_{j=1}^\infty \langle \Psi(e_j), \Phi(e_j) \rangle.$$

10.11 Suppose $\{e_j, j \geq 1\}$ is an orthonormal basis in $\mathcal{H}$. Show that the operators $e_j \otimes e_i$ defined by $(e_j \otimes e_i)(x) = \langle e_i, x \rangle e_j$ form an orthonormal basis in $\mathcal{S}$.

10.12 Show that if L is bounded then L^* is also bounded, and

$$\|L^*\|_{\mathcal{L}} = \|L\|_{\mathcal{L}}, \qquad \|L^*L\|_{\mathcal{L}} = \|L\|_{\mathcal{L}}^2.$$

10.13 Let $\{\lambda_j, j \geq 1\}$ be the set of nonzero eigenvalues of a Hilbert–Schmidt operator Ψ. Show that $\|\Psi\|_{\mathcal{S}}^2 = \sum_{j=1}^\infty \lambda_j^2$.

10.14 Prove Theorem 10.5.1.

10.15 Show that a Linear operator is bounded if it is bounded on an orthonormal basis $\{e_j\}$.

10.16 Show that if a sequence $x_n \to x$ in $\mathcal{H}$, then $x_n \otimes x_n \to x \otimes x$ in $\mathcal{S}$.

11

Random functions

In this chapter, we continue the exposition of the mathematical foundations of FDA we started in Chapter 10. We revisit many concepts introduced in Chapter 3. We rigorously define and explore in some depth: random functions in abstract metric space, expectation and covariance in a Hilbert space, Gaussian random functions and principal component analysis. This chapter provides background required to state certain mathematical results of previous chapters. It is also a concise introduction to the most important concepts of theoretical FDA. Familiarity with its contents will enable the readers to understand most research papers in the field of FDA. Results that have simple proofs are proven. By studying these proofs, the reader will be able to understand all concepts much better than by just reading the definitions and theorems.

This chapter requires some background in measure theory. However, no difficult results of measure theory are used. Basically, only familiarity with the concepts of the σ–algebra and a measurable map is required. Many textbooks provide such background. For example, Kallenberg (1997) or Rudin (1987) (the first few pages of Chapter 1 of both) provide more than enough background. These concepts are recalled in a very concise manner at the beginning of Section 11.1.

11.1   Random elements in metric spaces

Recall that a set $\mathcal{S}$ is a metric space if there is a function d defined on $\mathcal{S} \times \mathcal{S}$ such that $d(x, y) = d(y, x) > 0$ if $x \neq y$, $d(x, x) = 0$ and $d(x, y) \leq d(x, z) + d(z, y)$. A collection $\mathfrak{S}$ of subsets of $\mathcal{S}$ is called a σ–algebra if $\emptyset \in \mathfrak{S}$, $A \in \mathfrak{S}$ implies $A^c \in \mathfrak{S}$ and $A_j \in \mathfrak{S}, j \geq 1$, implies $\bigcup_{j=1}^{\infty} A_j \in \mathfrak{S}$. The smallest σ–algebra containing open sets is called the Borel σ–algebra, and its members the Borel sets. In the following, we denote the Borel σ–algebra by $\mathfrak{S}$. A function μ defined on $\mathfrak{S}$ is called a probability measure if it is nonnegative, $\mu(\bigcup_{j=1}^{\infty} A_j) = \sum_{j=1}^{\infty} \mu(A_j)$ if no two A_j have a nonempty intersection (the A_j are pairwise disjoint), and $\mu(\mathcal{S}) = 1$. In the following, *we assume that $\mathcal{S}$ is a separable metric space*, i.e. it contains a dense countable subset. The assumption of topological separability in needed to prove the results we cite in this section, see Chapter 1 of Billingsley (1968) for more details. All metric spaces we use in this book are separable.

Suppose Ω is a probability space with a σ–algebra $\mathfrak{O}$ and a probability measure P. Let $\mathcal{S}$ be a metric space. We say that a mapping $X : \Omega \to \mathcal{S}$ is a random element in $\mathcal{S}$ if $X^{-1}(A) \in \mathfrak{O}$ for every Borel set A. If $\mathcal{S}$ is the real line, we call X a random variable, if $\mathcal{S} = \mathbb{R}^k$, the k–dimensional Euclidean space, X is a random vector. Our interest will focus on settings in which $\mathcal{S}$ is a space of functions, in which case X is called a random function. Many convergence properties of random functions do not depend on the structure of the function space, and can be stated in the general context of a metric space. Every random element X induces a probability measure on $\mathcal{S}$ defined by

$$\mu(A) = P(X^{-1}(A)) = P(\omega \in \Omega : X(\omega) \in A) = P(X \in A).$$

The measure μ is called the *distribution* of X.

DEFINITION 11.1.1 [weak convergence] Suppose $\mu_n, n \geq 1$, and μ are probability measures on a metric space $\mathcal{S}$. We say that the sequence $\{\mu_n\}$ converges weakly to μ if

$$\int_{\mathcal{S}} f d\mu_n \to \int_{\mathcal{S}} f d\mu$$

for every bounded continuous real function on $\mathcal{S}$.

The integrals in Definition 11.1.1 are abstract integrals defined e.g. in Chapter 1 of Rudin (1987). Theorem 11.1.1 provides some intuition behind the concept of weak convergence. It is proven in Billingsley (1968) and in several other advanced textbooks. A set $A \in \mathfrak{S}$ is called a μ–continuity set if the measure μ of its topological boundary is zero.

THEOREM 11.1.1 *Suppose $\mu_n, n \geq 1$, and μ are probability measures on a metric space $\mathcal{S}$. The following conditions are equivalent.*

a) $\int_{\mathcal{S}} f d\mu_n \to \int_{\mathcal{S}} f d\mu$ *for every bounded continuous real function on $\mathcal{S}$.*

b) $\mu_n(A) \to \mu(A)$ *for every μ–continuity set A.*

Recall that if X_n and X are random variables, then the sequence $\{X_n\}$ converges in distribution to X if $F_n(t) \to F(t)$ at every continuity point t of F, where F_n and F are cumulative distribution functions. Using the theory of σ–algebras, this definition can be shown to be equivalent to condition 2 in Theorem 11.1.1, where μ_n and μ are probability measures generated by F_n and F, respectively. This motivates the following definition.

DEFINITION 11.1.2 [convergence in distribution] Suppose $X_n, n \geq 1$, and X are random elements in $\mathcal{S}$. We say that the sequence $\{X_n\}$ converges in distribution to X, denoted $X_n \overset{d}{\to} X$, if the distributions μ_n of X_n converge weakly to the distribution μ of X.

By condition 2 of Theorem 11.1.1, convergence in distribution is equivalent to $P(X_n \in A) \to P(X \in A)$ for every $A \in \mathfrak{S}$ such that $P(X \in \partial A) = 0$, where ∂A is the boundary of A.

Another concept that will be often used is the convergence in probability. To define it, we must assume that all random variables are defined on the same probability space to be able to consider functions like $d(X_n, X)$ or $d(X_n, Y_n)$ which are defined on Ω via $\omega \mapsto d(X_n(\omega), X(\omega))$ and $\omega \mapsto d(X_n(\omega), Y_n(\omega))$. *From now on we assume that all random variables are defined on the common probability space* $(\Omega, \mathfrak{D}, P)$. This assumption is not restrictive in any practical sense; probability spaces are abstract constructs that are not observed directly, and if random variables are not defined on the same probability space, one can construct a larger probability space on which all copies in distribution are defined. We will not concern ourselves with these issues.

DEFINITION 11.1.3 [convergence in probability] Suppose $X_n, n \geq 1$, and X are random elements in $\mathcal{S}$. We say that the sequence $\{X_n\}$ converges in probability to X, denoted $X_n \overset{P}{\to} X$, if for every $\epsilon > 0$,

$$P(\omega \in \Omega : d(X_n(\omega), X(\omega)) > \epsilon) = P(d(X_n, X) > \epsilon) \to 0.$$

Analogously, for two sequences, $\{X_n\}$ and $\{Y_n\}$, we say that $d(X_n, Y_n) \overset{P}{\to} 0$, if the random variable $d(X_n, Y_n)$ converges to zero, i.e. $P(d(X_n, Y_n) > \epsilon) \to 0$.

We will now state several theorems which are extremely useful in FDA, and, in fact, in most arguments of asymptotic statistics. These results are proven in Billingsley (1968).

THEOREM 11.1.2 *If* $X_n \overset{d}{\to} X$ *and* $d(X_n, Y_n) \overset{P}{\to} 0$, *then* $Y_n \overset{d}{\to} X$.

Taking $X_n = X$, we see that convergence in probability implies convergence in distribution.

In many arguments, establishing the convergence in distribution of $\{X_n\}$ to X may be difficult, but there may be a truncation type parameter u such that for every fixed u $X_n(u) \overset{d}{\to} X(u)$, and $X(u) \overset{d}{\to} X$, as u approaches some limit. The question is then: when can we conclude that $X_n \overset{d}{\to} X$? The following result is often used.

THEOREM 11.1.3 *Suppose that for each* u, $X_n(u) \overset{d}{\to} X(u)$ $(n \to \infty)$, *and* $X(u) \overset{d}{\to} X$ $(u \to \infty)$. *If*

$$\lim_{u \to \infty} \limsup_{n \to \infty} P(d(X_n(u), X_n) > \epsilon) = 0,$$

then $X_n \overset{d}{\to} X$.

The next result is known as the *continuous mapping theorem*. Suppose $h : \mathcal{S} \to \mathcal{S}'$, where $\mathcal{S}'$ is another separable metric space. Denote by D_h the set of points in $\mathcal{S}$ at which h is not continuous.

THEOREM 11.1.4 [continuous mapping theorem] *Suppose $X_n, n \geq 1$, and X are random elements in $\mathcal{S}$ and $P(X \in D_h) = 0$. If $X_n \overset{d}{\to} X$, then $h(X_n) \overset{d}{\to} h(X)$.*

In many settings, we work with two sequences, $\{X_n\}$ and $\{Y_n\}$, and are interested in the convergence in distribution of the sequence $\{(X_n, Y_n), n \geq 1\}$, which is a random element in the Cartesian product $\mathcal{S} \times \mathcal{S}$. Since the projections are continuous functions, it follows from Theorem 11.1.4 that the convergence $(X_n, Y_n) \overset{d}{\to} (X, Y)$ implies $X_n \overset{d}{\to} X$ and $Y_n \overset{d}{\to} Y$. The converse is not true. The following result is used very often.

THEOREM 11.1.5 [Slutsky's theorem] *Suppose $X_n \overset{d}{\to} X$ and $Y_n \overset{P}{\to} a$, where a is a point in $\mathcal{S}$. Then $(X_n, Y_n) \overset{d}{\to} (X, a)$.*

11.2 Expectation and covariance in a Hilbert space

In order to extend the concepts of the expected value and variance, and establish extensions of results like the law of large numbers and the central limit theorem, we must assume that the random elements take values in a vector space. A number of profound results can be established in Banach spaces assuming they have some additional properties. This was a very active area of research in the 1970's, an interested reader is referred to Linde (1986) and Vakhaniia *et al.* (1987), a concise introduction is given in Chapter 7 of Laha and Roghatgi (1979). Chapters 1 and 2 of Bosq (2000) also provide a relevant summary. We introduce in this section the concepts and results used in the remainder of the book in the setting of a separable Hilbert space. We assume that all random functions are defined on a common probability space Ω and take values in a separable Hilbert space $\mathcal{H}$ with the distance defined by the norm.

DEFINITION 11.2.1 A random function X is said to be integrable if $E\|X\| < \infty$. It is said to be weakly integrable if, for any $y \in \mathcal{H}$, $E|\langle X, y \rangle| < \infty$ and if there is $e \in \mathcal{H}$ such that $E[\langle X, y \rangle] = \langle e, y \rangle$.

Integrable functions are also called strongly integrable, and weakly integrable vectors are called integrable. It is not difficult to show, Problem 11.6, that integrability implies weak integrability. The function e in Definition 11.2.1 is unique because if $\langle e_1, y \rangle = \langle e_2, y \rangle$ for any $y \in \mathcal{H}$, then $e_1 = e_2$. We denote this unique function by EX, and call it the expected value of X. By definition, we thus have

$$E[\langle X, y \rangle] = \langle EX, y \rangle, \quad \forall y \in \mathcal{H}. \tag{11.1}$$

EXAMPLE 11.2.1 Suppose $\mathcal{H} = L^2$ is the Hilbert space of square integrable functions introduced in Definition 10.1.4. The random function X is integrable if

$$E \left\{ \int X^2(t)dt \right\}^{1/2} < \infty.$$

Suppose X is integrable and denote by e its expected value EX; e is a function in L^2. We will verify that $e(t) = E[X(t)]$ for almost all $t \in [0, 1]$.

By the Cauchy–Schwarz inequality in L^2,

$$E \int |X(t)y(t)|dt \leq \left\{ \int y^2(t)dt \right\}^{1/2} E \left\{ \int X^2(t)dt \right\}^{1/2} < \infty.$$

The absolute integrability of $X(\omega, t)y(\omega, t)$ on $\Omega \times [0, 1]$ allows us to change the order of E and $\int$ (Fubini's theorem). Therefore, for any $y \in \mathcal{H}$,

$$E \int X(t)y(t)dt = \int E[X(t)]y(t)dt = \int e(t)y(t)dt.$$

The functions $t \mapsto E[X(t)]$ and $e(t)$ are thus equal as elements of L^2, and so for almost all $t \in [0, 1]$. $\qquad \square$

The following theorem provides two useful properties of the expected value.

THEOREM 11.2.1 *Suppose X is integrable.*

a) $\|EX\| \leq E \|X\|$ *(contractive property);*

b) *For any bounded operator L, $L(X)$ is integrable and $E[L(X)] = L(EX)$.*

PROOF: Set $e = EX$. Then

$$|\langle e, y \rangle| = |E \langle X, y \rangle| \leq E|\langle X, y \rangle| \leq \|y\| E \|X\|.$$

Taking $y = e$, we obtain $\|e\|^2 \leq \|e\| E \|X\|$, so the first claim follows.

To prove the second claim, first note that $L(X)$ is integrable because

$$E \|L(X)\| \leq E \|L\|_{\mathcal{L}} \|X\| = \|L\|_{\mathcal{L}} E \|X\| < \infty.$$

Next, denote $e' = L(e)$. The equality $e' = L(e)$ will follow if we can show that for every $y \in \mathcal{H}$, $\langle e', y \rangle = \langle L(e), y \rangle$. Recall the definition of the adjoint operator L^*, cf. Definition 10.3.1. Then

$$\langle L(e), y \rangle = \langle e, L^*(y) \rangle = E \langle X, L^*(y) \rangle = E \langle L(X), y \rangle = \langle e', y \rangle.$$

■

The conclusions of Theorem 11.2.1 remain valid if L maps $\mathcal{H}$ into another separable Hilbert space $\mathcal{H}'$; the proof requires more general definitions of the operator norm and of the adjoint operator. In particular, if $\mathcal{H}' = \mathbb{R}$, the Riesz representation theorem, Theorem 10.2.3, shows that claim 2 reduces to (11.1).

The second order properties of Hilbert space valued random functions are described by the covariance operator.

DEFINITION 11.2.2 A random function X is said to be square integrable if $E\|X\|^2 < \infty$. The covariance operator of a square integrable random function is defined by

$$\mathcal{H} \ni y \mapsto C(y) = E[\langle X - EX, y \rangle (X - EX)] \in \mathcal{H}.$$

In the remainder of this section, to lighten the formulas, we will often assume that $EX = 0$ so that $C(y) = E[\langle X, y \rangle X]$. If $\mathcal{H} = \mathbb{R}$, then $C(y) = E[XyX] = \mathrm{Var}[X]y$, so the action of C reduces to the multiplication by $\mathrm{Var}[X]$, and so C can be identified with the variance. If $\mathcal{H} = \mathbb{R}^d$, C can be similarly identified with the covariance matrix, Problem 11.7. If $\mathcal{H} = L^2$, as in Example 11.2.1, then $C(y)$ is a function in L^2 whose value at $t \in [0, 1]$ is

$$C(y)(t) = E\left[\left(\int X(s)y(s)ds\right)X(t)\right] = \int c(t, s)y(s)ds, \qquad (11.2)$$

where

$$c(t, s) = E[X(t)X(s)], \quad s, t \in [0, 1]. \qquad (11.3)$$

We see that the covariance operator C is an integral operator with kernel (11.3), called the *covariance function*. Equation (11.2) shows that the covariance operator can be identified with the corresponding covariance function (11.3), which is actually more often used in FDA. The covariance operator leads to a more elegant description of the second order structure in general Banach spaces. In theoretical arguments of FDA, it is more convenient to treat the covariance operator, and its sample counterparts, as elements of the Hilbert space $\mathcal{S}$ of Hilbert–Schmidt operators. Before we elaborate on this point, we state a theorem which characterizes covariance operators. We refer to Section 10.4 for the required definitions.

THEOREM 11.2.2 *Suppose $C : \mathcal{H} \to \mathcal{H}$ is a linear operator. The operator C is the covariance operator of some square integrable random function X if and only if*

a) *C is symmetric;*

b) *C is nonnegative-definite;*

c) *the eigenvalues of C satisfy $\sum_{j=1}^{\infty} \lambda_j < \infty$.*

PROOF: We will prove only that every covariance operator has properties 1–3. The converse statement follows by showing that if conditions 1–3 hold, then $\varphi(y) = \exp\{-\langle C(y), y\rangle/2\}$ is the characteristic functional of a square integrable random function in $\mathcal{H}$ with covariance operator C, cf. Section 11.3. This is established by using Bochner's theorem in a Hilbert space, the Minlos–Sazonov theorem, e.g. Section 7.6.2 of Laha and Roghatgi (1979). The details are too extensive to be presented here.

Verification of condition 1: By (11.1),

$$\langle C(y), z\rangle = \langle E[\langle X, y\rangle X], z\rangle = E\langle[\langle X, y\rangle X], z\rangle = E[\langle X, y\rangle \langle X, z\rangle]. \quad (11.4)$$

This shows that $\langle C(y), z\rangle = \langle y, C(z)\rangle$.
Verification of condition 2: By (11.4), $\langle C(y), y\rangle = E[\langle X, y\rangle^2] \geq 0$.
Verification of condition 3: Suppose $v_j, j \geq 1$, are the eigenfunctions of C. By (11.4),

$$E[\langle X, v_j\rangle^2] = \langle C(v_j), v_j\rangle = \langle \lambda_j v_j, v_j\rangle = \lambda_j.$$

By Parseval's identity (10.4),

$$\sum_{j=1}^{\infty} \lambda_j = \sum_{j=1}^{\infty} E[\langle X, v_j\rangle^2] = E\sum_{j=1}^{\infty} \langle X, v_j\rangle^2 = E\|X\|^2 < \infty.$$

∎

Operators for which $\sum_j |\lambda_j| < \infty$ are called *nuclear*. Theorem 11.2.2 thus states that an operator is a covariance operator if and only if it is symmetric, nonnegative–definite and nuclear. One can show that every covariance operator is Hilbert–Schmidt.

11.3 Gaussian functions and limit theorems

In this section, we continue to assume that all random functions take values in a separable Hilbert space. Just as in the scalar case, the expected value and the covariance operator do not determine the distribution. The appropriate extension of the characteristic function is the *characteristic functional*.

DEFINITION 11.3.1 The characteristic functional of a random function X in a separable Hilbert space is defined by

$$\mathcal{H} \ni y \mapsto \varphi_X(y) = E\exp\{i\langle y, X\rangle\}.$$

One can show that if $\varphi_X = \varphi_Y$, then X and Y have the same distribution, and that if $X_n \xrightarrow{d} X$, then $\varphi_{X_n}(y) \to \varphi_X(y)$ for every $y \in \mathcal{H}$. In finite dimensional spaces, the pointwise convergence of characteristic functions implies convergence in distribution. In infinite dimensional spaces this is no longer true, Problem 11.13, an additional assumption of *tightness* is needed. An interested reader is referred to Billingsley (1968) or Chapter 7 of Laha and Roghatgi (1979). Some other properties of the characteristic functional are stated in Problem 11.12.

DEFINITION 11.3.2 *A random function X is said to be Gaussian if its characteristic functional has the form*

$$\varphi_X(y) = \exp\left\{ i\langle \mu, y \rangle - \frac{1}{2}\langle C(y), y \rangle \right\},$$

where $\mu \in \mathcal{H}$ and C is a covariance operator.

It can be shown, e.g. Section 7.6 of Laha and Roghatgi (1979), that if X is Gaussian, then $E\|X\|^2 < \infty$, $EX = \mu$ and C is the covariance operator of X. Definition 11.3.2 shows that for every $y \in \mathcal{H}$, $\langle y, X \rangle$ is a normal random variable with expected value $\langle y, \mu \rangle$ and variance $\langle C(y), y \rangle$. The converse statement is true. The following theorem provides a way of verifying that a random function in $\mathcal{H}$ is Gaussian.

THEOREM 11.3.1 *A random function X is Gaussian (with $EX = 0$) if and only if for any $y \in \mathcal{H}$, $\langle y, X \rangle$ is a normal random variable (with mean zero).*

A proof is given in Chapter 7 of Laha and Roghatgi (1979), Proposition 7.6.5.
  The central limit theorem in a Hilbert space is stated as Theorem 11.3.2. It is proven e.g. in Bosq (2000), Theorem 2.7.

THEOREM 11.3.2 *Suppose $X_i, i = 1, 2, \ldots$ are iid square integrable with expected value μ and covariance operator C. Then*

$$N^{-1/2}\sum_{i=1}^{N}(X_i - \mu) \overset{d}{\to} Z,$$

where Z is normal with $EX = 0$ and covariance operator C.

The law of large numbers also takes a familiar form.

THEOREM 11.3.3 *Suppose $X_i, i = 1, 2, \ldots$ are iid integrable with expected value μ. Then*

$$N^{-1}\sum_{i=1}^{N}X_i \overset{a.s.}{\to} \mu.$$

A proof is given in Section 7.2 of Laha and Roghatgi (1979), Theorem 7.2.1.
  The iid assumption in Theorems 11.3.2 and 11.3.3 can be relaxed in many ways. In most applications, it is useful to retain the assumption of the identical distribution, but allow some weak dependence between the X_i. This point will be revisited in the following chapters.

11.4   Functional principal components

As we have seen in Chapter 1, a fundamental approach of FDA is to expand observed functions $X_n(t), t \in [0,1]$, as $X_n(t) = \sum_{m=1}^{M} c_{nm} B_m(t)$, where the B_m are some basis functions. The basis functions can be deterministic, e.g. Fourier or spline functions, but one can ask if there are some basis functions which are in some sense optimal for an observed data set $X_1, X_2, \ldots, X_N$. We will address this problem in Chapter 12. In this section, we study it in the context of a *random* function X taking values in L^2. We thus define population parameters which can be estimated from a sample $X_1, X_2, \ldots, X_N$.

Fix $p \geq 1$ and suppose $u_1, u_2, \ldots, u_p$ is any orthonormal system in L^2. The projection of a random function X on the subspace spanned by it is $\sum_{k=1}^{p} \langle X, u_k \rangle u_k$. The system $u_1, u_2, \ldots, u_p$ is optimal in the sense of minimizing the expected distance between X and its span if it minimizes

$$S(u_1, u_2, \ldots, u_p) = E \left\| X - \sum_{k=1}^{p} \langle X, u_k \rangle u_k \right\|^2 \tag{11.5}$$

over all orthonormal systems consisting of p functions. To find such a system observe that, by the orthonormality of the u_k,

$$S(u_1, u_2, \ldots, u_p) = E \left\langle X - \sum_{k=1}^{p} \langle X, u_k \rangle u_k, X - \sum_{k=1}^{p} \langle X, u_k \rangle u_k \right\rangle$$

$$= E \|X\|^2 - 2E \sum_{k=1}^{p} \langle X, u_k \rangle^2 + E \sum_{k=1}^{p} \langle X, u_k \rangle^2$$

$$= E \|X\|^2 - \sum_{k=1}^{p} E \langle X, u_k \rangle^2 .$$

Since, by (11.1),

$$E \langle X, u_k \rangle^2 = E \langle \langle X, u_k \rangle X, u_k \rangle = \langle C(u_k), u_k \rangle ,$$

minimizing $S(u_1, u_2, \ldots, u_p)$ is equivalent to maximizing $\sum_{k=1}^{p} \langle C(u_k), u_k \rangle$. We now assume that condition (10.11) holds. By Theorem 10.4.5, the largest possible value of $\langle C(u_1), u_1 \rangle$ is λ_1 and it is attained if $u_1 = v_1$ (up to a sign). Then, since u_2 is orthogonal to $u_1 = v_1$, the largest possible value of $\langle C(u_2), u_2 \rangle$ is λ_2 and it is attained at $u_2 = v_2$. Continuing in this manner, establishes the following result.

THEOREM 11.4.1 *Suppose X is square integrable and the eigenvalues of its covariance operator C satisfy (10.11). For any fixed $p \geq 1$, the expected error (11.5) is minimized if $u_j = v_j$, where the v_j are the eigenfunctions of C with unit norm.*

Setting $\mu = EX$, observe that

$$X - \mu = \sum_{j=1}^{\infty} \xi_j v_j, \qquad \xi_j = \langle X - \mu, v_j \rangle. \tag{11.6}$$

Decomposition (11.6) is called the *Karhunen–Loéve expansion*. The deterministic functions v_j are called the *functional principal components*. Throughout the book, we will use the abbreviation FPC's. The FPC's are thus the eigenfunctions of the covariance operator normalized to have unit norm. They are defined only up to a sign. The random variables ξ_j are called their *scores*. Observe that

$$E\xi_j = 0, \quad E\xi_j^2 = \lambda_j \quad E[\xi_j \xi_i] = 0, \text{ if } i \neq j. \tag{11.7}$$

Therefore, by Parseval's identity,

$$E \|X - \mu\|^2 = \sum_{j=1}^{\infty} \lambda_j. \tag{11.8}$$

Identity (11.8) has the following interpretation: The variance of X is equal to the sum of the variances of the projections of X onto the FPC's. This decomposition, or analysis, of variance is used very often in FDA.

In the space L^2, decompositions (11.6) and (11.8) take the form

$$X(t) = \mu(t) + \sum_{j=1}^{\infty} \xi_j v_j(t), \qquad E \int (X(t) - \mu(t))^2 dt = \sum_{j=1}^{\infty} \lambda_j.$$

We see that every square integrable random function is the sum of the mean function, $\mu(\cdot)$, and a random error function whose expected value is zero. The error function is the sum of the FPC's $v_j(\cdot)$ with mean zero random weights ξ_j. The importance of each FPC v_j is quantified by the standard deviation of its weight ξ_j, which is equal to $\sqrt{\lambda_j}$. The expected integrated squared difference between a random function and its mean function is equal to the sum of the eigenvalues of the covariance operator.

EXAMPLE 11.4.1 [Gaussian random function] If X is a Gaussian function in L^2, then by Theorem 11.3.1, $\sum_{j=1}^{d} a_j \xi_j = \left\langle X - \mu, \sum_{j=1}^{d} a_j v_j \right\rangle$ is a normal random variable for any d and any numbers $a_1, a_2, \ldots, a_d$. This means that the ξ_j are jointly normal. By (11.7) they are uncorrelated, so they are, in fact, independent with variances λ_j. We conclude that a Gaussian random function in L^2 admits the decomposition

$$X(t) = \mu(t) + \sum_{j=1}^{\infty} \sqrt{\lambda_j} N_j v_j(t),$$

where the N_j are independent standard normal and

$$\lambda_j = E\left\{\int (X(t) - \mu(t))v_j(t)dt\right\}^2.$$

$\square$

Example 11.4.1 shows that the distribution a Gaussian random function on $[0, 1]$ is fully characterized by the eigenfunctions $v_j(\cdot)$ and and the eigenvalues λ_j. The two Gaussian random functions occurring most often in various settings are the Wiener process, also called the Brownian motion, and the Brownian bridge. We will use them extensively, so we now define them.

DEFINITION 11.4.1 [Wiener process and Brownian bridge] A random function $\{W(t), t \in [0, 1]\}$ is called the *Wiener process* if the following conditions hold

a) $W(0) = 0$ a.s..

b) If $0 \le s < t \le 1$, then $W(t) - W(s)$ is normal with mean zero and variance $t - s$.

c) For any $0 \le t_0 < t_1 < \dots < t_k \le 1$, the random variables $W(t_j) - W(t_{j-1}), 1 \le j \le k$, are independent.

If $\{W(t), t \in [0, 1]\}$ is the Wiener process, then

$$B(t) = W(t) - tW(1), \quad t \in [0, 1]$$

is called the *Brownian bridge*.

Based on Definition 11.4.1, the usual way of approximating a Wiener process is based on the observation that for $0 \le k \le K$

$$W\left(\frac{k}{K}\right) - W\left(\frac{k-1}{K}\right) \stackrel{d}{=} N\left(0, \frac{1}{K}\right) \stackrel{d}{=} \frac{1}{\sqrt{K}}N_k,$$

so, with independent standard normal N_k,

$$W\left(\frac{k}{K}\right) \stackrel{d}{=} \frac{1}{\sqrt{K}}\sum_{i=1}^{k}N_k.$$

The Karhunen–Loéve expansion allows us to view the Wiener process and the Brownian bridge from a different angle. We begin with Proposition 11.4.1 whose proof is assigned as Problem 11.16.

PROPOSITION 11.4.1 *The covariance function of the Wiener process is* $c_W(t, s) = \min(t, s)$. *The covariance function of the Brownian bridge is* $c_B(t, s) = \min(t, s) - ts$.

THEOREM 11.4.2 *The functional principal components and the corresponding eigenvalues of the Wiener process are*

$$v_j(t) = \sqrt{2}\sin\left(\left(j - \frac{1}{2}\right)\pi t\right), \quad \lambda_j = \frac{1}{\left(j - \frac{1}{2}\right)^2 \pi^2}, \quad j = 1, 2, \ldots.$$

For the Brownian bridge, these are

$$v_j(t) = \sqrt{2}\sin(\pi j t), \quad \lambda_j = \frac{1}{j^2\pi^2}, \quad j = 1, 2, \ldots.$$

PROOF: We will verify the claim for the Wiener process. The verification for the Brownian bridge is similar.

By Proposition 11.4.1, we must solve the integral equation $\int_0^1 \min(s,t)v(s)ds = \lambda v(t)$, which can be be written as

$$\int_0^t sv(s)ds + t\int_t^1 v(s)ds = \lambda v(t).$$

Differentiating with respect to t yields

$$\int_t^1 v(s)ds = \lambda v'(t). \tag{11.9}$$

Differentiating again, we obtain the differential equation

$$-v(t) = \lambda v''(t).$$

A general solution to this equation has the form

$$v(t) = A\sin\left(\frac{t}{\sqrt{\lambda}}\right) + B\cos\left(\frac{t}{\sqrt{\lambda}}\right).$$

The constants A and B are determined by boundary conditions. Since $\int_0^0 = 0$, $v(0) = 0$, so $B = 0$. Setting $t = 1$ in (11.9), gives $v'(1) = 0$, which implies

$$\cos\left(\frac{1}{\sqrt{\lambda}}\right) = 0.$$

This means that $1/\sqrt{\lambda} = -\pi/2 + j\pi, j \geq 1$, so all possible eigenvalues are as specified in the claim, and the corresponding eigenfunctions are $A\sin(\pi(j - 1/2)t)$. The condition $\int v_j^2(t)dt = 1$, yields $A = \sqrt{2}$. ∎

Theorem 11.4.2 combined with Example 11.4.1 lead to the following Corollary.

COROLLARY 11.4.1 *Let $N_j, j \geq 1$, be independent standard normal random variables.*

The Karhunen–Loéve expansion of the Wiener process is

$$W(t) = \sum_{j=1}^{\infty} \frac{\sqrt{2}}{\left(j - \frac{1}{2}\right)\pi} N_j \sin\left(\left(j - \frac{1}{2}\right)\pi t\right).$$

For the Brownian bridge, this expansion is

$$B(t) = \sum_{j=1}^{\infty} \frac{\sqrt{2}}{j\pi} N_j \sin\left(j\pi t\right).$$

We conclude this section by explaining how functional data with specified FPC's can be simulated. Set

$$X_n(t) = \sum_{j=1}^{p} a_j Z_{jn} e_j(t), \tag{11.10}$$

where a_j are real numbers, for every n, the Z_{jn} are iid mean zero random variables with unit variance, and the e_j are orthogonal functions with unit norm. Denote by X a random function with the same distribution as each X_n, i.e. $X(t) = \sum_{j=1}^{p} a_j Z_j e_j(t)$. Then the covariance operator of X acting on x is equal to

$$C(x)(t) = E\left[\left(\int X(s)x(s)ds\right)X(t)\right] = \int E[X(t)X(s)]x(s)ds.$$

By the independence of the Z_j, the covariance function is equal to

$$E[X(t)X(s)] = E\left[\sum_{j=1}^{p} a_j Z_j e_j(t) \sum_{i=1}^{p} a_i Z_i e_i(s)\right] = \sum_{j=1}^{p} a_j^2 e_j(t) e_j(s).$$

Therefore,

$$C(x)(t) = \sum_{j=1}^{p} a_j^2 \left(\int e_j(s)x(s)ds\right) e_j(t).$$

It follows that the FPC's of the X_n are the e_j, and the eigenvalues are $\lambda_j = a_j^2$.

11.5   Chapter 11 problems

11.1 Slutsky's theorem is often known as the following implications for random variables ($S = \mathbb{R}$):

a) If $X_n \overset{d}{\to} X$ and $Y_n \overset{P}{\to} a$, then $X_n + Y_n \overset{d}{\to} X + a$ and $Y_n X_n \overset{d}{\to} aX$;

b) If $X_n \overset{d}{\to} X$ and $Y_n \overset{P}{\to} a \neq 0$, then $X_n/Y_n \overset{d}{\to} X/a$.

Show how these results follow from Theorems 11.1.4 and 11.1.5.

11.2 A sequence $\{X_n\}$ of random variables is bounded in probability ($X_n = O_P(1)$) if

$$\lim_{M \to \infty} \limsup_{n \to \infty} P(|X_n| > M) = 0.$$

This means that for every $\epsilon > 0$, there is $M > 0$ such that $P(|X_n| > M) < \epsilon$ for sufficiently large n (those greater than some n_0). Show that if $X_n \overset{d}{\to} X$, for some random variable X, then $X_n = O_P(1)$.
Hint: Show that the cdf F of X can have at most countably many discontinuity points.

11.3 (a) Suppose $g : \mathbb{R} \to \mathbb{R}$ is continuous at a. Show that $X_n \overset{P}{\to} a$ implies $g(X_n) \overset{P}{\to} g(a)$.

(b) Suppose $f : \mathbb{R} \to \mathbb{R}$ has a derivative at a. Show that if $X_n \overset{P}{\to} a$, then

$$f(X_n) = f(a) + f'(a)(X_n - a) + (X_n - a)Z_n,$$

with $Z_n \overset{P}{\to} 0$.

11.4 Suppose for each $k = 1, 2, \ldots, M$, $Y_{k,n}, Y_k$ are random variables such that $Y_{k,n} \overset{P}{\to} Y_k$.

(a) Show that for any numbers w_k, $\sum_{k=1}^{M} w_k Y_{k,n} \overset{P}{\to} \sum_{k=1}^{M} w_k Y_k$.

(b) Show that

$$[Y_{1,n}, Y_{2,n}, \ldots, Y_{M,n}]^T \overset{P}{\to} [Y_1, Y_2, \ldots, Y_M]^T$$

in the Euclidean space $\mathbb{R}^M$.

11.5 Suppose $Y_{k,n}, Y_k$ are random variables such that for every $M \geq 1$,

$$[Y_{1,n}, Y_{2,n}, \ldots, Y_{M,n}]^T \overset{d}{\to} [Y_1, Y_2, \ldots, Y_M]^T$$

in the Euclidean space $\mathbb{R}^M$. Suppose $\{w_k, k \geq 1\}$ is a sequence of numbers such that

$$\sum_{k=1}^{\infty} |w_k| E|Y_k| < \infty \quad \text{and} \quad \sum_{k=1}^{\infty} |w_k| \sup_{n \geq 1} E|Y_{k,n} - Y_k| < \infty.$$

Using Theorem 11.1.3, show that $\sum_{k=1}^{\infty} w_k Y_{k,n} \overset{d}{\to} \sum_{k=1}^{\infty} w_k Y_k$.

11.6 Show that if X is integrable, then it is weakly integrable, cf. Definition 11.2.1.

11.7 Suppose $\mathcal{H} = \mathbb{R}^d$ and X is square integrable. Show how the covariance operator of X can be identified with the covariance matrix of X.

11.8 Suppose $c(t, s)$, $t, s \in [0, 1]$ is the kernel of a covariance operator of X. Show that if $E\,\|X\|^4 < \infty$, then

$$\iint c^2(t, s)dt\,ds \le \left(E\|X - \mu\|^2\right)^2.$$

Without using Theorem 11.2.2, conclude that every covariance operator of such X is Hilbert–Schmidt.

11.9 Suppose $\mathcal{H}$ is an infinite dimensional separable Hilbert space and $\{e_j, j \ge 1\}$ is an orthonormal system. Define the operator Ψ by

$$\Psi(x) = \sum_{j=1}^{\infty} j^{-1} \langle x, e_j \rangle\, e_j.$$

Show that Ψ is bounded, symmetric and nonnegative definite, but it is not a covariance operator.

11.10 Suppose $\Psi : L^2 \to L^2$ is a symmetric Hilbert–Schmidt operator with kernel $\psi(t, s), t, s \in [0, 1]$. Show that if Ψ is nonnegative definite, then the kernel $\psi(\cdot, \cdot)$ is a nonnegative definite function in the sense that

$$\sum_{i,k=1}^{d} \psi(t_i, t_k) z_i \bar{z}_k \ge 0, \tag{11.11}$$

for any integer d, any $t_1, t_2, \ldots, t_d \in [0, 1]$ and any complex numbers $z_1, z_2, \ldots, z_d$.

11.11 Construct a kernel $\psi(t, s)$, $t, s \in [0, 1]$, which is symmetric and non-negative definite in the sense of (11.11), but which is not a covariance function of any square integrable $X \in L^2$.

11.12 Using Definition 11.3.1, prove the following properties:

a) $|\varphi_X(y)| \le 1$, $\varphi_X(0) = 1$.

b) The functional φ_X is nonnegative definite in the sense that

$$\sum_{i,k=1}^{d} \varphi_X(y_i - y_k) z_i \bar{z}_k \ge 0,$$

for any integer d, any $y_1, y_2, \ldots, y_d \in \mathcal{H}$ and any complex numbers $z_1, z_2, \ldots, z_d$.

c) φ_X is uniformly continuous.

11.13 Consider the space ℓ^2 of Example 10.1.1. Define (constant) random variables $X_n, n \geq 1$, by

$$X_1(\omega) = \{1, 0, 0, 0, \ldots\}$$
$$X_2(\omega) = \{0, 1, 0, 0, \ldots\}$$
$$X_3(\omega) = \{0, 0, 1, 0, \ldots\}$$
$$\vdots \quad = \quad \vdots$$

For every $y = \{y_1, y_2, y_3, \ldots\} \in \ell^2$ find the limit of $\varphi_{X_n}(y)$. Observe that the limit is the value at y of the characteristic functional of a random element X in ℓ^2, i.e. $\varphi_{X_n}(y) \to \varphi_X(y)$. Show that X_n does not converge in distribution to X.

11.14 Suppose X satisfies Definition 11.3.2 and let L be a bounded operator. Show that $L(X)$ is Gausian; find its expected value and covariance operator.

11.15 Verify the identities in (11.7).

11.16 Prove Proposition 11.4.1.

12

Inference from a random sample

In this chapter, we assume that we observe a simple random sample consisting of N curves, $X_1, X_2, \ldots X_N$, which have been converted into functional objects using, for example, one of the methods described in Section 1.1. We thus assume that the curves are completely observed. We view each curve as a realization of a random function X which is a random element of L^2 with the same distribution as each X_i. For ease of reference, we state the following assumption.

ASSUMPTION 12.0.1 *The functions $X_1, X_2, \ldots X_N$ are iid in L^2, and have the same distribution as X, which is assumed to be square integrable.*

In many theoretical results, we will replace L^2 by an arbitrary separable Hilbert space and impose assumption on the moments of the X_i.

The Chapter is organized as follows. In Section 12.1, we provide asymptotic justifications for the estimators introduced in Chapter 1, the sample mean and the sample covariance function, utilizing the operator and tensor formulations. We show that the estimators converge to their population counterparts. i.e. that they are consistent. Section 12.2 studies in detail the estimated (or empirical) functional principal components and shows that they are consistent estimators of the population functional principal components. Section 12.3 elaborates on the results of Sections 12.1 and 12.2 by establishing asymptotic distributions and expansions. The results of Section 12.3 are used in Sections 12.4 and 12.5 to develop, respectively, testing an confidence band construction procedures. We use testing and confidence band estimation of the mean function to illustrate chief challenges encountered in many inferential procedures of FDA. Section 12.6 provides an illustration of these procedures using the data set of Bank of America (BOA) cumulative intraday returns introduced in Chapter 1.

12.1 Consistency of sample mean and covariance functions

The following functional parameters have been defined in Chapters 3 and 11:

$$\mu(t) = E[X(t)] \qquad \text{(mean function)};$$
$$c(t,s) = E[(X(t) - \mu(t))(X(s) - \mu(s))] \qquad \text{(covariance function)};$$
$$C = E[(X - \mu) \otimes (X - \mu)] \qquad \text{(covariance tensor)};$$
$$C(\cdot) = E[\langle (X - \mu), \cdot \rangle (X - \mu)] \qquad \text{(covariance operator)}.$$

The mean function μ is estimated by the sample mean function

$$\hat{\mu}(t) = N^{-1} \sum_{n=1}^{N} X_n(t)$$

and the covariance function by its sample counterpart

$$\hat{c}(t,s) = N^{-1} \sum_{n=1}^{N} (X_n(t) - \hat{\mu}(t))(X_n(s) - \hat{\mu}(s)).$$

The sample covariance operator is defined by

$$\widehat{C}(x) = N^{-1} \sum_{n=1}^{N} \langle X_n - \hat{\mu}, x \rangle (X_n - \hat{\mu}), \quad x \in L^2.$$

In this chapter, we use the normalization with N, rather than the $N - 1$ used in Chapter 1. Asymptotic results, as $N \to \infty$, are identical in both cases.

Note that $\widehat{C}$ maps L^2 into a finite dimensional subspace spanned by X_1, $X_2, \ldots, X_N$. This illustrates a limitation of statistical inference for functional observations; a finite sample can recover only a projection of an unknown functional parameter onto a finite dimensional subspace. The estimators do, however, have the usual desirable properties in the asymptotic setting, as $N \to \infty$.

Theorem 12.1.1 states that $\hat{\mu}$ is an unbiased, L^2–consistent estimator of μ. It implies that it is consistent in probability: $\|\hat{\mu} - \mu\| \xrightarrow{P} 0$. The theorem and its proof parallel analogous results for the average of scalar observations. In the functional case, we need the following simple lemma whose proof is assigned as Problem 12.1.

LEMMA 12.1.1 *If $X_1, X_2 \in L^2$ are independent, square integrable and $EX_1 = 0$, then $E[\langle X_1, X_2 \rangle] = 0$.*

THEOREM 12.1.1 *If Assumption 12.0.1 holds, then $E\hat{\mu} = \mu$ and $E\|\hat{\mu} - \mu\|^2 = O(N^{-1})$.*

PROOF: For every n, for almost all $t \in [0, 1]$, $EX_n(t) = \mu(t)$, so it follows that $E\hat{\mu} = \mu$ in L^2. Next, observe that by Lemma 12.1.1,

$$E\|\hat{\mu} - \mu\|^2 = N^{-2} \sum_{n,m=1}^{N} E[\langle(X_n - \mu), (X_m - \mu)\rangle]$$

$$= N^{-2} \sum_{n=1}^{N} E\|X_n - \mu\|^2 = N^{-1} E\|X - \mu\|^2.$$

∎

Replacing μ by $\hat{\mu}$ has an asymptotically negligible effect on the second order properties. We will therefore often assume in theoretical work that $\mu = 0$. This simplifies the appearance of many formulas. When applying such results to real data, it is important to remember to first subtract the sample mean function $\hat{\mu}$ from functional observations. From now on, we thus assume that the functions have zero expectation, so that

$$\hat{c}(t, s) = N^{-1} \sum_{n=1}^{N} X_n(t) X_n(s); \quad \widehat{C}(x) = N^{-1} \sum_{n=1}^{N} \langle X_n, x \rangle X_n$$

and

$$\widehat{C}(x)(t) = \int \widehat{C}(t, s) x(s) ds, \quad x \in L^2. \tag{12.1}$$

The sample covariance operator $\widehat{C}$ is, almost surely, a covariance operator (see Definition 11.2.2 and consider X which takes values $X_1, X_2, \ldots, X_N$ with probabilities N^{-1}). It is therefore a Hilbert–Schmidt operator, and it follows from (10.10) that $\widehat{C}(\cdot, \cdot) \in L^2([0, 1] \times [0, 1])$. Theorem 12.1.2 shows that $E\|X\|^4 < \infty$ implies $E\|\widehat{C}\|_{\mathcal{S}}^2 < \infty$, where $\|\cdot\|_{\mathcal{S}}$ is the Hilbert–Schmidt norm.

THEOREM 12.1.2 *If $E\|X\|^4 < \infty$, $(EX = 0)$, and Assumption 12.0.1 holds, then*

$$E\|\widehat{C}\|_{\mathcal{S}}^2 \leq E\|X\|^4.$$

PROOF: By definition, cf. (10.9),

$$E\|\widehat{C}\|_{\mathcal{S}}^2 = \sum_{i=1}^{\infty} E\langle \widehat{C}(e_i), \widehat{C}(e_i) \rangle.$$

Plugging in the expression for $\widehat{C}$, the above can be expressed as

$$E\|\widehat{C}\|_{\mathcal{S}}^2 = \sum_{i=1}^{\infty} \sum_{n=1}^{N} \sum_{m=1}^{N} N^{-2} E\langle \langle X_n, e_i \rangle X_n, \langle X_m, e_i \rangle X_m \rangle$$

$$= N^{-2} \sum_{n=1}^{N} \sum_{m=1}^{N} \sum_{i=1}^{\infty} E\langle \langle X_n, e_i \rangle X_n, \langle X_m, e_i \rangle X_m \rangle.$$

Applying Parceval's identity (10.8), we obtain

$$\sum_{i=1}^{\infty} E\left\langle\langle X_n, e_i\rangle X_n, \langle X_m, e_i\rangle X_m\right\rangle = E\left\langle X_n, X_m\right\rangle \sum_{i=1}^{\infty}\langle X_n, e_i\rangle\langle e_i, X_m\rangle = E\langle X_n, X_m\rangle^2.$$

Therefore,

$$E\|\widehat{C}\|_{\mathcal{S}}^2 = N^{-2}\sum_{n=1}^{N}\sum_{m=1}^{N}E\langle X_n, X_m\rangle^2.$$

Applying the Cauchy-Schwarz inequality twice, we get

$$E\|\widehat{C}\|_{\mathcal{S}}^2 \le N^{-2}\sum_{n=1}^{N}\sum_{m=1}^{N}(E\|X_n\|^4)^{1/2}(E\|X_m\|^4)^{1/2} = E\|X\|^4,$$

completing the proof.

■

We conclude this section by establishing in Theorem 12.1.4 the consistency of the sample covariance operator. By (10.10), the conclusion of Theorem 12.1.4 can be equivalently stated as

$$E\iint[\hat{c}(t, s) - c(t, s)]^2 dt\, ds \le N^{-1}E\|X\|^4.$$

This implies that $\hat{c}(t, s)$ is an L^2–consistent estimator of the covariance function $c(t, s)$.

Theorem 12.1.4 can be proven using the approach found in the proof of Theorem 12.1.2. We thus use this as an opportunity to demonstrate how to use tensors in such proofs. We first state the following relationship which the reader is asked to verify in Problem 12.6.

THEOREM 12.1.3 *Let* $x, y \in \mathcal{H}$, *then*

$$\|x \otimes y\|_{\mathcal{H}\otimes\mathcal{H}} = \|\langle y, \cdot\rangle x\|_{\mathcal{S}}.$$

Notice that this implies that the equivalence between the two norms can be extended to the entirety of both spaces. We can now state the desired convergence result.

THEOREM 12.1.4 *If* $E\|X\|^4 < \infty$, $EX = 0$, *and Assumption 12.0.1 holds, then*

$$E\|\widehat{C} - C\|_{\mathcal{S}}^2 \le N^{-1}E\|X\|^4.$$

PROOF: By Theorem 12.1.3

$$E\|\widehat{C}-C\|_{\mathcal{S}}^2 = E\|\widehat{C}-C\|_{\mathcal{H}\otimes\mathcal{H}}^2 = N^{-2}\sum_n^N\sum_m^N E\langle X_n\otimes X_n-C, X_m\otimes X_m-C\rangle_{\mathcal{H}\otimes\mathcal{H}}.$$

Since the sample is iid all cross terms in the sum are zero, thus the above simplifies to

$$N^{-2}\sum_n^N E\langle X_n\otimes X_n - C, X_n\otimes X_n - C\rangle_{\mathcal{H}\otimes\mathcal{H}}$$
$$= N^{-1}E\langle X\otimes X - C, X\otimes X - C\rangle_{\mathcal{H}\otimes\mathcal{H}}$$
$$= N^{-1}(E\|X\|^4 - \|C\|_{\mathcal{S}}^2)$$
$$\leq N^{-1}E\|X\|^4.$$

∎

12.2 Estimated functional principal components

The functional principal components (FPC's) are defined in Section 11.4 as orthonormal eigenfunctions of the covariance operator C. Their interpretation as parameters, and their estimation, must be approached with care. The eigenfunctions v_j are defined by $C(v_j) = \lambda_j v_j$, so if v_j is an eigenfunction, then so is av_j, for any nonzero scalar a (by definition, eigenfunctions are nonzero functions). The FPC's v_j are normalized (orthonormal by definition), so $\|v_j\| = 1$, but this does not determine the sign of v_j. Thus if $\hat{v}_j$ is an estimate computed from the data, we can only hope that $\hat{c}_j\hat{v}_j$ is close to v_j, where

$$\hat{c}_j = \text{sign}(\langle\hat{v}_j, v_j\rangle).$$

When discussing properties of the $\hat{v}_j$ we will always implicitly assume that this sign has been taken into account,

$$\hat{v}_j \equiv \hat{c}_j\hat{v}_j,$$

and will omit writing it out from here on. Note that $\hat{c}_j$ cannot be computed form the data, so it must be ensured that statistics we want to work with do not depend on the $\hat{c}_j$. To ensure orthogonality, we must assume that $\lambda_1 > \lambda_2 > \cdots$, see Theorem 10.4.3. In practice, we can estimate only the p largest eigenvalues, and assume that $\lambda_1 > \lambda_2 > \cdots > \lambda_p > \lambda_{p+1}$, which implies that the first p eigenvalues are nonzero and their eigendirections are uniquely determined and orthogonal. This assumption is stated as condition (10.11).

With these preliminaries in mind, we define the estimated functional principal components (EFPC's) by

$$\int \hat{c}(t,s)\hat{v}_j(s)ds = \hat{\lambda}_j\hat{v}_j(t), \quad j = 1,2,\ldots,N. \tag{12.2}$$

The $\hat{v}_j$ are the principal components discussed in Section 1.3. Theorem 12.2.1 shows that the EFPC's consistently estimate the FPC, and the same is true for the eigenvalues. Moreover, if the functions are completely observable, the suitable distance between the parameters and their estimators is of the order $O_P(N^{-1/2})$, the usual parametric rate.

THEOREM 12.2.1 *Suppose* $E\|X\|^4 < \infty$, *Assumption 12.0.1 and condition* (10.11) *hold. Then, for each* $1 \le j \le p$,

$$\limsup_{N\to\infty} NE\left[\|\hat{v}_j - v_j\|^2\right] < \infty, \quad \limsup_{N\to\infty} NE\left[|\lambda_j - \hat{\lambda}_j|^2\right] < \infty. \tag{12.3}$$

The proof of Theorem 12.2.1 is presented in Section 12.7.

As a final reminder, in Theorem 12.2.1 consistency of $\hat{v}_j$ requires that the sign be chosen such that $\hat{v}_j$ points in the direction of v_j (as opposed to $-v_j$). If Assumption (10.11) is replaced by $\lambda_j > \lambda_{j+1} > 0$ for every $j \ge 1$, then (12.3) holds for every $j \ge 1$. Theorem 12.2.1 and its extensions are used to derive asymptotic distributions of statistics based on the EFPC's. If the assumptions of Theorem 12.2.1 do not hold, the direction of the $\hat{v}_k$ may not be close to the v_k. Examples of this type, with many references, are discussed in Johnstone and Lu (2009). However, relations (12.3) continue to hold if the X_i follow a stationary functional time series model such that the dependence between the observations decays sufficiently fast with time separation, see Section 16.2 of Horváth and Kokoszka (2012).

We now turn to the interpretation of the EFPC's $\hat{v}_j$; an analogous discussion for the v_j is provided in Section 11.4. EFPC's can be computed for any collection of observed functions, and the assumptions that the X_i are draws from the same population and their distribution satisfies regularity conditions are needed only to establish the convergence of the $\hat{v}_j$ to the v_j. Suppose then that we observe some functions $x_1, x_2, \ldots, x_N$. Fix an integer $p < N$. We think of p as being much smaller than N, typically a single digit number. We want to find an orthonormal basis $u_1, u_2, \ldots, u_p$ such that

$$\hat{S}^2 = \sum_{i=1}^{N}\left\|x_i - \sum_{k=1}^{p}\langle x_i, u_k\rangle u_k\right\|^2$$

is minimum. Once such a basis is found, we can replace each curve x_i by $\sum_{k=1}^{p}\langle x_i, u_k\rangle u_k$, to a good approximation. For the p we have chosen, this approximation is optimal in the sense of minimizing $\hat{S}^2$. This means that instead of working with infinitely dimensional curves x_i, we can work with p–dimensional vectors

$$\mathbf{x}_i = [\langle x_i, u_1 \rangle, \langle x_i, u_2 \rangle, \ldots, \langle x_i, u_p \rangle]^T.$$

This is the same fundamental idea as using spline or Fourier basis expansions; the difference is that the u_k are determined by the observed functions. For this reason, the functions u_j are called the *optimal empirical orthonormal basis* or *natural orthonormal components*, the words "empirical" and "natural" emphasizing that they are computed from the observed data.

The functions $u_1, u_2, \ldots, u_p$ minimizing $\hat{S}^2$ are equal (up to a sign) to the normalized eigenfunctions of the sample covariance operator $\widehat{C}$, i.e. to the EFPC's $\hat{v}_j$. To see this, suppose first that $p = 1$, i.e. we want to find u with $\|u\| = 1$ which minimizes

$$\sum_{i=1}^{N} \|x_i - \langle x_i, u \rangle u\|^2 = \sum_{i=1}^{N} \|x_i\|^2 - 2 \sum_{i=1}^{N} \langle x_i, u \rangle^2 + \sum_{i=1}^{N} \langle x_i, u \rangle^2 \|u\|^2$$

$$= \sum_{i=1}^{N} \|x_i\|^2 - \sum_{i=1}^{N} \langle x_i, u \rangle^2,$$

i.e. maximizes $\sum_{i=1}^{N} \langle x_i, u \rangle^2 = \langle \widehat{C}(u), u \rangle$. By Theorem 10.4.5, we conclude that $u = \hat{v}_1$. The general case is treated analogously. Since

$$\hat{S}^2 = \sum_{i=1}^{N} \|x_i\|^2 - \sum_{i=1}^{N} \sum_{k=1}^{p} \langle x_i, u_k \rangle^2,$$

we need to maximize

$$\sum_{k=1}^{p} \sum_{i=1}^{N} \langle x_i, u_k \rangle^2 = \sum_{k=1}^{p} \langle \widehat{C}(u_k), u_k \rangle$$

$$= \sum_{j=1}^{\infty} \hat{\lambda}_j \langle u_1, \hat{v}_j \rangle^2 + \sum_{j=1}^{\infty} \hat{\lambda}_j \langle u_2, \hat{v}_j \rangle^2 + \cdots + \sum_{j=1}^{\infty} \hat{\lambda}_j \langle u_p, \hat{v}_j \rangle^2.$$

By Theorem 10.4.5, the sum cannot exceed $\sum_{k=1}^{p} \hat{\lambda}_k$, and this maximum is attained if $u_1 = \hat{v}_1, u_2 = \hat{v}_2, \ldots, u_p = \hat{v}_p$.

We now discuss the sample variance decomposition analogous to the population decomposition (11.8). Suppose the functions are centered, i.e. assume that $\bar{x}_N = 0$. The statistic

$$\frac{1}{N} \sum_{i=1}^{N} \langle x_i, x \rangle^2 = \langle \widehat{C}(x), x \rangle$$

can be viewed as the sample variance of the data "in the direction" of the

function x. If we are interested in finding the function x which is "most correlated" with the variability of the data (away from the mean if the data are
not centered), we must thus find x which maximizes $\langle \widehat{C}(x), x \rangle$. Clearly, we
must impose a restriction on the norm of x, so if we require that $\|x\| = 1$,
we see from Theorem 10.4.5 that $x = \hat{v}_1$, the first EFPC. Next, we want to
find a second direction, orthogonal to $\hat{v}_1$, which is "most correlated" with the
variability of the data. By Theorem 10.4.5, this direction is $\hat{v}_2$. Observe that
since

$$\operatorname{span}\{x_1, \ldots, x_N\} = \operatorname{span}\{\hat{v}_1, \ldots, \hat{v}_N\}$$

we have that

$$\frac{1}{N} \sum_{i=1}^{N} \|x_i\|^2 = \frac{1}{N} \sum_{i=1}^{N} \sum_{j=1}^{N} \langle x_i, \hat{v}_j \rangle^2 = \sum_{j=1}^{N} \frac{1}{N} \sum_{i=1}^{N} \langle x_i, \hat{v}_j \rangle^2 = \sum_{j=1}^{N} \hat{\lambda}_j. \quad (12.4)$$

Thus, we may say that the variance in the direction $\hat{v}_j$ is $\hat{\lambda}_j$, or that $\hat{v}_j$ explains
the fraction of the total sample variance equal to $\hat{\lambda}_j / (\sum_{k=1}^{N} \hat{\lambda}_k)$. Recall that
the corresponding population analysis of variance (assuming $EX = 0$) is

$$E\|X\|^2 = \sum_{j=1}^{\infty} E[\langle X, v_j \rangle^2] = \sum_{j=1}^{\infty} \langle C(v_j), v_j \rangle = \sum_{j=1}^{\infty} \lambda_j.$$

In most applications, it is important to determine a value of p such that the
actual data can be replaced by the approximation $\sum_{i=1}^{p} \langle \hat{v}_j, X_n \rangle \hat{v}_j$ reasonably
well. A popular method is the *scree plot*. This is a graphical method proposed,
in a different context, by Cattell (1966). To apply it, one plots the successive
eigenvalues $\hat{\lambda}_j$ against j, as shown in Figure 12.1. The method recommends to
find j where the decrease of the eigenvalues appears to level off. This point is
used as the selected value of p; to the right of it, one finds only the "factorial
scree" ("scree" is a geological term referring to the debris that collects on the
lower part of a rocky slope). The scree plot approach works well in the right
panel of Figure 12.1, but its application in the left panel is more controversial.

Another approach that uses the magnitude of the $\hat{\lambda}_j$ is the *CPV method*
defined as follows. According to (12.4), the cumulative percentage of total
variance (CPV) explained by the first p EFPC's is

$$CPV(p) = \frac{\sum_{k=1}^{p} \hat{\lambda}_k}{\sum_{k=1}^{N} \hat{\lambda}_k}.$$

We choose p for which $CPV(p)$ exceeds a desired level, 85% is the recommended value for most applications. It is the approach we will apply most
often. Other methods, known as *pseudo–AIC* and *cross–validation* have also
been proposed. They are described and implemented in the MATLAB package
PACE developed at the University of California at Davis. It is often useful
to complement the methods described above by examining the shapes of the
EFPC's. An irregular white noise type shape indicates that such a $\hat{v}_j$ does not
contribute useful information to the data.

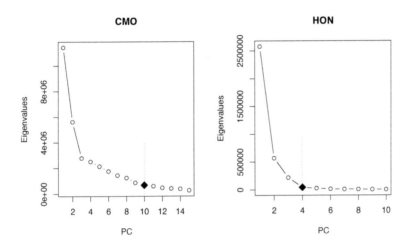

FIGURE 12.1: Eigenvalues $\hat{\lambda}_j$ as functions of j for space physics data analyzed by Kokoszka *et al.* (2008). Left panel uses daily curves observed at College (CMO), Alaska; right panel analogous curves observed at Honolulu (HON), Hawaii. The black diamond denotes the number of the most important principal components selected by the scree test. The selection is fairly obvious for HON, but is difficult for CMO. This illustrates the limitations of this approach.

12.3   Asymptotic normality

In the previous two sections, we have seen that expected values $E \|\hat{\mu} - \mu\|^2$, $E\|\widehat{C} - C\|^2$, $E \|\hat{v}_j - v_j\|^2$ and $E(\hat{\lambda}_j - \lambda_j)^2$ are all $O(N^{-1})$. By Chebyshev's inequality, this implies that the estimators converge to the corresponding population parameters in probability. To develop significance tests, it is useful to find the asymptotic distributions of these estimators. This is the subject of this section. It turns our that under Assumption 12.0.1, all these estimators are asymptotically normal. The results of this section thus also imply consistency of the estimators (convergence in probability), but we emphasize that convergence in distribution, in general does not imply convergence of moments, so the bounds stated in Sections 12.1 and 12.2 are of value. In this section we will be a bit more general and assume that the data come from a real separable Hilbert space $\mathcal{H}$. We begin with the following result whose proof is based on the Hilbert space CLT, Theorem 11.3.2.

THEOREM 12.3.1 *Let $X_1, \ldots, X_N$ be an iid sequence in $\mathcal{H}$ with $E\|X_n\|^2 < \infty$, then*

$$\sqrt{N}(\hat{\mu} - \mu) \xrightarrow{d} \mathcal{N}(0, C), \ in \ \mathcal{H},$$

where $C = E[(X_1 - \mu) \otimes (X_1 - \mu)]$. If, in addition, $E\|X\|^4 < \infty$, then

$$\sqrt{N}(\widehat{C} - C) \xrightarrow{d} \mathcal{N}(0, \Gamma), \ in \ \mathcal{S},$$

where $\Gamma = E[[(X_1 - \mu) \otimes (X_1 - \mu) - C] \otimes [(X_1 - \mu) \otimes (X_1 - \mu) - C]] \in \mathcal{S} \otimes \mathcal{S}$.

PROOF: The asymptotic normality of the mean is a direct application of the Hilbert space CLT. We thus only need to show the result for the covariance operator. Notice that the result would follow from the CLT if we weren't estimating μ. So let

$$\tilde{C} = \frac{1}{N} \sum_{n=1}^{N} (X_n - \mu) \otimes (X_n - \mu).$$

Then by the CLT we have that

$$\sqrt{N}(\tilde{C} - C) \xrightarrow{d} \mathcal{N}(0, \Gamma).$$

The CLT for $\widehat{C}$ will then follow if we can show that

$$\sqrt{N}(\tilde{C} - \widehat{C}) \xrightarrow{P} 0.$$

Writing it out and doing a number of simple manipulations, we have that

$$\sqrt{N}(\tilde{C} - \hat{C}) = \frac{1}{\sqrt{N}} \sum_{n=1}^{N} [(X_n - \mu) \otimes (X_n - \mu) - (X_n - \hat{\mu}) \otimes (X_n - \hat{\mu})]$$

$$= \frac{1}{\sqrt{N}} \sum_{n=1}^{N} [-\mu \otimes X_n - X_n \otimes \mu + \mu \otimes \mu + \hat{\mu} \otimes X_n + X_n \otimes \hat{\mu} - \hat{\mu} \otimes \hat{\mu}]$$

$$= \frac{1}{\sqrt{N}} [-N\mu \otimes \hat{\mu} - N\hat{\mu} \otimes \mu + N\mu \otimes \mu + N\hat{\mu} \otimes \hat{\mu} + N\hat{\mu} \otimes \hat{\mu} - N\hat{\mu} \otimes \mu]$$

$$= \sqrt{N} [-\mu \otimes \hat{\mu} - \hat{\mu} \otimes \mu + \mu \otimes \mu + \hat{\mu} \otimes \hat{\mu}]$$

$$= \frac{1}{\sqrt{N}} \left[\sqrt{N}(\hat{\mu} - \mu) \otimes \sqrt{N}(\hat{\mu} - \mu) \right]$$

$$= O_P(N^{-1/2}) = o_P(1).$$

Thus the difference between the two is asymptotically negligible and the claim holds.

We now turn to the asymptotic normality of the empirical eigenvalues and eigenfunctions. While more technical, one can nearly always tie their asymptotic properties to that of the empirical covariance operator. If the empirical covariance operator is asymptotically normal (and the eigenvalues are distinct), then the empirical eigenvalues/functions will also be asymptotical normal. First assume that the first p eigenvalues of C are distinct:

$$\lambda_1 > \lambda_2 > \cdots > \lambda_p > \lambda_{p+1} \geq \ldots .$$

It will be helpful to define the following quantities,

$$Z_N = \sqrt{N}(\hat{C} - C) \in \mathcal{S},$$

and

$$T_{jN} = \sum_{k \neq j} (\lambda_j - \lambda_k)^{-1} \langle Z_N, v_k \otimes v_j \rangle v_k \in \mathcal{H}.$$

We then have the following.

THEOREM 12.3.2 *Let $X_1, \ldots, X_N$ be iid elements of $\mathcal{H}$ with $E\|X\|^4 < \infty$. Assume that the first p eigenvalues of C are distinct. Then we have for each $j = 1, \ldots, p$,*

$$N^{1/2}(\hat{\lambda}_j - \lambda_j) = \langle Z_N, v_j \otimes v_j \rangle + o_P(1),$$

and

$$N^{1/2}(\hat{v}_j - v_j) = T_{jN} + o_P(1).$$

The proof of theorem 12.3.2 can be found in Kokoszka and Reimherr (2013), who generalize the result of Hall and Hosseini-Nasab (2006). Its conclusion is actually a bit stronger than asymptotic normality. It gives asymptotic expressions for the estimates up to an order of $o_P(1)$. Interestingly, one can actually relax the independence assumption and show that the above still holds, see Kokoszka and Reimherr (2013). Using Theorem 12.3.2, one can also compute the covariances between the $\hat{\lambda}_j$ or the cross covariance operators between the v_j if desired. The general formulas are complex, but if the processes are Gaussian, then they can be simplified, see Kokoszka and Reimherr (2013). The following result is useful when working with Gaussian functions.

THEOREM 12.3.3 *Let $X_1, \ldots, X_N$ be iid Gaussian elements of $\mathcal{H}$, then*

$$\langle \Gamma, v_k \otimes v_j \otimes v_l \otimes v_j \rangle = \lambda_j \lambda_k 1_{k=l} + \lambda_j^2 1_{j=k=l}$$

PROOF: To lighten notation, assume that $\mu = 0$. We can write

$$\Gamma = E[(X_1 \otimes X_1) \otimes (X_1 \otimes X_1)] - E[X_1 \otimes X_1] \otimes E[X_1 \otimes X_1].$$

We then have

$$\langle \Gamma, v_k \otimes v_j \otimes v_l \otimes v_j \rangle = E[\langle X_1, v_k \rangle \langle X_1, v_j \rangle \langle X_1, v_l \rangle \langle X_1, v_j \rangle] - \langle C, v_k \otimes v_j \rangle \langle C, v_l \otimes v_j \rangle$$
$$= E[\langle X_1, v_k \rangle \langle X_1, v_j \rangle \langle X_1, v_l \rangle \langle X_1, v_j \rangle] - \lambda_j^2 1_{j=k=l}.$$

Applying Isserlis' theorem, the above is

$$E[\langle X_1, v_j \rangle \langle X_1, v_j \rangle] E[\langle X_1, v_k \rangle \langle X_1, v_l \rangle] + E[\langle X_1, v_j \rangle \langle X_1, v_k \rangle] E[\langle X_1, v_j \rangle \langle X_1, v_l \rangle]$$
$$+ E[\langle X_1, v_j \rangle \langle X_1, v_l \rangle] E[\langle X_1, v_j \rangle \langle X_1, v_k \rangle] - \lambda_j^2 1_{j=k=l}$$
$$= \lambda_j \lambda_k 1_{k=l} + \lambda_j^2 1_{j=k=l} + \lambda_j^2 1_{j=k=l} - \lambda_j^2 1_{j=k=l},$$
$$= \lambda_j \lambda_k 1_{k=l} + \lambda_j^2 1_{j=k=l},$$

which completes the proof.

∎

We conclude this section with two corollaries which follow fairly directly from the above results, and whose proof is requested in Problem 12.8.

COROLLARY 12.3.1 *Under the assumptions of Theorem 12.3.2,*

$$N^{1/2}(\hat{\lambda}_j - \lambda_j) \overset{d}{\to} \mathcal{N}(0, \langle \Gamma, v_j \otimes v_j \otimes v_j \otimes v_j \rangle),$$

and

$$N^{1/2}(\hat{v}_j - v_j) \overset{d}{\to} \mathcal{N}(0, C_j),$$

where

$$C_j = \sum_{k \neq j} \sum_{l \neq j} (\lambda_j - \lambda_k)^{-1} (\lambda_j - \lambda_l)^{-1} \langle \Gamma, v_k \otimes v_j \otimes v_l \otimes v_j \rangle (v_k \otimes v_l).$$

COROLLARY 12.3.2 *Let $X_1, \ldots, X_N$ be iid Gaussian elements of $\mathcal{H}$. Assume that the first p eigenvalues of C are distinct. Then, for each $j = 1, \ldots, p$,*

$$N^{1/2}(\hat{\lambda}_j - \lambda_j) \xrightarrow{d} \mathcal{N}(0, 2\lambda_j^2),$$

and

$$N^{1/2}(\hat{v}_j - v_j) \xrightarrow{d} \mathcal{N}(0, C_j),$$

where

$$C_j = \sum_{k \neq j} \frac{\lambda_k \lambda_j}{(\lambda_j - \lambda_k)^2} (v_k \otimes v_k).$$

12.4 Hypothesis testing about the mean

While inference for the mean function is important in its own right, it also serves the purpose of demonstrating the key points concerning hypothesis testing on functional parameters. The goal is to construct test statistics and calculate p-values, which allow us to compare the following hypotheses:

$$H_0 : E[X_n] = \mu_0 \qquad \text{versus} \qquad H_A : E[X_n] \neq \mu_0,$$

keeping in mind that X_n is a realization of a random function X and μ_0 is a specific deterministic function. While the above problem is fairly simple, it already illustrates many aspects commonly encountered in FDA, but not in the scalar or vector settings. Many more complex testing problems are studied in Horváth and Kokoszka (2012).

As usual, H_0 denotes the null hypothesis, and H_A the alternative hypothesis. A test statistic is then some function of the data $X_1, X_2, \ldots, X_N$ whose observed value researchers use as either evidence to support the null, or evidence to reject the null in favor of the alternative. As for scalar data, it is natural to base the test statistic on the difference $\hat{\mu} - \mu_0$, where $\hat{\mu}$ is the sample mean function. However, since this is an infinite dimensional object, there is no "best" way to go about doing this. We discuss two approaches: the norm approach and the PC approach. In the norm approach, one uses the norm of the function space to construct a test statistic, while using the PC approach, one uses FPCA to first reduce the dimension of the data, and then constructs a multivariate test statistic.

The norm approach

The norm approach is useful as it takes advantage of the natural distance in the underlying function space, and can be tailored by moving to spaces other than L^2 to emphasize different properties of the data. In our setting, the test statistic is given by

$$T_{\text{Norm}} = N \|\hat{\mu} - \mu_0\|^2.$$

One can then show the following.

THEOREM 12.4.1 *Suppose Assumption 12.0.1 holds. Under H_0,*

$$T_{\text{Norm}} \xrightarrow{d} T_\infty := \sum_{i=1}^{\infty} \lambda_i \chi_i^2(1),$$

where the $\chi_i^2(1)$ are iid chi-square random variables with one degree of freedom, and the λ_i are the eigenvalues of the covariance operator of X_1.

Under H_A, with $\mu^\star$ denoting the true parameter,

$$T_{\text{Norm}} = N\|\mu^\star - \mu_0\|^2 + O_P(N^{1/2}) \xrightarrow{P} \infty.$$

PROOF: We begin by working under H_0. By Theorem 12.3.1, $\sqrt{N}(\hat{\mu} - \mu) \xrightarrow{d} \mathcal{N}(0, C)$. Thus, by the continuous mapping theorem,

$$T_{\text{Norm}} \xrightarrow{d} \|Z\|^2,$$

where $Z \sim \mathcal{N}(0, C)$. Problem 12.9 asks for the verification of the following commonly used identity:

$$\|Z\|^2 = \sum_{i=1}^{\infty} \lambda_i N_i^2, \tag{12.5}$$

where the N_i are iid standard normals. Since the square of a standard normal is $\chi^2(1)$, the claims holds.

Turning to H_A, observe that by Theorem 12.3.1, $N^{1/2}(\hat{\mu} - \mu^\star)$ converges in dostribution (to a normal limit), and so is bounded in probability. Therefore, letting $\Delta = \mu^\star - \mu_0$, we obtain

$$
\begin{aligned}
N\|\hat{\mu} - \mu_0\|^2 &= N \langle \hat{\mu} - \mu_0, \hat{\mu} - \mu_0 \rangle \\
&= N \langle \hat{\mu} - \mu^\star + \Delta, \hat{\mu} - \mu^\star + \Delta \rangle \\
&= N \left[\|\Delta\|^2 + 2 \langle \hat{\mu} - \mu^\star, \Delta \rangle + \langle \hat{\mu} - \mu^\star, \hat{\mu} - \mu^\star \rangle \right] \\
&= N\|\Delta\|^2 + N^{1/2} 2 \left\langle N^{1/2}(\hat{\mu} - \mu^\star), \Delta \right\rangle + \left\langle N^{1/2}(\hat{\mu} - \mu^\star), N^{1/2}(\hat{\mu} - \mu^\star) \right\rangle \\
&= N\|\Delta\|^2 + O_P(N^{1/2}) + O_P(1),
\end{aligned}
$$

and the claim holds.

∎

We now provide some comments on Theorem 12.4.1. The asymptotic distribution under H_0 depends on the unknown eigenvalues λ_i. Even if the λ_i were known, there is no closed form formula for the cdf or the density. We encountered a similar problem in Section 8.6. We first replace the λ_i by the $\hat{\lambda}_i$. Then one can use truncation of the infinite sum and simulation to find the empirical distribution, or some other numerical approach to compute the critical values or P–values. The R function `imhof()` in the `CompQuadForm` package can be used. While there are other methods, for statistical purposes it is important that they provide good approximations in the extreme tail (since that is how one obtains P–values), which `imhof` does, as we will see in examples that follow. One way or the other, to perform the test at a fixed significance level α, we compute the upper quantile, say $c_N(\alpha)$, of an approximation to T_∞. The subscript N emphasizes that this quantile is an estimate obtained from the sample. We then reject H_0 if $T_{\text{Norm}} > c_N(\alpha)$.

The asymptotic behavior under H_A is used to show that the test is *consistent*. A test is consistent, by definition, if

$$\lim_{N\to\infty} P(\text{reject } H_0|H_A \text{ is true}) = 1. \tag{12.6}$$

Since, $T_{\text{Norm}} \xrightarrow{P} \infty$, as $N \to \infty$, we see that for any *fixed c*,

$$\lim_{N\to\infty} P(T_{\text{Norm}} > c|H_A \text{ is true}) = 1. \tag{12.7}$$

We cannot however automatically conclude that the test is consistent. We would need to show that (12.7) holds with c replaced by $c_N(\alpha)$. This is in fact true for any reasonable method of computing $c_N(\alpha)$. For example, suppose we approximate the distribution of T_∞ by the empirical distribution of R replications of $\widehat{T}_p = \sum_{i=1}^p \widehat{\lambda}_i N_i^2$, with some fixed truncation level p. Since in the definition of $\widehat{C}$, the sample mean $\widehat{\mu}$ is subtracted from each X_i, we see that that the second relation in (12.3) holds also under H_A. (The $\widehat{\lambda}_i$ are computed under H_A, while the λ_i are defined under H_0.) It follows that $c_N(\alpha) \xrightarrow{P} c_p(\alpha)$, the upper αth quantile of $T_p = \sum_{i=1}^p \lambda_i N_i^2$, which does not depend on N. Thus, conditionally on H_A being true, $P(T_{\text{Norm}} > c_N(\alpha)) \to P(T_{\text{Norm}} > c_p(\alpha)) \to 1$, by (12.7). Since simulation is used to approximate the distribution of $\widehat{T}_p$, one finally must show that, essentially by the law of large numbers, the empirical quantile $c_p^{(R)}(\alpha)$ obtained from R replications of $\widehat{T}_p$ converges in probability to $c_p(\alpha)$, as $R \to \infty$ and $N \to \infty$. Detailed probabilistic arguments require some care and go slightly beyond the scope of this textbook.

In more complex testing problems, establishing consistency is more difficult, but it can often be done in the spirit of the above argument. Relation (12.6) is often verified by simulation. It is shown that for some representative alternatives, $P(\text{reject } H_0|H_A \text{ is true})$ approaches 1, as the sample size increases.

The PC approach

In the norm approach, the test statistic is based on a natural distance between the postulated mean function μ_0 and its estimate $\widehat{\mu}$. While such an approach is intuitive, as we have seen, its disadvantage is that even the limit distribution of the test statistic depends on unknown parameters, the eigenvalues λ_i.

The PC approach allows us to derive a test statistic whose asymptotic distribution is the standard chi–square distribution, which does not depend on any parameters, but involves a truncation level which turns out to be the number of its degrees of freedom. Before we present the details, we briefly discuss the analogous multivariate testing problem.

Suppose we observe a sample $\mathbf{X}_1, \ldots, \mathbf{X}_N$ of column vectors in $\mathbb{R}^p$ with the mean vector $\boldsymbol{\mu} = E\mathbf{X}_i$. We want to test the null hypothesis $H_0 : \boldsymbol{\mu} = \boldsymbol{\mu}_0$ for a specific vector $\boldsymbol{\mu}_0$. In the multivariate setting, the sample mean and variance are defined, respectively, by

$$\bar{\mathbf{X}} = \frac{1}{N} \sum_{k=1}^{N} \mathbf{X}_k, \quad \mathbf{S} = \frac{1}{N-1} \sum_{k=1}^{N} (\mathbf{X}_k - \bar{\mathbf{X}})(\mathbf{X}_k - \bar{\mathbf{X}})^{\top}.$$

An analog of the usual univariate t–statistic (its square to be precise) is Hotelling's one sample T^2 statistic defined by

$$T^2 = N(\bar{\mathbf{X}} - \boldsymbol{\mu}_0)^{\top} \mathbf{S}^{-1} (\bar{\mathbf{X}} - \boldsymbol{\mu}_0).$$

It is discussed in many multivariate statistics textbook, e.g. in Chapter 5 of Johnson and Wichern (2007). If the $\mathbf{X}_i$ are multivariate normal, its null distribution is a scaled F–distribution. Without the normality assumption, the distribution of T^2 converges to the χ_p^2 distribution, as $N \to \infty$.

To construct a functional analog we would need to be able to construct $\hat{C}^{-1}$, the inverse of the empirical covariance operator. However, the range of $\hat{C}$ is an N-dimensional subspace of $\mathcal{H}$, thus the inverse does not exist and the T^2 does not have a direct generalization to functional data. The PC approach utilizes FPCA to first reduce the dimension of the data, and then invokes a T^2 test. Since the covariance operator of $\hat{\mu}$ is $N^{-1}C$, we obtain (Problem 12.10),

$$\text{Cov}(\langle \hat{\mu}, v_i \rangle, \langle \hat{\mu}, v_j \rangle) = N^{-1} \lambda_i 1_{i=j}. \tag{12.8}$$

In other words, when using the FPC's as a basis, one ends up with uncorrelated projections whose variances are given by the eigenvalues of C. So to carry out the hypothesis test, we can use the projections $\langle \hat{\mu}, \hat{v}_i \rangle$ to construct the test statistic:

$$T_{\text{PC}} = N \sum_{i=1}^{p} \frac{\langle \hat{\mu} - \mu_0, \hat{v}_i \rangle^2}{\hat{\lambda}_i}.$$

Notice that the above statistic can be viewed as a quadratic form (like T^2), but since the projections are uncorrelated, we do not have any cross terms, and it simplifies in the above way. Additionally, the sum must end at some finite p. To see this, notice that we can construct the following operator

$$\Gamma_p(x) = \sum_{i=1}^{p} \lambda_i^{-1/2} \langle x, v_i \rangle v_i.$$

This operator projects onto the first p FPC's and normalizes by the corresponding eigenvalues. It does not converge, as $p \to \infty$, either in $\mathcal{S}$ or in $\mathcal{L}$. This is because the weights $\lambda_i^{-1/2}$ grow rapidly as $i \to \infty$, since the λ_i are summable. Furthermore, when one starts replacing the FPCs with their estimates, then Γ_p can become very unstable for large p due to the difficulty in estimating small eigenfunctions. In practice, it is therefore recommended to take p relatively small, generally following the 85% of explained variance rule.

Returning to the testing problem, we have the following result.

THEOREM 12.4.2 *Let $X_1, \ldots, X_N$ be iid square integrable elements of $\mathcal{H}$. Furthermore, assume that the first p eigenvalues are distinct. Then under H_0 we have that*

$$T_{\mathrm{PC}} \overset{d}{\to} \chi^2(p),$$

and under H_A,

$$T_{\mathrm{PC}} = N\|\Gamma_p(\Delta)\|^2 + O_P(N^{1/2}) \overset{P}{\to} \infty,$$

as long as $\langle \mu^\star, v_i \rangle \neq 0$ for some $i \leq p$.

PROOF: We begin with the behavior under H_0. Denote by $Z \sim \mathcal{N}(0, C)$ the limit of $\sqrt{N}(\hat{\mu} - \mu)$ in Theorem 12.3.1. Applying, in addition, Theorem 12.2.1, Slutsky's theorem and the continuous mapping theorem, we have that

$$T_{\mathrm{PC}} \overset{d}{\to} \sum_{i=1}^{p} \frac{\langle Z, v_i \rangle^2}{\lambda_i} \sim \chi^2(p).$$

Turning to the alternative, we observe that

$$\begin{aligned}
T_{\mathrm{PC}} &= \sum_{i=1}^{p} \frac{N \langle \hat{\mu} - \mu_0, \hat{v}_i \rangle^2}{\hat{\lambda}_i} \\
&= \sum_{i=1}^{p} \frac{N \langle \hat{\mu} - \mu^\star + \Delta, \hat{v}_i \rangle^2}{\hat{\lambda}_i} \\
&= \sum_{i=1}^{p} \frac{N[\langle \hat{\mu} - \mu^\star, \hat{v}_i \rangle^2 + 2 \langle \hat{\mu} - \mu^\star, \hat{v}_i \rangle \langle \Delta, \hat{v}_i \rangle + \langle \Delta, \hat{v}_i \rangle^2]}{\hat{\lambda}_i} \\
&= \sum_{i=1}^{p} \frac{N \langle \Delta, v_i \rangle^2}{\lambda_i} + O_P(N^{1/2}) \\
&= N\|\Gamma_p(\Delta)\|^2 + O_P(N^{1/2}),
\end{aligned}$$

which completes the argument. The verification of the penultimate equality is requested in Problem 12.11.

∎

Notice that the power of the norm approach is driven by $\|\Delta\|$ while the power of the PC approach is driven by $\|\Gamma_p(\Delta)\|$. Unfortunately, it is very difficult to predict which of these two quantities is going to be larger, and therefore lead to a more powerful procedure. If one can imagine the types of likely alternatives and examine the FPC's, then it might be possible to anticipate the relative powers of the two approaches. In particular, the FPC's and the rate of decay of the eigenvalues are often (though not always) related to smoothness of the data. If one has a relatively smooth alternative, often the PC approach will work well. We therefore recommend, in general, that the PC approach be used when a very small number of PCs explain a larger proportion of variability, and to use the norm approach otherwise. A simulation study that studies alternatives of interests is often useful in comparing the norm and the PC approaches, as we will see in the following.

We end this section by discussing two alternative, though equivalent, formulations of the PC approach. The first uses data projections (as opposed to projecting the parameter estimate), while the second uses generalized inverses. Notice that

$$\langle \hat{\mu} - \mu_0, \hat{v}_i \rangle = \left(N^{-1} \sum_{n=1}^{N} \langle X_n, \hat{v}_i \rangle \right) - \langle \mu_0, \hat{v}_i \rangle.$$

So, the PC approach can also be viewed as projecting the data onto $\{\hat{v}_1, \ldots, \hat{v}_p\}$ and seeing if the mean of the projections is close to $\langle \mu_0, \hat{v}_1 \rangle, \ldots, \langle \mu_0, \hat{v}_p \rangle$.

To directly generalize the T^2 test, one needs to be able to construct an inverse of $\widehat{C}$. One way of doing this is via the Moore-Penrose generalized inverse, where one uses the spectral representation of $\widehat{C}$ and inverts it by inverting the nonzero eigenvalues (or at least the larger eigenvalues). Recall that

$$\widehat{C} = \sum_{j=1}^{N} \hat{\lambda}_j \hat{v}_j \otimes \hat{v}_j.$$

We could then construct a generalized inverse by only inverting the first $p < N$ eigenvalues, and setting the rest to zero:

$$\widehat{C}_p^+ := \sum_{j=1}^{p} \hat{\lambda}_j^{-1} \hat{v}_j \otimes \hat{v}_j.$$

The generalization of the T^2 test statistic would then be

$$N \langle \widehat{C}_p^+ (\hat{\mu} - \mu_0), (\hat{\mu} - \mu_0) \rangle = N \sum_{j=1}^{p} \frac{\langle \hat{\mu} - \mu_0, \hat{v}_j \rangle^2}{\hat{\lambda}_j},$$

which is exactly T_{PC}. This provides an alternative perspective on the PC approach, which is a solution to the limitation that one cannot invert covariances like in the low dimensional setting.

A simulation study

We present a small simulation study to demonstrate that, in general, neither the PC nor the norm approach can claim to be superior; it will always depend on the alternative. The parameters of the simulation are as follows: $N = 100$ with 50 evenly spaced time points over $[0, 1]$. The data are simulated as

$$X_n(t) = \mu(t) + \varepsilon_n(t),$$

with the ε_n taken to be iid Matérn processes (see Problem 1.5 and Example 2.1.1) with variance 1, scale $1/4$, and smoothness $5/2$. We take the mean function μ to be

$$\mu(t) = c_1 \sqrt{2} \sin((k - 1/2)\pi t).$$

The parameter c_1 controls the size of the alternative, $|c_1| = \|\mu\|$, and the k controls the frequency behavior of the mean, with larger k resulting in more chaotic mean functions. We conduct 1000 repetitions of each scenario with a type one error rate of $\alpha = 0.1$. Notice that with this number of repetitions, the SE of the rejection rate (at least under the null) should be around $\sqrt{0.9 \times 0.1/1000} \approx 0.01$. The results are given in Table 12.1.

c_1	k	Norm	PC-1	PC-2	PC-3	PC-4	PC-5	PC-6
0	·	0.114	0.105	0.114	0.128	0.129	0.135	0.146
0.1	1	0.371	0.339	0.368	0.346	0.328	0.314	0.313
0.1	3	0.382	0.099	0.164	0.540	0.817	0.812	0.806
0.1	5	0.392	0.095	0.107	0.137	0.267	0.965	1.000
0.2	1	0.862	0.797	0.852	0.819	0.833	0.815	0.806
0.2	3	0.989	0.108	0.363	0.971	1.000	1.000	1.000
0.2	5	1.000	0.106	0.130	0.209	0.647	1.000	1.000

TABLE 12.1: Empirical rejection rates for the test of $H_0 : \mu = 0$ at the nominal significance level $\alpha = 0.10$. The row $c_1 = 0$ corresponds to the empirical size.

There are a few interesting points which help illustrate the challenges of choosing between different functional procedures. We begin by noticing that the calibration of the tests (i.e. type 1 error) is acceptable for the norm and lower PC tests, but not for the higher PC tests. We see here that, given the sample size, the distortion in type 1 error increases substantially as p increases. Using up to 4 FPC's might be reasonable, but anything higher and calibration becomes a substantial concern. As we discussed above, this is intuitively due to the fact that T_{PC} involves division by the $\hat{\lambda}_i$. If the unknown population eigenvalues λ_i are small, and this will happen if we take p too large, then their estimates $\hat{\lambda}_i$ may be almost zero with a large probability.

Turning to the power of the tests, we see that, as expected, the power of the norm approach is primarily driven by the L^2 norm of the mean function. The frequency of the mean function (i.e. k) plays a much smaller role. When comparing to the PC approach, we see that when $k = 1$, the two tests are very similar. The mean function has a very low frequency which is captured well by the first one or two FPC's. When we move to $k = 3$ or 5, we see the beginning of a frustrating dynamic. Clearly, if one chooses the right number of FPC's, then the power of the PC approach is superior to the norm approach. This is because the power of the PC approach comes from the later FPC's which capture the variability of the mean function. So, by moving the signal to the later FPC's, while keeping its magnitude the same, the inclusion of the $\hat{v}_i$ with small $\hat{\lambda}_i$ results in a substantial increase in power as compared to $k = 1$. However, if one uses a smaller number of FPC's, which is a reasonable choice given the calibration of the test, then one can have essentially no power.

This illustrates the challenges of functional testing. There are multiple approaches to testing, and no single procedure can claim to be optimal. We will, however, make the following general recommendation. If one has a clear guidance on the number of FPC's to use, then we recommend PC test. In the absence of such guidance, we recommend the norm approach. Of course, this is just a simple rule of thumb and should be treated as such. Substantial power gains with the PC approach can be obtained if any of the later FPC's capture the signal well. If the signal can be captured by the first 1–2 FPC's, then the norm and PC approaches will not substantially differ.

Below we provide an example code illustrating how to carry out the two hypothesis tests for the discussed simulation scheme. The first section of code mainly sets up the simulation parameters for the data. The primary points of interest are the two code chunks within the for loop. Both use the eigenfunctions and eigenvalues from the `pca.fd`. In the PCA approach, one projects the estimated mean onto some number of FPCS (in this case 3), normalizes by the eigenvalues, and then computes the vector norm. The norm approach is nearly the same, but one does not normalize by the eigenvalues and uses all of the FPCs (that have a nonzero eigenvalue). The eigenvalues are then used in the `imhof` function to compute p-values.

```
library(fda); library(RandomFields); library(CompQuadForm)
# RandomFields is used to simulate thefunctions.
# CompQuadForm is used to find the p-values for the norm
    approach.
# First we set an option for RandomFields so that matrices are
    output.
RFoptions(spConform=FALSE) # makes output a matrix
# Then we set the parameters of the Matern error
nu<-3/2; var = 1; scale = 1/4
Mat_model<-RMmatern(nu = nu , var = 1 , scale = scale)
N<-10; m<-50;

# Next we generate the mean function
times<-seq(0,1, length=m)
c1<-0; k<-1
mu<-function(x){c1*sqrt(2)*sin((k-1/2)*pi*x)}
mu_vec<-mu(times)
```

```
# Now we simulate the data and tests.
reps<-10
TPC3_pval<-numeric(reps)
Tnorm_pval<-numeric(reps)
for(i in 1:reps){
    # Step 1 - Simulate Data
    #Prevent RFsimulate from printing
    capture.output(eps<-RFsimulate(Mat_model,times,n=N))
    X_mat<-scale(t(eps),center=-mu_vec,scale=FALSE)
    X.f<-Data2fd(times,t(X_mat))

    # Step 2 - Estimate parameters
    muhat<-mean.fd(X.f)
    X.pca<-pca.fd(X.f,nharm=20)
    lambda<-X.pca$values
    scores<-X.pca$scores
    v<-X.pca$harmonics

    # Step 3 - Compute tests and p-values
    # PCA test with 3 PCs
    TPC3<-N*sum(inprod(v[1:3],muhat)^2/lambda[1:3])
    TPC3_pval[i]<-pchisq(TPC3,3,lower.tail=FALSE)
    # Norm test
    Tnorm<-N*sum(inprod(v[lambda[1:20]>0],muhat)^2)
    Tnorm_pval[i]<-imhof(Tnorm,lambda[lambda[1:20]>0])[[1]]}
```

12.5   Confidence bands for the mean

Here we describe how to construct a simultaneous confidence band for the mean function μ. As with Section 12.4, these ideas can be readily generalized to other parameter estimates. We first explain what the term *simultaneous confidence band* means.

From Theorem 12.3.1 we have that for every fixed t, $\hat{\mu}(t)$ is asymptotically normal with mean $\mu(t)$ and variance $N^{-1}c(t,t)$. The above actually holds only for almost all t since we are working in L^2. It follows that the $1 - \alpha$ confidence interval for $\mu(t)$ is

$$\hat{\mu}(t) \pm z_{1-\alpha/2} N^{-1/2} \sqrt{\hat{c}(t,t)},$$

where $z_{1-\alpha/2}$ is the upper $\alpha/2$ quantile of the standard normal distribution. In other words, for each t,

$$P\left(|\hat{\mu}(t) - \mu(t)| \le z_{1-\alpha/2} N^{-1/2} \sqrt{\hat{c}(t,t)}\right) \approx 1 - \alpha.$$

A simulataneous confidence band is defined by the requirement

$$P\left(\operatorname*{ess\,sup}_{t\in[0,1]} \frac{|\hat{\mu}(t) - \mu(t)|}{\sqrt{\hat{c}(t,t)}} \leq c_N(\alpha)\right) \approx 1 - \alpha.$$

We use the *essential supremum* rather than the supremum because we can neglect a set of t's which has Lebesque measure zero. For continuous functions, the essential supremum coincides with the supremum. In practice, it means that we want to find $c_N(\alpha)$ such that the probability that $|\hat{\mu}(t) - \mu(t)| \leq c_N(\alpha)\sqrt{\hat{c}(t,t)}$, *for all* t, is approximately $1 - \alpha$. The event

$$\left\{\forall\ t,\ |\hat{\mu}(t) - \mu(t)| \leq z_{1-\alpha/2} N^{-1/2} \sqrt{\hat{c}(t,t)}\right\}$$

will have a very small probability, so $c_N(\alpha)$ must be much larger than $z_{1-\alpha/2} N^{-1/2}$. In the following sections, we explain how it can be determined. We will describe two general approaches to constructing confidence bands. The first is simulation based using either a parametric or nonparametric bootstrap. The second approach is analytic and can produce closed form bands, but is conservative, resulting in bands that can be wider than needed.

Bootstrap

If we knew the distribution of

$$D_N := \sup_{t\in[0,1]} \frac{|\hat{\mu}(t) - \mu(t)|}{\sqrt{\hat{c}(t,t)}},$$

we would take $c_N(\alpha)$ as its αth upper quantile. Bootstrap methods approximate the distribution of D_N using various forms of sampling and simulation. A bootstrap procedure generally refers to any method which (1) estimates the population distribution, (2) creates resamples from that estimated distribution, and (3) uses those resamples to estimate the relevant aspects of the population distribution. If the population distribution is estimated using a parametric family, such the normal/Gaussian distribution, then we say the bootstrap is *parametric*. If the empirical distribution is used, for example resampling residuals, then we say the bootstrap is *nonparametric*, or *residual* in this specific case.

Using a parametric bootstrap, the distribution of the X_n is estimated as $\mathcal{N}(\hat{\mu}, \widehat{C})$. We then randomly generate, say, B iid samples (each of size N) from this distribution. For each bootstrap sample, we get an estimate of the mean $\hat{\mu}_b(t)$. Next we compute

$$D_b = \sup_{t\in[0,1]} \frac{|\hat{\mu}_b(t) - \hat{\mu}(t)|}{\sqrt{\widehat{C}(t,t)}},$$

and then we take $c_N(\alpha) = c_N(\alpha, B)$ to be the upper αth upper quantile of $\{D_1, \ldots, D_B\}$.

A nonparametric bootstrap version is nearly the same, but constructs the resamples using the residuals. In particular, suppose $\{\hat{\varepsilon}_n(t)\}$ are the residuals. Then one bootstrap sample, $\{X_n^*(t)\}$, is generated by randomly selecting N of the $\{\hat{\varepsilon}_n(t)\}$ with replacement, call them $\{\varepsilon_n^*(t)\}$, and setting $X_n^*(t) = \hat{\mu}(t) + \hat{\varepsilon}_n^*(t)$. Repeating this B times will result in B bootstrap samples and then one proceeds the same as in the parametric case.

The advantage of the bootstrap approach is that the band it gives is nearly exact, and it generalizes quite readily to more complicated models. The downside is that it is computationally quite expensive and thus may be prohibitive in certain settings. Another consideration is that, after taking into account the point–wise variance, the same threshold is used across all time points. Intuitively, one might prefer a band which somehow exploits the within curve dependence more effectively, but currently this is not available and remains an area of ongoing research.

Modified Scheffe's method

The methodology discussed here was introduced in Choi and Reimherr (2016), and we refer the interested reader to that work for a deeper exposition. An analytic expression for a confidence band can be used by carefully applying a variant of Scheffe's method from multivariate statistics. In the FDA setting, this method relies heavily on the Karhunen–Loéve expansion for $\hat{\mu}(t)$:

$$\sqrt{N}(\hat{\mu}(t) - \mu(t)) = \sqrt{\hat{c}(t,t)} \sum_{j=1}^{\infty} \hat{\xi}_j \sqrt{\hat{\tau}_j} \hat{u}_j(t),$$

where $\hat{\tau}_j$ and $\hat{u}_j(t)$ are the eigenvalues/eigenfunctions of the correlation function

$$\hat{\rho}(t,s) = \hat{c}(t,s)/\sqrt{\hat{c}(t,t)\hat{c}(s,s)}.$$

Here we normalize by the point-wise variance as it often leads to tighter confidence bands. If one can find positive numbers $r_1, r_2, \ldots$ and a threshold $c_N(\alpha)$ such that

$$P\left(\sum_{j=1}^{\infty} \frac{\hat{\xi}_j^2 \hat{\tau}_j}{r_j^2} < c_N(\alpha)\right) = 1 - \alpha,$$

then a confidence band which has at least $1 - \alpha$ confidence can be constructed as

$$\hat{\mu}(t) \pm b(t) \qquad \text{where} \qquad b(t) := N^{-1/2}\sqrt{\hat{c}(t,t)}\sqrt{c_N(\alpha)\sum r_j^2 \hat{u}_j^2(t)}. \quad (12.9)$$

There are in fact infinitely many choices of r_j that will work, but in practice using $r_j = \hat{\tau}_j^{1/4}$ works well and, in some sense, leads to the narrowest possible band (see Problem 12.14). This choice can also be viewed as in-between the PCA approach and the norm approach from Section 12.4. Once the r_j are

selected, one can find $c_N(\alpha)$ using a numerical approximation such as the
`imof` function in R. To see why this gives a proper band notice that $\mu(t)$ falls
in the band if and only if

$$N(\hat{\mu}(t) - \mu(t))^2 \hat{c}(t,t)^{-1} \le c_N(\alpha) \sum r_j^2 \hat{u}_j(t)^2.$$

We can re-express the left side using the KL expansion and follow it up with
a careful application of the Cauchy–Schwarz inequality

$$\left(\sum_{j=1}^{\infty} \hat{\xi}_j \sqrt{\hat{\tau}_j} \hat{u}_j(t) \right)^2 = \left(\sum_{j=1}^{\infty} \frac{\hat{\xi}_j \sqrt{\hat{\tau}_j}}{r_j} r_j \hat{u}_j(t) \right)^2 \le \sum_{j=1}^{\infty} \frac{\hat{\xi}_j^2 \hat{\tau}_j}{r_j^2} \sum_{j=1}^{\infty} r_j^2 \hat{u}_j^2(t).$$

So then we have that

$$P\left(\hat{\mu}(t) - b(t) \le \mu(t) \le \hat{\mu}(t) + b(t); \text{ for almost all } t \right)$$

$$\ge P\left(\sum_{j=1}^{\infty} \frac{\hat{\xi}_j^2 \hat{\lambda}_j}{r_j^2} \le c_N(\alpha) \right) = 1 - \alpha.$$

Since we are working in $L^2[0,1]$, the confidence band holds for almost all t, i.e.
except a possible set of measure zero. If one switches to using a Reproducing
Kernel Hilbert Space, then this claim can be made to hold for all t, however
this is not usually much of a concern in practice. (Choi and Reimherr, 2016)
recommend taking $r_j = \lambda_j^{1/4}$, and showed that this leads to the narrowest
band in terms of average squared width. A minor technical point is that, in
that case, $c_N(\alpha)$ will be finite if and only if $\sum \lambda_j^{1/2} < \infty$. This property is not
guaranteed to hold (for example, it does not hold for Brownian motion), but
in many applications it will. Lastly, normalizing by the variance $c(t,t)$ is not
absolutely necessary, but can be very helpful when the point-wise variances
change greatly.

12.6 Application to BOA cumulative returns

We return to the Bank of America example in Section 1.4. There we saw
that the mean function looked slightly positive. Using point-wise confidence
intervals, we saw that this positive mean was likely real, but to verify this,
we use the hypothesis testing procedures developed in Section 12.4. We thus
assume that the daily CIDR curves are iid. Table 12.2 reports the P–values,
along with the explained variance (for the PC approach). We see that all tests
reject the null hypothesis of zero mean function. The p-values for the PC
approach are substantially lower for $p \ge 2$. However, the first FPC already

	Norm	PC-1	PC-2	PC-3	PC-4
P–value	8.9e-04	1.6e-03	8.6e-05	4.9e-05	1.3e-05
Explained variance	·	0.851	0.924	0.949	0.962

TABLE 12.2: P–values for the test of $H_0 : \mu = 0$ for the BOA cumulative intraday returns.

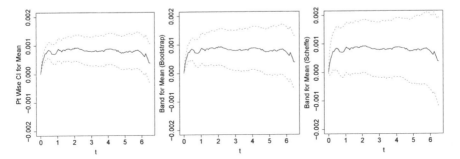

FIGURE 12.2: 95% confidence bands for the mean CIDR for BoA. The left panel plots point-wise confidence intervals with no global coverage. The middle and right panels give simultaneous bands using parametric bootstrap and a modified Scheffe's method respectively.

explains 85% of the variability, so if one follows the 85% of explained variance rule, one already gets a rejection.

We finish this application by constructing confidence bands. In Figure 12.2 we plot the raw mean estimates with three forms of bands. In the left panel, we plot point-wise 95% CI, which does not guarantee a global type 1 error rate of 5%. In the middle and right panels we plot 95% simultaneous confidence bands which do give the proper global type 1 error rate. The middle panel uses a parametric bootstrap while the right panel uses a modified Scheffe's method. Notice that the parametric bootstrap gives a narrower band. However, it also takes substantially longer to compute. For example, here we used 1000 bootstrap samples and a grid of 100 time points. Since the sample size was 2510, this means we had to generate 2.51 million multivariate normals, each of dimension 100. This is still computationally manageable, but it is clear that the computational burden can quickly compound.

To illustrate the role of smoothing on the bands, we also provide confidence bands where the means are estimated using 10 cubic B–splines (with equally spaced knots), see Figure 12.3. Here we see that the estimates are noticeably smoother, while still maintaining the main features of the mean function. Examining the bands, we can see that while the point-wise and bootstrap bands are still fairly similar, Scheffe's method now produces a notably narrower band than before. Furthermore, the bootstrap and Scheffe's method now produce bands of very similar widths. In general, for smoother

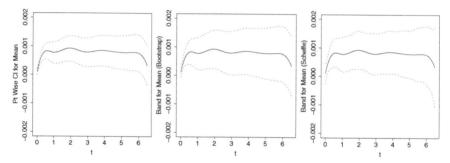

FIGURE 12.3: 95% confidence bands for the smoothed mean CIDR for BoA. The left panel plots point-wise confidence intervals with no global coverage. The middle and left panels give simultaneous bands using parametric bootstrap and a modified Scheffe's method.

estimates (which are very common in FDA), Scheffe's method works well and can be computed very quickly. For estimates that are, for whatever reason, rougher, a bootstrap approach will result in a noticeable improvement.

12.7 Proof of Theorem 12.2.1

The proof of Theorem 12.2.1 is based on Lemmas 12.7.1 and 12.7.2. These lemmas have wider applicability, and will be used in subsequent chapters. They state that if the operators are close, then their eigenvalues and eigenfunctions (adjusted for the sign) are also close.

We begin with a very useful Lemma 12.7.1, which can be stated for operators in a general class of *compact* operators, which are also called *completely continuous* operators. There are several equivalent definition of these operators and an interested reader is referred to Chapter 4 of Debnath and Mikusinski (2005). The definition that emphasizes the complete continuity is as follows: T is completely continuous if $\langle x_n, y \rangle \to \langle x, y \rangle$ for every $y \in \mathcal{H}$ implies $\|T(x_n) - T(x)\| \to 0$. *Every compact operator is bounded. Every Hilbert–Schmidt operator is a compact operator.* In fact, in Theorem 10.4.4 and Corollary 10.4.1, the assumption that $\Psi \in \mathcal{S}$ can be replaced by the assumption that Ψ is compact. If the assumption of symmetry is dropped, the representation in Corollary 10.4.1 must be replaced by a more general representation $\Psi(x) = \sum_{j=1}^{\infty} \lambda_j \langle x, v_j \rangle u_j$, called the *singular value decomposition*, where $\{u_j, j \geq 1\}$ is an orthonormal basis which may be different from the basis $\{v_j, j \geq 1\}$. The eigenvalues of a compact operator satisfy $\lim_{j \to \infty} \lambda_j = 0$, they do not have to be nonnegative.

Consider thus two compact operators, C and K, with singular value decompositions

$$C(x) = \sum_{j=1}^{\infty} \lambda_j \langle x, v_j \rangle f_j, \quad K(x) = \sum_{j=1}^{\infty} \gamma_j \langle x, u_j \rangle g_j. \tag{12.10}$$

The following Lemma is proven in Section VI.1 of Gohberg *et al.* (1990), see their Corollary 1.6 on p. 99.

LEMMA 12.7.1 *Suppose* $C, K \in \mathcal{L}$ *are two compact operators with singular value decompositions (12.10). Then, for each* $j \geq 1$, $|\gamma_j - \lambda_j| \leq \|K - C\|_{\mathcal{L}}$.

If we replace K, by the zero operator, Lemma 12.7.1 formally reduces to $|\lambda_j| \leq \|C\|_{\mathcal{L}}$. This special case is very easy to verify, see Problem 12.2. In order to prove Theorem 12.2.1, we will apply Lemma 12.7.1 with C being the covariance operator and K the sample covariance operator $\widehat{C}$. We therefore now tighten the conditions on the operator C by assuming that it is symmetric and $C(v_j) = \lambda_j v_j$, i.e. $f_j = v_j$ in (12.10); any covariance operator C satisfies these conditions. To lighten the notation, we also define

$$v'_j = c_j v_j, \quad c_j = \text{sign}(\langle u_j, v_j \rangle).$$

LEMMA 12.7.2 *Suppose* $C, K \in \mathcal{L}$ *are two symmetric compact operators with singular value decompositions (12.10)* $(f_j = v_j$ *and* $g_j = u_j)$. *If* C's *eigenvalues satisfy (10.11), then*

$$\|u_j - v'_j\| \leq \frac{2\sqrt{2}}{\alpha_j} \|K - C\|_{\mathcal{L}}, \quad 1 \leq j \leq p,$$

where $\alpha_1 = \lambda_1 - \lambda_2$ *and* $\alpha_j = \min(\lambda_{j-1} - \lambda_j, \lambda_j - \lambda_{j+1})$, $2 \leq j \leq p$.

PROOF: For a fixed $1 \leq j \leq p$, introduce the following quantities

$$D_j = \|C(u_j) - \lambda_j u_j\|, \quad S_j = \sum_{k \neq j} \langle u_j, v_k \rangle^2.$$

The claim will follow, once we have established that

$$\|u_j - v'_j\|^2 \leq 2S_j, \tag{12.11}$$

$$\alpha_j^2 S_j \leq D_j^2, \tag{12.12}$$

and

$$D_j \leq 2\|K - C\|_{\mathcal{L}}. \tag{12.13}$$

Verification of (12.11): By Parseval's identity

$$\|u_j - v'_j\|^2 = \sum_{k=1}^{\infty} (\langle u_j, v_k \rangle - c_j \langle v_j, v_k \rangle)^2 = (\langle u_j, v_j \rangle - c_j)^2 + S_j.$$

If $c_j = 0$, then (12.11) clearly holds, since in this case $\|u_j - v'_j\| = \|u_j\| = 1$ and $S_j = \|u_j\|^2 - \langle u_j, v_j \rangle = 1$.

If $|c_j| = 1$, then $(\langle u_j, v_j \rangle - c_j)^2 = (1 - |\langle u_j, v_j \rangle|)^2$, and using the identity $\sum_k \langle u_j, v_k \rangle^2 = 1$, we obtain

$$(1 - |\langle u_j, v_j \rangle|)^2 = \sum_{k=1}^{\infty} \langle u_j, v_k \rangle^2 - 2|\langle u_j, v_j \rangle| + \langle u_j, v_j \rangle^2.$$

Thus, if $|c_j| = 1$,

$$(\langle u_j, v_j \rangle - c_j)^2 = S_j + 2\left(\langle u_j, v_j \rangle^2 - |\langle u_j, v_j \rangle|\right) \leq S_j.$$

Verification of (12.12): By Parseval's identity

$$D_j^2 = \sum_{k=1}^{\infty} \left(\langle C(u_j), v_k \rangle - \lambda_j \langle u_j, v_k \rangle\right)^2.$$

Since C is *symmetric* and $C(v_j) = \lambda_j v_j$, $\langle C(u_j), v_k \rangle = \lambda_k \langle u_j, v_k \rangle$. Therefore

$$D_j^2 = \sum_{k \neq j} (\lambda_k - \lambda_j)^2 \langle u_j, v_k \rangle^2 \geq S_j \min_{k \neq j}(\lambda_k - \lambda_j)^2.$$

Verification of (12.13): Observe that since K is symmetric

$$C(u_j) - \lambda_j u_j = (C - K)(u_j) + (\gamma_j - \lambda_j)u_j.$$

Therefore, by Lemma 12.7.1,

$$D_j \leq \|C - K\|_{\mathcal{L}}\|u_j\| + |\gamma_j - \lambda_j|\|u_j\| \leq 2\|K - C\|_{\mathcal{L}}. \tag{12.14}$$

■

PROOF OF THEOREM 12.2.1: By Theorem 10.3.1, $\|Z_N\|_{\mathcal{L}} \leq \|Z_N\|_{\mathcal{S}}$, where Z_N is the operator with the kernel

$$z_N(t, s) = N^{1/2} \sum_{n=1}^{N} [\hat{c}(t, s) - c(t, s)].$$

By Theorem 12.1.4, $E\|Z_N\|_{\mathcal{S}}^2 = O(N)$, so the result follows from Lemmas 12.7.1 and 12.7.2.

12.8   Chapter 12 problems

12.1 Prove Lemma 12.1.1.

12.2 Show that for every eigenvalue λ of a bounded operator L, we have $|\lambda| \leq \|L\|_{\mathcal{L}}$.

12.3 Recall the simulation scheme given in Problem 1.5. For the three ν scenarios, increase N to 1000.

(a) Plot the estimates for the first 10 eigenvalues and include confidence intervals. Include everything in one plot if you can (but use no more than three plots).

(b) How many of the CI's did not include 0? Suppose you used that to determine the number of FPC's; how does that compare to the explained variance approach?

(c) Using the formula for the CI's and your estimated eigenvalues, approximately how many observations would you need to conclude that the 10th eigenvalue is not zero?

12.4 Suppose that $X_1, \ldots, X_N$ are iid and normally distributed.

(a) What is the largest that $\|\hat{v}_i - v_i\|$ could possibly be?

(b) Find an estimate for $E\|\hat{v}_i - v_i\|^2$ in terms of the estimated eigenvalues.

(c) Part (b) gives you a way of quantifying the variability of a an eigenfunction estimate. Construct a plot of these estimates for the first 10 FPC's for the simulation scheme given in Problem 12.3. What can you conclude?

12.5 Assume that $X_1, \ldots, X_N$ are iid elements of $L^2[0,1]$ with $E\|X_n\|^4 < \infty$ and whose first p eigenvalues are distinct. Prove that

$$|N\langle \hat{v}_j - v_j, v_j \rangle| = O_P(1), \quad \text{for } j = 1, \ldots p.$$

Why is this a seemingly unusual convergence rate? (Hint: start by showing that $|\langle \hat{v}_j - v_j, v_j \rangle| = \frac{1}{2}\|\hat{v}_j - v_j\|^2$)

12.6 Prove Theorem 12.1.3.

12.7 Suppose that the data $\{X_n(t) : t \in [0,1], 1 \leq n \leq N\}$ are expressed using an orthonormal basis $e_1, \ldots, e_J$:

$$X_n(t) = \sum_{j=1}^{J} x_{nj} e_j(t).$$

In this case, the EFPC's, $\hat{v}_i(t)$, can also be expressed as

$$\hat{v}_i(t) = \sum_{j=1}^{J} \hat{v}_{ij} e_j(t).$$

Explain how to obtain the coefficients $\hat{v}_{ij}$ from the x_{nj}. Justify your answer.

12.8 Prove Corollaries 12.3.1 and 12.3.2.

12.9 Using the expansion in Example 11.4.1 and (11.7), verify identity (12.5).

12.10 Suppose Assumption 12.0.1 holds. Show that the covariance operator of the sample mean function $\hat{\mu} = \sum_{i=1}^{N} X_i$ is $N^{-1}C$ and then establish (12.8).

12.11 Referring to the proof of Theorem 12.4.2, verify in detail that

$$\frac{N[\langle \hat{\mu} - \mu^\star, \hat{v}_i \rangle^2 + 2\langle \hat{\mu} - \mu^\star, \hat{v}_i \rangle \langle \Delta, \hat{v}_i \rangle + \langle \Delta, \hat{v}_i \rangle^2]}{\hat{\lambda}_i} = \frac{N\langle \Delta, v_i \rangle^2}{\lambda_i} + O_P(N^{1/2}).$$

12.12 Under the same assumptions as in Problem 12.5 show that, for $j \neq k$, and $1 \leq j \leq p$,

$$\langle \hat{v}_j - v_j, v_k \rangle = \frac{\langle \hat{C} - C, \hat{v}_j \otimes v_k \rangle}{\hat{\lambda}_j - \lambda_k}.$$

What can you conclude about the asymptotic distribution of $N^{1/2}\langle \hat{v}_j - v_j, v_k \rangle$? (justify your answer).

12.13 Under the same assumptions as in Problem 12.5 consider, for $j \leq p$, the mappings $G_{jN} : \mathcal{S} \to \mathcal{H}$ defined by

$$G_{jN}(\cdot) = \sum_{k \neq j} \frac{\langle \cdot, \hat{v}_j \otimes v_k \rangle}{\hat{\lambda}_j - \lambda_k} v_k.$$

(a) Show that G_{jN} is linear and continuous.
(b) Find a nontrivial Hilbert space which contains G_{jN}. What is the inner product?
(c) Show that there exists mappings G_j such that $G_{jN} \xrightarrow{P} G_j$ with respect to the operator norm.

12.14 Consider the confidence band in (12.9). Finding the r_j which lead to the narrowest band is, in general, a very difficult task. However, we can readily solve a slight variant of it. Namely, we can find the r_j which minimize

$$\int \sum r_j^2 v_j(t)^2 \, dt$$

subject to the constraint that

$$\sum \frac{\lambda_j}{r_j^2} = 1.$$

Intuitively, this can be thought of replacing the quantile $c_N(\alpha)$ with its mean. Find the r_j.

References

Akhiezier, N. I. and Glazman, I. M. (1993). *Theory of Linear Operators in Hilbert Space*. Dover, New York.

Apostol, T. M. (1957). *Mathematical Analysis*. Addison-Wesley.

Aston, J., Pigoli, D. and Tavakoli, S. (2016). Tests for separability in non-parametric covariance operators of random surfaces. *The Annals of Applied Statistics*, **6**, 1906–1948.

Aue, A., Norinho, D. D. and Hörmann, S. (2015). On the prediction of stationary functional time series. *Journal of the American Statistical Association*, **110**, 378–392.

Banerjee, S., Carlin, B. P. and Gelfand, A. E. (2004). *Hierarchical Modeling and Analysis for Spatial Data*. Chapman & Hall/CRC.

Beirlant, J., Goegebeur, Y., Segers, J. and Teugels, J. (2006). *Statistics of Extremes: Theory and Applications*. John Wiley & Sons.

Berlinet, A. and Thomas-Agnan, C. (2011). *Reproducing kernel hilbert spaces in probability and statistics*. Springer Science & Business Media.

Billingsley, P. (1968). *Convergence of Probability Measures*. Wiley, New York.

Booth, H., Hyndman, R. J. and Tickle, L. (2014). Prospective life tables. In *Computational Actuarial Science with R* (ed. A. Charpentier.), The R Series, chapter 8, pp. 323–348. Chapman & Hall/CRC.

Bosq, D. (2000). *Linear Processes in Function Spaces*. Springer.

Brockwell, P. J. and Davis, R. A. (1991). *Time Series: Theory and Methods*. Springer, New York.

Brockwell, P. J. and Davis, R. A. (2002). *Introduction to Time Series and Forecasting*, Second edn. Springer.

Caballero, W., Giraldo, R. and Mateu, J. (2013). A universal kriging approach for spatial functional data. *Stochastic Environmental Research and Risk Assessment*, **27**, 1553–1563.

Casella, G. and Berger, R. L. (2002). *Statistical Inference*. Duxbury.

Cattell, R. B. (1966). The scree test for the number of factors. *Journal of Multivariate Behavioral Research*, **1**, 245–276.

Chiou, J-M. and Müller, H-G. (2007). Diagnostics for functional regression via residual processes. *Computational Statistics and Data Analysis*, **15**, 4849–4863.

Choi, H. and Reimherr, M. (2016). A geometric approach to confidence regions and bands for functional data. Technical Report. Pennsylvania State University.

Constantinou, P., Kokoszka, P. and Reimherr, M. (2015). Testing separability of space–time functional processes. Technical Report. Penn State University.

Constantinou, P., Kokoszka, P. and Reimherr, M. (2016). Testing separability of functional time series. Technical Report. Penn State University.

Cressie, N. and Wikle, C. K. (2011). *Statistics for Spatio-Temporal Data*. Wiley.

Cressie, N. A. C. (1993). *Statistics for Spatial Data*. Wiley.

Csörgő, M. and Horváth, L. (1997). *Limit Theorems in Change-Point Analysis*. Wiley.

Cuevas, Antonio (2014). A partial overview of the theory of statistics with functional data. *Journal of Statistical Planning and Inference*, **147**, 1–23.

Dai, Xiongtao, Müller, Hans-Georg and Yao, Fang (2016). Optimal bayes classifiers for functional data and density ratios. *arXiv preprint arXiv:1605.03707*.

Debnath, L. and Mikusinski, P. (2005). *Introduction to Hilbert Spaces with Applications*. Elsevier.

Delaigle, A. and Hall, P. (2010). Defining probability density function for a distribution of random functions. *The Annals of Statistics*, **38**, 1171–1193.

Delaigle, Aurore and Hall, Peter (2012). Achieving near perfect classification for functional data. *Journal of the Royal Statistical Society: Series B (Statistical Methodology)*, **74**, number 2, 267–286.

Delicado, P., Giraldo, R., Comas, C. and Mateu, J. (2010). Statistics for spatial functional data: some recent contributions. *Environmetrics*, **21**, 224–239.

Earls, Cecilia and Hooker, Giles (2016). Adapted variational bayes for functional data registration, smoothing, and prediction. *Bayesian Analysis*.

Ettinger, B., Perotto, S. and Sangalli, L. (2016). Spatial regression models over two-dimensional manifolds. *Biometrika*, **103**, 71–88.

Fan, J. and Gijbels, I. (1996). *Local Polynomial Modelling and its Applications.* Chapman & Hall/CRC.

Faraway, J. J. (1997). Regression analysis for a functional response. *Technometrics*, **39,** 254–261.

Faraway, J. J. (2009). *Linear Models with R.* Taylor & Francis.

Ferraty, F. (2011) (ed.). *Recent Advances in Functional Data Analysis and Related Topics.* Physica–Verlag.

Ferraty, F. and Romain, Y. (2011) (eds). *The Oxford Handbook of Functional Data Analysis.* Oxford University Press.

Ferraty, F. and Vieu, P. (2006). *Nonparametric Functional Data Analysis: Theory and Practice.* Springer.

French, J., Kokoszka, P., Stoev, S. and Hall, L. (2016). Quantifying the risk of extreme heat waves over North America using climate model forecasts. Technical Report. Colorado State University.

Gabrys, R., Horváth, L. and Kokoszka, P. (2010). Tests for error correlation in the functional linear model. *Journal of the American Statistical Association*, **105,** 1113–1125.

Gelfand, A. E., Diggle, P. J., Fuentes, M. and Guttorp, P. (2010) (eds). *Handbook of Spatial Statistics.* CRC Press.

Gohberg, I., Golberg, S. and Kaashoek, M. A. (1990). *Classes of Linear Operators.* Operator Theory: Advances and Applications, volume 49. Birkhaüser.

Goldsmith, J., Bobb, J., Crainiceanu, C., Caffo, B. and Reich, D. (2011). Penalized functional regression. *Journal of Computational and Graphical Statistics*, **20,** 830–851.

Goldsmith, J., Bobb, J., Crainiceanu, C., Caffo, B. and Reich, D. (2012a). Penalized functional regression. *Journal of Computational and Graphical Statistics*, **20,** number 4, 830–851.

Goldsmith, J., Crainiceanu, C., Caffo, B. and Reich, D. (2012b). Longitudinal penalized functional regression for cognitive outcomes on neuronal tract measurements. *Journal of the Royal Statistical Society: Series C (Applied Statistics)*, **61,** number 3, 453–469.

Goldsmith, J. and Scheipl, F. (2014). Estimator selection and combination in scalar–on–function regression. *Computational Statistics and Data Analysis*, **70,** 362–372.

Górecki, T., Krzyśko, M. and Wolyński, W. (2015). Classification problems based on regression models for multidimensional functional data. *Statistics in Transition*, **16,** 97–110.

Gromenko, O. and Kokoszka, P. (2012). Testing the equality of mean functions of spatially distributed curves. *Journal of the Royal Statistical Society (C)*, **61**, 715–731.

Gromenko, O. and Kokoszka, P. (2013). Nonparametric inference in small data sets of spatially indexed curves with application to ionospheric trend determination. *Computational Statistics and Data Analysis*, **59**, 82–94.

Gromenko, O., Kokoszka, P. and Reimherr, M. (2016). Detection of change in the spatiotemporal mean function. *Journal of the Royal Statistical Society (B)*, **000**, 000–000; Forthcoming.

Gromenko, O., Kokoszka, P. and Sojka, J. (2016). Evaluation of the global cooling trend in the ionosphere using functional regression models with incomplete curves. Technical Report. Colorado State University.

Hackbusch, W. (2012). *Tensor Spaces and Numerical Tensor Calculus*. Springer.

Hall, P. and Hosseini-Nasab, M. (2006). On properties of functional principal components. *Journal of the Royal Statistical Society (B)*, **68**, 109–126.

Hamilton, J. D. (1994). *Time Series Analysis*. Princeton University Press, Princeton, NJ.

Hörmann, S., Kidziński, L. and Hallin, M. (2015). Dynamic functional principal components. *Journal of the Royal Statistical Society(B)*, **77**, 319–348.

Horváth, L. and Kokoszka, P. (2012). *Inference for Functional Data with Applications*. Springer.

Horváth, L., Kokoszka, P. and Reeder, R. (2013). Estimation of the mean of functional time series and a two sample problem. *Journal of the Royal Statistical Society (B)*, **75**, 103–122.

Horváth, L., Kokoszka, P. and Rice, G. (2014). Testing stationarity of functional time series. *Journal of Econometrics*, **179**, 66–82.

Hsing, T. and Eubank, R. (2015). *Theoretical Foundations of Functional Data Analysis, with an Introduction to Linear Operators*. Wiley.

Hunter, J. and Nachtergaele, B. (2001). *Applied Analysis*. World Scientific.

Hyndman, R. J. and Booth, H. (2008). Stochastic population forecasts using functional data models for mortality, fertility and migration. *International Journal of Forecasting*, **24**, 323–342.

Hyndman, R. J. and Shang, H. L. (2009). Forecasting functional time series (with discussion). *Journal of the Korean Statistical Society*, **38**, 199–221.

Hyndman, R. J. and Ullah, S. (2007). Robust forecasting of mortality and fertility rates: A functional data approach. *Computational Statistics and Data Analysis*, **51**, 4942–4956.

Ibragimov, I. and Rozanov, Y. (1978). *Gaussian random processes*, volume 9. Springer Science & Business Media.

Ivanescu, A. E., Staicu, A-M., Scheipl, F. and Greven, S. (2015). Penalized function–on–function regression. *Computational Statistics*, **30**, 539–568.

Jacques, J. and Preda, C. (2014). Functional data clustering: a survey. *Advances in Data Analysis and Classification*, **8**, 231–255.

Johnson, R. A. and Wichern, D. W. (2007). *Applied Multivariate Statistical Analysis*. Prentice Hall.

Johnstone, I. M. and Lu, A. Y. (2009). On consistency and sparcity for principal components analysis in high dimensions. *Journal of the Americal Statistical Association*, **104**, 682–693.

Kallenberg, O. (1997). *Foundations of Modern Probability*. Springer.

Kneip, A. and Ramsay, J. (2008). Combining registration and fitting for functional models. *Journal of the American Statistical Association*, **103**, 1155–1165.

Kokoszka, P., Maslova, I., Sojka, J. and Zhu, L. (2008). Testing for lack of dependence in the functional linear model. *Canadian Journal of Statistics*, **36**, 207–222.

Kokoszka, P. and Reimherr, M. (2013). Asymptotic normality of the principal components of functional time series. *Stochastic Processes and their Applications*, **123**, 1546–1562.

Kokoszka, P. and Young, G. (2016). KPSS test for functional time series. *Statistics*, **50**, 957–973.

Kraus, D. (2015). Components and completion of partially observed functional data. *Journal of the Royal Statistical Society (B)*, **77**, 1369–7412.

Krivobokova, T. and Kauermann, G. (2007). A note on penalized spline smoothing with correlated errors. *Journal of the American Statistical Association*, **102**, 1328–1337.

Laha, R. G. and Roghatgi, V. K. (1979). *Probability Theory*. Wiley.

Li, W. and Linde, W. (1999). Approximation, metric entropy and small ball estimates for Gaussian measures. *The Annals of Probability*, **27**, 1556–1578.

Liebl, D. (2013). Modeling and forecasting electricity prices: A functional data perspective. *The Annals of Applied Statistics*, **7**, 1562–1592.

Linde, W. (1986). *Probability in Banach Spaces - Stable and Infinitely Divisible Distributions*. Wiley.

Liu, C., Ray, S. and Hooker, G. (2014). Functional principal components analysis of spatially correlated data. Technical Report. Cornell University.

Lu, N. and Zimmerman, D. (2005). The likelihood ratio test for a separable covariance matrix. *Statistics & Probability Letters*, **73**, 449–457.

Manly, B. F. J. (1991). *Randomization, and Monte Carlo Methods in Biology*. Chapman and Hall.

Marron, J. S., Ramsay, J. O., Sangalli, L. M. and Srivastava, A. (2015). Functional data analysis of amplitude and phase variation. *Statistical Science*, **30**, 468–484.

McCullagh, P. and Nelder, J. (1989). *Generalized linear models*, volume 37. CRC press.

McLean, M. W., Hooker, G., Staicu, A-M., Schleip, F. and Ruppert, D. (2014). Functional generalized additive models. *Journal of Computational and Graphical Statistics*, **23**, 249–269.

Menafoglio, A., Secchi, P. and Rosa, M. D. (2013). A universal kriging predictor for spatially dependent functional data of a hilbert space. *Electronic Journal of Statistics*, **7**, 2209–2240.

Mitchell, M. W., Genton, M. G. and Gumpertz, M. L. (2006). A likelihood ratio test for separability of covariances. *Journal of Multivariate Analysis*, **97**, 1025–1043.

Morris, J. S. (2015). Functional regression. *Annual Review of Statistics and Its Applications*, **2**, 321–359.

Panaretos, V. M. and Tavakoli, S. (2013a). Cramér–Karhunen–Loève representation and harmonic principal component analysis of functional time series. *Stochastic Processes and their Applications*, **123**, 2779–2807.

Panaretos, V. M. and Tavakoli, S. (2013b). Fourier analysis of stationary time series in function space. *The Annals of Statistics*, **41**, 568–603.

Ramsay, J., Hooker, G. and Graves, S. (2009). *Functional Data Analysis with R and MATLAB*. Springer.

Ramsay, J. O. and Silverman, B. W. (2002). *Applied Functional Data Analysis*. Springer.

Ramsay, J. O. and Silverman, B. W. (2005). *Functional Data Analysis*. Springer.

Ramsay, Jim O, Hooker, G, Campbell, D and Cao, J (2007). Parameter estimation for differential equations: a generalized smoothing approach. *Journal of the Royal Statistical Society: Series B (Statistical Methodology)*, **69**, number 5, 741–796.

Reiss, P. T., Goldsmith, J., Shang, H. L. and Ogden, R. T. (2016). Methods for scalar–on–function regression. *International Statistical Review*, **00**,; DOI: 10.1111/insr.12163.

Reiss, P. T., Huang, L. and Mennes, M. (2010). Fast function–on–scalar regression with penalized basis expansions. *The International Journal of Biostatistics*, **6**, 1–28.

Rishbeth, H. (1990). A greenhouse effect in the ionosphere? *Planet. Space Sci.*, **38**, 945–948.

Roble, R. G. and Dickinson, R. E. (1989). How will changes in carbon dioxide and methane modify the mean structure of the mesosphere and thermosphere? *Geophys. Res. Lett.*, **16**, 1441–1444.

Rudin, W. (1976). *Principles of Mathematical Analysis*, Third edn. McGraw-Hill, Singapore.

Rudin, W. (1987). *Real and Complex Analysis*, International edn. McGraw-Hill, Singapore.

Ruppert, D., Wand, M. P. and Caroll, R. J. (2003). *Semiparametric Regression*. Cambridge.

Schabenberger, O. and Gotway, C. A. (2005). *Statistical Methods for Spatial Data Analysis*. Chapman & Hall/CRC.

Scheipl, F., Staicu, A-M. and Greven, S. (2015). Functional additive mixed models. Technical report. Ludwig Maximilians Universität München.

Seber, G. A. F. and Lee, A. J. (2003). *Linear Regression Analysis*. Wiley, New York.

Shang, H. L. (2013). ftsa: An R package for analyzing functional time series. *The R Journal*, **5**, 64–72.

Shang, H. L. (2017). Functional time series forecasting with dynamic updating: An application to intraday particulate matter concentration. *Econometrics and Statistics*; Forthcoming.

Sherman, M. (2011). *Spatial Statistics and Spatio–Temporal data: Covariance Functions and Directional Properties*. Wiley.

Shi, J. Q. and Choi, T. (2011). *Gaussian Process Regression Analysis for Functional Data*. CRC Press.

Shorack, G. R. and Wellner, J. A. (1986). *Empirical Processes with Applications to Statistics*. Wiley.

Shumway, R. H. and Stoffer, D. S. (2011). *Time Series Analysis and Its Applications with R Examples*. Springer.

Stein, M. L. (1999). *Interpolation of Spatial Data: Some Theory for Krigging*. Springer.

Sun, Y. and Genton, M. G. (2011). Functional boxplots. *Journal of Computational and Graphical Statistics*, **20**, 316–334.

Vakhaniia, N. N., Tarieladze, V. I. and Chobanian, S. A. (1987). *Probability Distributions on Banach Spaces*. Springer.

Wackernagel, H. (2003). *Multivariate Geostatistics*. Springer.

Wand, Matt P and Jones, M Chris (1994). *Kernel smoothing*. CRC Press.

Wang, J-L., Chiou, J-M. and Müller, H-G. (2016). Review of functional data analysis. *The Annual Review of Statistics and Its Application*, **3**, 257–295.

Wood, S. (2006). *Generalized additive models: An introduction with R*. CRC press.

Yao, F., Müller, H-G. and Wang, J-L. (2005). Functional data analysis for sparse longitudinal data. *Journal of the American Statistical Association*, **100**, 577–590.

Yao, F., Müller, H-G. and Wang, J-L. (2005b). Functional linear regression analysis for longitudinal data. *The Annals of Statistics*, **33**, 2873–2903.

Index

T - #0831 - 101024 - C306 - 234/156/14 - PB - 9781032096599 - Gloss Lamination